Confocal Scanning Optical Microscopy and Related Imaging Systems

Confocal Scanning Optical Microscopy and Related Imaging Systems

Timothy R. Corle
National Applied Science
Portland, Oregon

Gordon S. Kino
Ginzton Laboratory
Stanford University
Stanford, California

Academic Press
San Diego   London   Boston   New York   Sydney   Tokyo   Toronto

Cover photographs: The second image from the top is courtesy of Dr. G. Q. Xiao; all others are courtesy of Dr. Timothy R. Corle.

This book is printed on acid-free paper. ∞

Copyright © 1996 by ACADEMIC PRESS

All Rights Reserved.
No part of this publication may be reproduced or transmitted in any form or by any means, electronic or mechanical, including photocopy, recording, or any information storage and retrieval system, without permission in writing from the publisher.

Academic Press, Inc.
525 B Street, Suite 1900, San Diego, California 92101-4495, USA
http://www.apnet.com

Academic Press Limited
24-28 Oval Road, London NW1 7DX, UK
http://www.hbuk.co.uk/ap/

Library of Congress Cataloging-in-Publication Data

Corle, Timothy R.
   Confocal scanning optical microscopy and related imaging systems / by Timothy R. Corle, Gordon S. Kino.
       p.   cm.
   Includes bibliographical references and index.
   ISBN 0-12-408750-7 (alk. paper)
   1. Confocal microscopy.   2. Imaging systems.   I. Kino, Gordon S. II. Title.
QH224.C67   1996
502' .8'2--dc20   96-1999
                                                                            CIP

PRINTED IN THE UNITED STATES OF AMERICA
96 97 98 99 00 01 BC 9 8 7 6 5 4 3 2 1

*This book is dedicated to our wives,
Toni and Dorothy*

Contents

Preface   xiii

1   Introduction

1.1 Confocal and Interferometric Microscopy   1
1.2 The Standard Optical Microscope   7
    1.2.1 Principle of Operation   7
    1.2.2 The Point Spread Function   10
    1.2.3 Coherent and Incoherent Illumination   13
    1.2.4 The Coherent Transfer Function, Line Spread
           Function, and Spatial Frequencies   17
    1.2.5 The Optical Transfer Function   21
    1.2.6 The Rayleigh and Sparrow Two-Point Definitions 22
    1.2.7 Brightness of the Image   24
    1.2.8 Imaging Techniques with the Standard Optical
           Microscope   26
1.3 The Confocal Microscope   31
    1.3.1 Principle of Operation   31
    1.3.2 Scanning   33
    1.3.3 Depth Response   34
    1.3.4 The Point Spread Function and Two-Point
           Resolution   38
    1.3.5 History of the CSOM   41
1.4 Optical Interference Microscopes   44
    1.4.1 Principle of Operation   44
    1.4.2 Signal Processing Techniques   48
    1.4.3 Depth and Transverse Resolution   51

	1.5 Comparison of Scanning Optical Microscopes with Other Types of Scanning Microscopes	56
	References	63
2	**Instruments**	
	2.1 Introduction	67
	2.2 The Confocal Scanning Laser Microscope	68
	2.2.1 The Illumination Source	69
	2.2.2 The Objective Lens	71
	2.2.3 The Scanning Stage	73
	2.2.4 The Intermediate Optics	73
	2.2.5 The Pinhole	74
	2.2.6 The Detector and Electronics	74
	2.2.7 Beam Scanning Techniques	75
	2.2.8 Commercial Examples	79
	2.2.9 Fiber-Optic Scanning Microscopes	83
	2.3 Nipkow Disk Scanning Microscopes	84
	2.3.1 One-Sided and Two-Sided Designs	84
	2.3.2 The Nipkow Disk	86
	2.3.3 Illumination of the Disk	89
	2.3.4 The Tilted Disk and Optical Isolator	91
	2.3.5 The Field Lens, Tube Lens, and Objective Lens	93
	2.3.6 The Imaging Path	94
	2.3.7 Commercial Examples	95
	2.4 Slit Microscopes	97
	2.4.1 Ophthalmologic Slit Microscopes	97
	2.4.2 Bilateral Scanning Slit Microscopes	99
	2.4.3 Hybrid Slit Microscopes	101
	2.5 Confocal Transmission Microscopes	104
	2.6 Alternative Imaging Configurations	108
	2.7 Interference Microscopes	110
	2.7.1 Interference CSOMs	110
	2.7.2 The Michelson Interference Microscope	113
	2.7.3 The Linnik Interference Microscope	115
	2.7.4 The Mirau Interference Microscope	116
	2.7.5 The Tolanski Interference Microscope	119
	2.8 Near-Field Microscopy	120
	2.8.1 The Near-Field Scanning Optical Microscope	120
	2.8.2 Applications of the NSOM	131
	2.8.3 The Solid Immersion Microscope	133
	2.9 Conclusion	138
	References	139

3 Depth and Transverse Resolution

3.1	Introduction	147
3.2	Depth Response of the Confocal Microscope with Infinitesimal Pinholes and Slits	149
	3.2.1 Scalar Theory for a Plane Reflector	149
	3.2.2 Scalar Theory for Depth Response of a Point Reflector	154
	3.2.3 Scalar Theory for Fluorescent Reflectors	157
	3.2.4 Scalar Theory for Confocal Slit Microscopes	160
	3.2.5 The Effect of Sample and Lens Aberrations on the Depth Response	161
3.3	Depth Response of the Confocal Microscope with Finite-Sized Pinholes	165
	3.3.1 Approximate Theory for Optimum Pinhole Size	166
	3.3.2 Approximate Theory for the Range Resolution vs. Pinhole Size	167
	3.3.3 Exact Theory for the Range Resolution vs. Pinhole Size	169
3.4	Transverse Response of the Confocal Microscope	175
	3.4.1 Transverse Response for Infinitesimal Pinholes	175
	3.4.2 Two-Point Resolution	179
	3.4.3 Edge and Line Response	180
	3.4.4 The Effect of Finite Pinhole Size on the Transverse Resolution	183
3.5	Depth and Transverse Resolution of the Interferometric Microscope	189
	3.5.1 Scalar Theory for the Depth Response with a Plane Reflector	189
	3.5.2 Transverse Resolution	195
	3.5.3 The Effect of the Thin-Film Beamsplitter and Mirror Support of the MCM on Signal Levels, Range, and Transverse Resolution	196
3.6	The Near-Field Scanning Optical Microscope (NSOM)	206
	3.6.1 Attenuation in a Tapered Rod or Fiber	206
	3.6.2 The Fields outside the Pinhole	209
3.7	The Solid Immersion Microscope (SIM)	212
	3.7.1 The Transverse and Longitudinal Magnifications of the SIL	212
	3.7.2 The Depth Response of the SIM	214
	3.7.3 The Transverse Response of the SIM	216
3.8	Conclusion	220
	References	220

4 Phase Imaging

4.1 Introduction	225
4.2 Phase-Contrast Imaging in Conventional Microscopes	226
4.3 Phase-Contrast Imaging in the CSOM	229
4.3.1 Phase Imaging with an Interferometer	229
4.3.2 Electro-optic Phase Imaging	233
4.3.3 The ac Zernike Technique	234
4.3.4 Acousto-optic Phase Imaging	239
4.4 Differential Interference Contrast Imaging	247
4.4.1 The Basic Theory of Nomarski Imaging	248
4.4.2 Imaging Modes of a DIC Microscope	250
4.4.3 Polarization-Shifted DIC Imaging	252
4.4.4 Split Detector DIC Imaging	254
4.4.5 Differential Probe Beam DIC Imaging	260
4.4.6 Differential Imaging with an AO Modulator	261
4.4.7 Differential Imaging with an Optical Fiber CSOM	263
4.5 Phase Imaging with an Interference Microscope	266
4.5.1 The Integrating Bucket Technique	266
4.5.2 The Fourier Transform Technique	269
4.6 Conclusion	272
References	272

5 Applications

5.1 Introduction	277
5.2 Semiconductor Metrology	278
5.2.1 Microlithography Measurements	278
5.2.2 Precision, Linearity, and Accuracy in Semiconductor Metrology	279
5.2.3 Critical Dimension Measurements	280
5.2.4 Experimental Results	284
5.2.5 Polarization-Enhanced Imaging of Dense Arrays	286
5.2.6 Calibration	294
5.2.7 Overlay Misregistration Measurements	295
5.3 Film Thickness Measurements	300
5.3.1 CARIS and VAMFO	300
5.3.2 Film Thickness Measurements with the Mirau Interference Microscope	307

5.4	Biological Imaging	308
	5.4.1 Brightfield and Phase Imaging	308
	5.4.2 Fluorescence Imaging	311
	5.4.3 Two-Wavelength and Two-Photon Fluorescence Imaging	314
5.5	Conclusion	316
	References	317

Appendix A: Vector Field Theory for Depth and Transverse Resolution of a CSOM

A.1	The Depth Response	323
A.2	Transverse Response	326
	References	330

Index 331

Preface

The development and proliferation of the confocal microscope during the past few years has provided an important new imaging tool to scientists and engineers. These microscopes supply researchers with the capability of optically cross sectioning transparent samples without physically slicing them into thin sections. In addition, the ability to remove the glare from out-of-focus layers in a sample has yielded major improvements in biological imaging and extended the use of optical microscopes in many other areas of science.

The application of the confocal microscope to inspection and to the measurement of submicrometer features in semiconductors and other materials is less well known, but also of great importance. The Nipkow disk-based real time scanning optical microscope has proven to be particularly useful in this context. At the same time, other closely related microscopes based on interferometric principles and capable of quantitative measurements of phase and amplitude have been developed. These devices have many of the same capabilities as the confocal microscope and provide a great deal of information on surface profiles with submicrometer resolution.

In our own laboratory we have carried out research on Nipkow disk real-time scanning optical microscopes, on several types of interferometric microscopes, and have been interested in applications of these new instruments to semiconductor measurements and optical storage. The text we have written is based on this experience and describes the large variety of scanning microscopes that have been developed, the theory behind them, and how they are used in materials applications. We also describe the recently developed near-field scanning optical microscope and a related type of microscope developed in our own laboratory, the solid immersion microscope. These instruments are capable of resolutions well below the normal diffraction limits.

Chapter 1 introduces the reader to the standard optical microscope and the theoretical concepts that have been developed to describe its operation. We introduce the various modes of operation, the basic definitions of beam size that are used, and the concepts of coherence and incoherence. A description of the various types of confocal and interferometric microscopes is then given.

Chapter 2 concentrates on some of the practical aspects of building a confocal scanning optical microscope (CSOM) or optical interference microscope and introduces near-field optical microscopy. This chapter also explores some of the trade-offs made in designing a CSOM for various applications.

Chapter 3 covers the theory behind the confocal and interferometric microscopes. It contains an in-depth discussion of the origin of the shallow depth response and improved transverse resolution of the CSOM and of the interferometric microscope. Fourier optics is used without introduction, so the reader should be familiar with Fourier imaging theory before reading this chapter. Chapter 3 uses a simple point spread function approach in discussing imaging theory rather than the transfer function approach commonly used in other texts. The theory of near-field scanning microscopy is also dealt with in some detail at the end of the chapter.

Chapter 4 discusses phase imaging. This chapter illustrates how the principles of phase contrast and differential interference contrast imaging have been adapted to the CSOM. It also contains a detailed explanation of how phase and amplitude images are generated in optical interference microscopes.

The last chapter, Chapter 5, covers applications of the CSOM and interference microscope to both semiconductor measurements and biology. Biological imaging has been well covered in the literature. Most notably, a book edited by Pawley, *Handbook of Biological Confocal Microscopy, 3rd Edition*, gives an excellent review of the current state of the art. In contrast, the problems of semiconductor and materials measurements are not well covered in the literature and need an adequate treatment. We therefore concentrate in this book on some of the challenges associated with critical dimension and overlay measurements on semiconductor wafers, and we give a relatively short treatment in the last chapter of some of the techniques available for optimizing a CSOM for biological and fluorescent imaging.

We thank several of our colleagues for their help. Professor Calvin Quate provided us with helpful insights into near-field microscopy, much encouragement, and many useful comments on the text. David Dickensheets, Alice Liu, and Fang-Cheng Fang made insightful comments on the text. David and Fang-Cheng also provided us with experimental and theoretical results. We also drew heavily from the research results of Stanley Chim, Guoqing Xiao, Scott Mansfield, and the ConQuest development team at the Prometrix division of Tencor Instruments.

Putting this book together has been a great learning experience for both of us. We hope that the reader will find as much enjoyment and enlightenment in reading it as we have found in writing it.

Tim Corle
Gordon Kino

CHAPTER 1

Introduction

1.1 Confocal and Interferometric Microscopy

Optical microscopes have a ubiquitous presence in modern society. Virtually everyone has used one at some point in their life, if only to dissect a frog in school or observe the life hidden in a drop of pond water. They are used in laboratories and health clinics around the world. They have been developed into powerful measurement and observational tools with applications in geology, medicine, and manufacturing, to name a few areas. In the last few years, new types of optical microscopes have emerged. These microscopes enable researchers to visualize submicron structures, determine their surface profiles, and observe selected cross sections of transparent materials without cutting the sample into thin slices. In biology, fluorescence imaging has become increasingly important because biological activity can be traced by the fluorescence of markers associated with particular atomic or molecular species as they move through a cell. Application of microscopy principles in other fields, such as optical storage on compact disks, has also become important.

This book will concentrate on mainly two new types of optical microscopes: the *confocal scanning optical microscope* (CSOM) and the *optical interference microscope* (OIM). These instruments differ from the standard optical microscope because they have a shallow depth of focus and hence are capable of accurate height and thickness measurements and of obtaining cross-sectional images. There will also be a discussion of the *near-field scanning optical microscope* (NSOM), which is capable of obtaining definition well below the normal diffraction limits of optics.

In a standard optical microscope, when an image is defocused, its features blur so that the edges become less sharp, but the average light intensity does not change. With the CSOM, some types of OIMs, and NSOMs, on the other hand, a defocused image disappears rather than

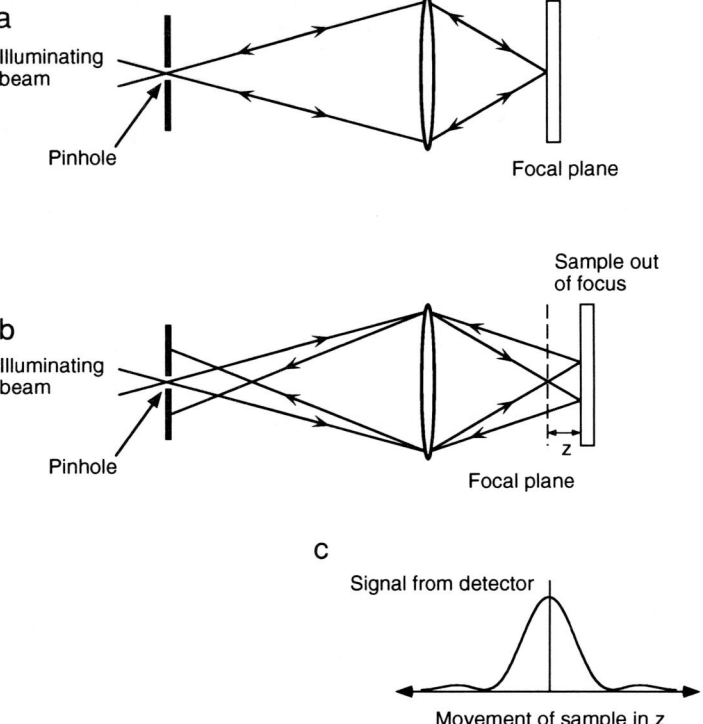

Figure 1.1 Simplified schematic of a confocal scanning optical microscope showing the sample (a) in the focal plane of the objective and (b) out of focus. (c) The form of the signal output from the detector as a function of sample defocus.

blurring. Put another way, the image intensity decreases as the image is defocused. This property enables structures which differ in height by as little as one optical wavelength to be independently imaged by these microscopes. As a result, quantitative measurements of height, surface profiles, and three-dimensional image reconstructions can be made.[1,2] The resulting images also tend to have more contrast, leading to better edge definition, than those obtained using a standard microscope.

Confocal Microscopy The basic principle of the confocal microscope, illustrated in Fig. 1.1(a), is to illuminate only one spot on the sample at a time through a pinhole. The light reflected from the sample is imaged by the objective back to the pinhole. By scanning the spot or the sample

1.1 Confocal and Interferometric Microscopy

in a raster pattern a complete image can be formed. If the sample moves out of focus, as shown in Fig. 1.1(b), the reflected light is defocused at the pinhole and hence does not pass through it to a detector located on the other side. The result is that the image of the defocused plane disappears. The signal output from a detector located behind the pinhole as the sample is moved in the focus direction is illustrated in Fig. 1.1(c).

Figure 1.2 shows three images of an integrated circuit taken at three different focus levels. Only the layers of the circuit which are within approximately ±0.25 μm of the focal plane appear in each of the photographs. For comparison, a conventional microscope image of the same sample is shown in Fig. 1.3(a). When the microscope is badly defocused, as in Fig. 1.3(b), the image blurs rather than disappearing.

Interference Microscopy Interference microscopes form an interference pattern with light reflected by the sample and a reference surface. If the reference surface is kept in a fixed position, as in the Michelson interferometer system illustrated in Fig. 1.4, interference fringes of each pixel in the image are formed as the reflecting sample is moved through focus. The contrast of the interference fringes falls off rapidly as the object is defocused.

By electronically processing the stored interference pattern, the envelope of the interference pattern for each pixel is determined. The shape of this envelope is very similar to the depth response of the confocal microscope. The cross-sectional images produced by the interference microscope are also very similar to those taken with the CSOM. Electronic processing of the fringe pattern allows measurement of not only the amplitude but also the phase of the reflected light. Since phase can be measured to an accuracy of a few degrees, it is possible to measure height or surface roughness to accuracies of a small fraction of an optical wavelength.

Near-Field Microscopy In this book, we will also discuss the NSOM. This device passes light through a small pinhole at the end of an optical fiber or a tapered aperture to illuminate or receive light from the sample, Fig. 1.5. If the pinhole is placed sufficiently close to the sample, the resolution is determined by the size of the pinhole rather than the wavelength of the light. Since the definition is not limited by diffraction as in other types of optical microscopes, it can be a very small fraction of a wavelength. An image is formed by scanning the pinhole in a raster pattern.[3,4,5] The microscope is called a near-field microscope because the evanescent fields just outside of the pinhole that are used for imaging require a close spacing between the sample and the probe.

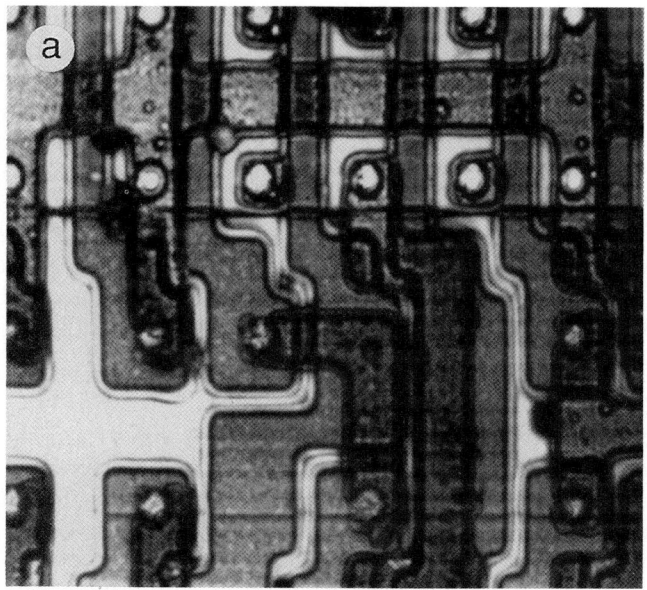

Figure 1.2 Three images of an integrated circuit taken at three different focus positions: (a) substrate (0 μm); (b) 1.2 μm defocused; (c) 2.4 μm defocused.

A related type of near-field microscope, the *solid immersion microscope* (SIM), focuses the light from a standard or confocal microscope beam onto the lower surface of a high-refractive-index solid transparent material, called a *solid immersion lens* (SIL), shown in Fig. 1.6. This additional lens element reduces the effective wavelength by the refractive index of the lens material improving the definition of the microscope.[6] Because of total internal reflection of rays making a large angle to the axis, this enhanced definition can be obtained only if the sample is placed close to the SIL. Therefore, the device is also a near-field microscope.

In order to put the performance of these microscopes into the proper context, we will discuss the operating principles of the standard optical microscope in the next section. Much of the basic mathematical theory is covered in the texts by Hecht and Zajac[7] and Born and Wolf.[8] We will not rederive all the results here but will quote from them in order to give a basis of comparison for the principles of the confocal and interferometric microscopes.

Section 1.2 will also introduce definitions and notation which will be used throughout this book. This introduction is followed by a discussion

1.1 Confocal and Interferometric Microscopy

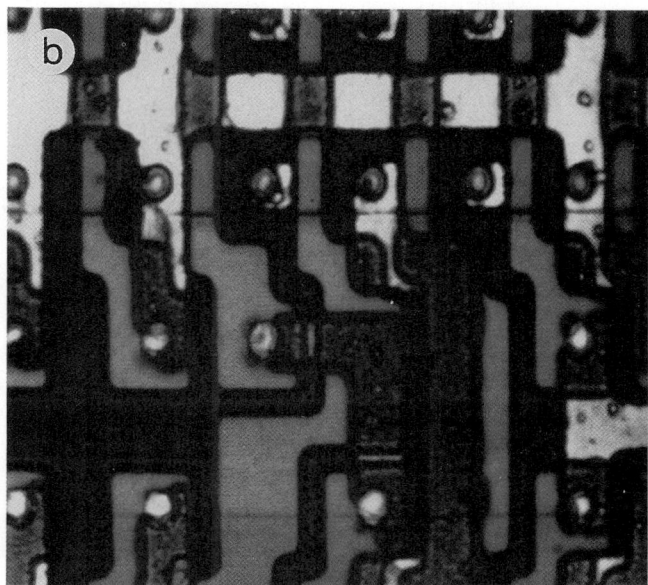

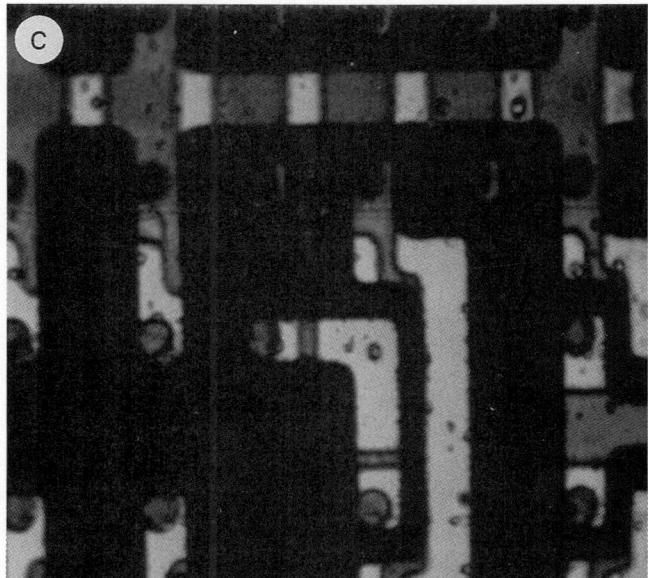

Figure 1.2—*Continued*

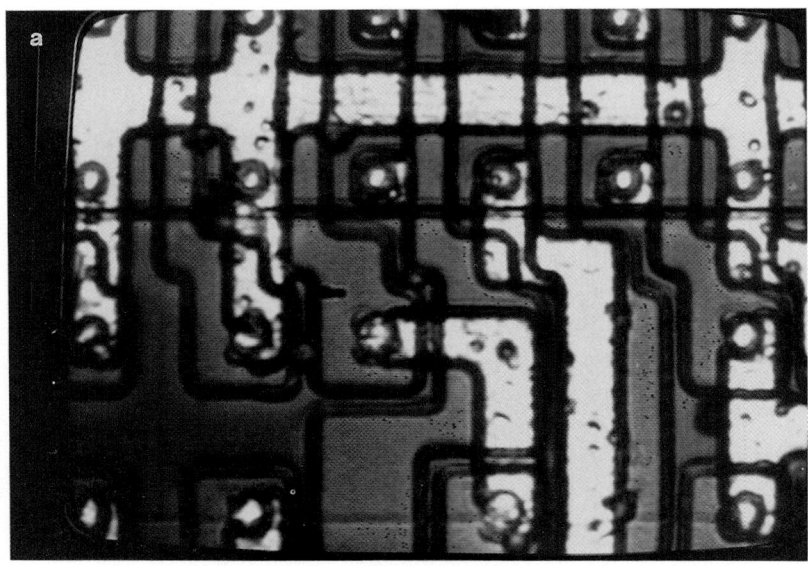

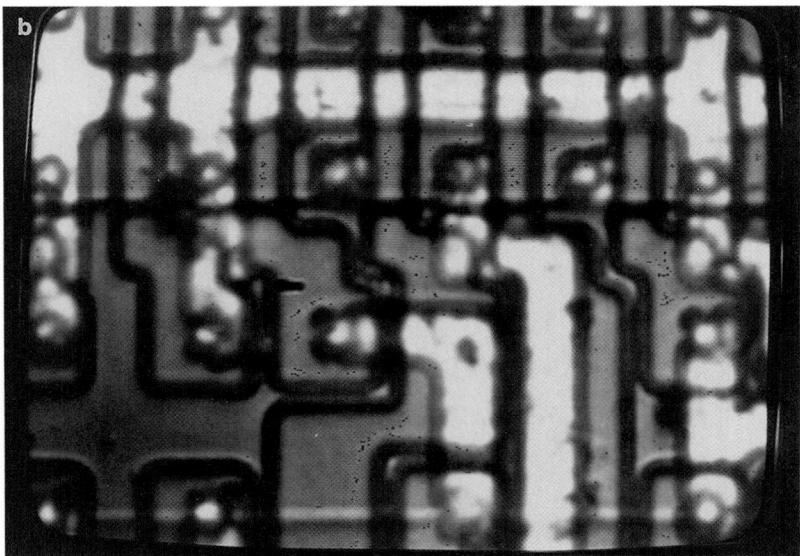

Figure 1.3 A standard microscope image of an integrated circuit: (a) when the sample is in focus; (b) when it is defocused by 2 μm.

1.2 The Standard Optical Microscope

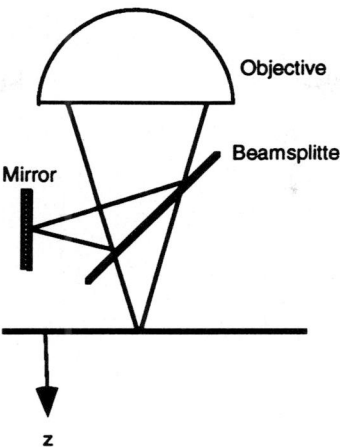

Figure 1.4 Simplified schematic of an interferometric microscope based on a Michelson interferometer.

of the different imaging techniques commonly used in a standard optical microscope. These imaging techniques are also available for use with the CSOM and thus form an important foundation for many of the later chapters in the book. Following the section on the standard optical microscope, the CSOM and the optical interference microscope will be described, with emphasis on the CSOM because of its simplicity of use and its wide range of applications. Chapter 1 concludes with a discussion comparing scanning optical microscopes with other types of scanning microscopes, such as the tunneling and force microscopes, the scanning electron microscope, and the scanning acoustic microscope. Our purpose is to illustrate the benefits and drawbacks of optical microscopy relative to other non-optical microscopy techniques.

1.2 The Standard Optical Microscope
1.2.1 Principle of Operation

Many of the design principles of the CSOM and interference microscopes are based on the standard microscope. A simplified form of a standard optical reflection microscope is shown schematically in Fig. 1.7. In this instrument, the sample is uniformly illuminated through the objective lens by a filament lamp or other bright incoherent light source such as a mercury vapor lamp. The objective lens forms a real inverted image

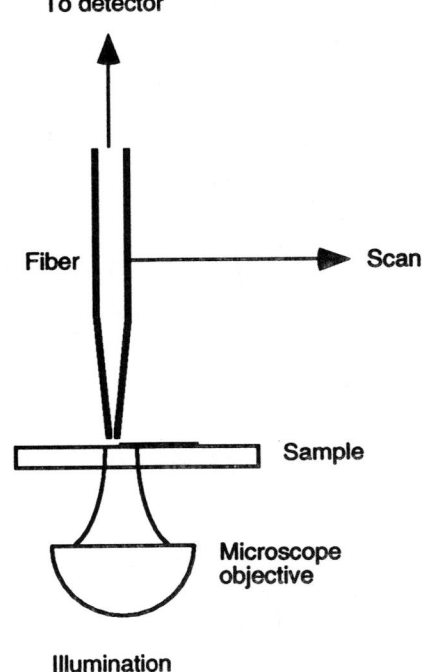

Figure 1.5 Simplified schematic of a near-field scanning optical microscope.

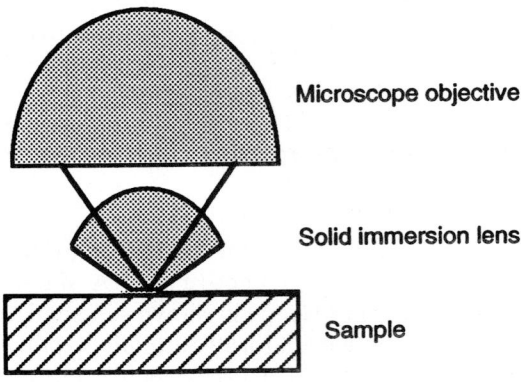

Figure 1.6 Basic component of a solid immersion microscope.

1.2 The Standard Optical Microscope

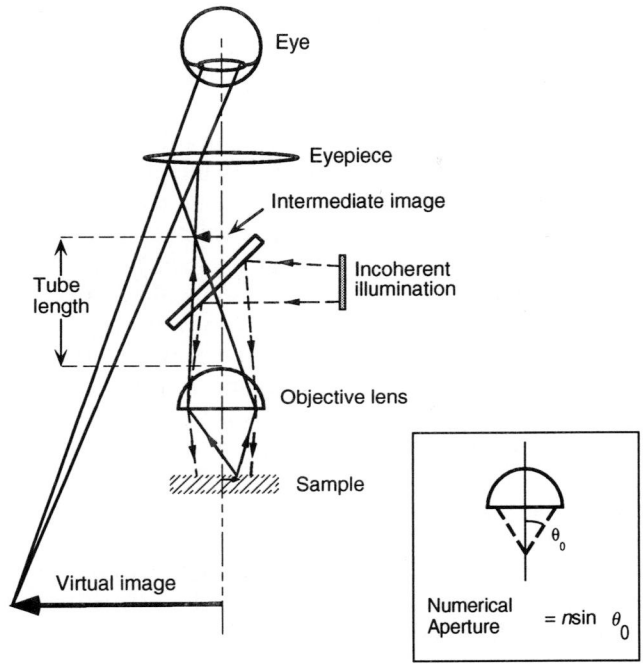

Figure 1.7 Simplified schematic of a standard optical microscope.

of the object at the intermediate image plane of the microscope. The distance of the intermediate image plane from the back focal plane of the objective is called the *tube length*.

This image is viewed through the eyepiece, which provides additional magnification. The eyepiece forms a virtual image of the object at a comfortable viewing distance from the eye, normally 250 mm for a "standard observer." With this arrangement, the total magnification is the product of the objective and eyepiece magnifications, so that high total values, up to 2000×, can be achieved in a small space. Several parameters are used to describe the performance of an optical microscope. Among these are the *magnification* (M) and the *numerical aperture* (N.A.). These parameters are defined below.

Magnification The magnification determines the size of the image at the detector. For a simple lens the magnification in the transverse direction is given by the negative ratio of the image distance to the object distance, $M_T = -d_i/d_o$. The negative sign accounts for the inversion of the image by a simple lens. This formula applies to the objective lens in a standard optical microscope with the image distance given by the tube length of the lens. It also follows from ray optics that the longitudinal magnification, or the magnification in the axial direction, is the square of the transverse magnification $M_L = -M_T^2$. The total magnification, M_r of the image in an optical microscope is given by the product of the objective and eyepiece magnifications. The eyepiece magnification M_E is commonly defined as the ratio of the size of the retinal image as seen through the instrument to the size of the retinal image as seen by the unaided eye at a normal viewing distance of 250 mm.[7] It is calculated from the formula $M_E = 250$ mm/f, where f is the focal length of the eyepiece lens. There is no negative sign in this formula because the eyepiece produces a virtual image. Common values of eyepiece magnification are 2.5-10.0×. Thus, as the magnification of the objective can vary from 1.5 to 200×, a wide range of total magnifications $M_r = M_T M_E$ can be obtained.

Numerical Aperture The magnification, by itself, does not determine the resolution of the microscope. To determine the resolution, the numerical aperture of the objective, defined by the relationship $N.A. = n \sin \theta_0$, must also be known. In this expression n is the refractive index of the medium between the lens and the sample, and θ_0 is the half-angle subtended by the lens at its focus, as illustrated in Fig. 1.7. The larger the angle, the more light that can be collected; hence, the numerical aperture is a measure of both the resolution and light gathering ability of the lens. Formulae for both the resolution and brightness of an optical microscope are given below.

1.2.2 The Point Spread Function

The Point Spread Function and Pupil Function The performance of an imaging system can be quantified by calculating its *point spread function* (PSF). The amplitude PSF, $h(x,y)$, of a lens is defined as the transverse spatial variation of the amplitude of the image received at the detector plane when the lens is illuminated by a perfect point source. Diffraction coupled with aberrations in the lens will cause the image of a perfect point to be smeared out into a blur spot occupying a finite area of the image plane. For a simple lens the concept of reciprocity applies, so that the function $h(x,y)$ also denotes the amplitude variation at the focal plane when the lens is illuminated using a point source. In the same way, the intensity PSF, $I_h(x,y) = |h(x,y)|^2$, of an objective is defined as the spatial variation of the intensity of the image received at the detector plane when the lens is illuminated by a perfect point source.

1.2 The Standard Optical Microscope

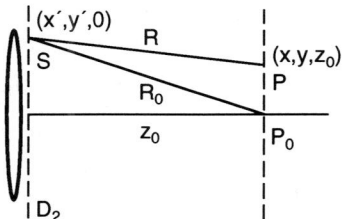

Figure 1.8 Schematic of the variables used in calculation of PSF in a standard optical microscope.

The definition of the PSF has it origin in linear systems theory. Its properties include the principle that for a linear spatially invariant imaging system, the image can be calculated by convolving a function characterizing the transmission $t(x,y)$ or reflectivity $r(x,y)$ of the sample with the PSF of the system. The analogy in electrical circuit theory is the convolution of an arbitrary signal with the response of a circuit. The PSF is the two-dimensional optical analog of the electrical impulse response of a circuit to a delta function input or infinitesimally narrow pulse.

The PSF is given in the paraxial approximation by the Fourier transform of the pupil function of the lens. To illustrate this mathematical relationship, we consider a beam of unit amplitude passing through the objective lens which is focused to a point P_0 at $(0,0,z_0)$ on the axis, of the lens as illustrated in Fig. 1.8. The pupil function of the objective, $P(x',y')$, is defined as the amplitude attenuation of the beam passing through the lens at a point $(x',y',0)$, on the plane D_2 just in front of the lens. The distance from the point S to the point P at (x,y,z_0) on the focal plane is

$$R = \sqrt{(x'-x)^2 + (y'-y)^2 + z_0^2}. \quad (1.1)$$

Similarly, the distance from S to the focal point P_0 is

$$R_0 = \sqrt{x'^2 + y'^2 + z_0^2}. \quad (1.2)$$

All the rays reaching the point P_0 will be in phase if the lens introduces a phase delay $\phi_0 = A - knR_0$ at each point $(x',y',0)$, where $k = 2\pi/\lambda$. In these expressions, λ is the free-space wavelength, n the refractive index of the medium between the lens and the point P_0, and A a constant. In this case the phase change along the ray of length R is $\phi = knR$.

Rayleigh-Sommerfeld diffraction theory can be used to calculate the scalar potential of the beam at the point (x,y,z_0). For a simple lens the scalar potential is just the amplitude PSF of the lens,[8,9]

$$h(x,y) = B \iint P(x',y') \frac{e^{-j(\phi-\phi_0)}}{R} dx' \, dy', \qquad (1.3)$$

where B is a constant.

To express Eq. (1.3) as a Fourier transform relation, the phase term is expanded to second order in x/z_0, y/z_0, x'/z_0, and y'/z_0 using the paraxial approximation, $z_0 \gg x'$, $z_0 \gg y'$, $z_0 \gg x$, and $z_0 \gg y$. Setting $R \approx z_0$ in the denominator of the integrand, the equation becomes

$$h(x,y) = De^{-jk(x^2+y^2)/2z_0} \iint P(x',y') e^{jkn(xx'+yy')/z_0} dx' \, dy'. \qquad (1.4)$$

In Eq. (1.4) D is a constant that is normally chosen to make the maximum value of $h(x,y)$ equal to unity.

Finally, assuming that the spot size is small, the exponential term in front of the integral is unity so that

$$h(x,y) = D \iint P(x',y') e^{jkn(xx'+yy')/z_0} dx' \, dy'. \qquad (1.5)$$

It is clear from Eq. (1.5) that the amplitude PSF of a simple lens at the focus is proportional to the Fourier transform of the pupil function.

Point Spread Function of a Spherical Lens In this book we will use the notation $h(r)$ for the radial variation of the amplitude PSF of a circularly symmetric aberration-free lens, where r is the distance from the center point of the image. If the pupil function is uniform, it can be shown from Eq. (1.5) in the paraxial approximation that $h(r)$ has the form of the Airy function[8,10]

$$h(r) = \frac{2J_1(v)}{v}. \qquad (1.6)$$

In Eq. (1.6), the normalized distance from the optical axis of the system is defined as $v = krn \sin \theta_0 = kr(N.A.)$, where $k = 2\pi/\lambda$ is the wave number, λ is the free-space wavelength, and $J_1(v)$ is a Bessel function of the first order and the first kind. The amplitude and intensity of the Airy function are plotted in Fig. 1.9. It will be noted that the amplitude is maximum at $v = 0$ and that there are subsidiary minima and maxima or sidelobes. The first zero of the response is located at $v = 3.832$ or $r = 0.61\lambda/n \sin \theta_0$. The first sidelobe or maximum in the amplitude response is at $v = 5.136$ or $r = 0.82\lambda/n \sin \theta_0$ and is reduced in amplitude by 0.132 or -17.6 dBs from the amplitude at the center of the main lobe. The amplitude PSF is related directly to the electric field at the sample, whereas the intensity PSF is related to the power per unit area or the square of the electric field.

1.2 The Standard Optical Microscope

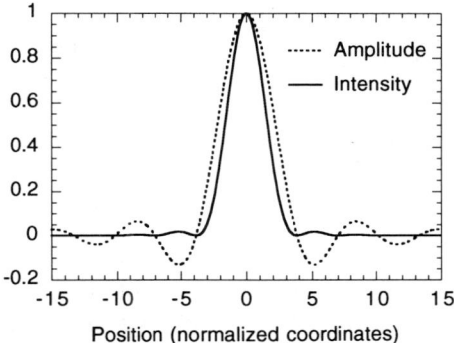

Figure 1.9 The amplitude variation (dotted line) and intensity (solid line) for the PSF of a simple lens.

The width between the half-power points of the main lobe, $d_r(3 \text{ dB})$, in the intensity response is known as *the full width at half-maximum* (FWHM) or 3-dB width and is given by the formula

$$d_r(3 \text{ dB}) = \frac{0.51\lambda}{n \sin \theta_0} = \frac{0.51\lambda}{N.A.}. \qquad (1.7)$$

This formula for the width of the image of a point object is also called *the single point resolution* of the standard optical microscope.

The resolution of an optical microscope can be improved by using liquid immersion objectives. These lenses use a high-refractive-index liquid between the sample and the microscope objective. Surrounding the sample with a high-index material reduces the effective wavelength in the medium, thereby improving the resolution. Two commonly used immersion fluids are water ($n = 1.33$) and immersion oil ($n = 1.52$). Water is often used for biological imaging when no cover glass is present on the sample. Oil is used for most other samples because its index most closely matches that of the cover glass, thus minimizing reflections from this surface and avoiding aberrations induced by the glass. Throughout this text, as is the common practice, objective lenses will be specified by both their magnification and their numerical aperture, for example, $80\times/0.95$ *N.A.*

1.2.3 Coherent and Incoherent Illumination

Throughout this book, we will primarily be concerned with two types of illumination: spatially coherent and spatially incoherent. The CSOM uses spatially coherent illumination, while the standard optical microscope generally uses spatially incoherent illumination. The interference microscope is a special case. Although the illumination source is incoherent, the interference of the sample and reference beams makes it effectively coherent, as will be discussed in Section 1.4 and Chapter 3. Studying the difference between these two types of illumination is an important foundation for understanding the differences between the various types of microscopes.

Following Goodman, we define *spatially coherent illumination* as having the property that the phasor amplitudes of all illuminated points vary in unison.[10] Thus, while the fields at any two spatially separated object points may have different *relative* phases, their absolute phases vary identically as the phase of the illumination changes. Therefore the phase difference between any two points remains constant with time.

The two most common examples of spatially coherent illumination are a point source, such as the light passing from an arc lamp through a small pinhole, and a laser source. An ideal single-frequency laser source has the additional property that its light is *temporally coherent*, which means that there is a definite phase relationship between the fields at any one point after a time delay T. In practice, real lasers maintain this phase relationship for only a finite time, called the *coherence time*, or when multiplied by the speed of light for a finite distance called the *coherence length* of the laser. We will revisit the concept of coherence length when we discuss the depth response of the interference microscope. With coherent illumination, the complex amplitudes of the image fields from several points in an object are added linearly to form the total field.

With *spatially incoherent illumination*, the phase relations between the fields from neighboring points are statistically random. In this case only the intensities, not the amplitudes, of two neighboring points add at the detector. Common examples of incoherent sources are filament lamps, fluorescent lamps, and gas discharge lamps.

In summary, with spatially coherent illumination the amplitude impulse responses are added to form the final image, while with spatially incoherent illumination the intensities of the impulse responses from separate points must be added to form the final image. If not carefully considered, these properties can lead to confusion about the imaging characteristics of a particular instrument. Optical systems can be linear in either intensity or amplitude depending on the light source and illumination path. In the first case the intensity image of a point, or the intensity PSF, must be convolved with the intensity object function to obtain the image. The intensity PSF can then be Fourier transformed to generate the spatial frequency response of the system as discussed in Section 1.2.4. For a

1.2 The Standard Optical Microscope

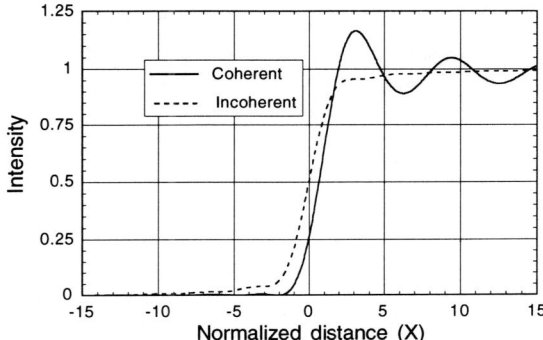

Figure 1.10 Theoretical response of a coherent and an incoherent imaging system to a knife edge.

system which is linear in amplitude the amplitude PSF must be used in the convolution integral. From now on, we shall refer to a spatially coherent source as simply "coherent" and spatially incoherent sources as "incoherent."

Speckle The presence of sidelobes in a coherently illuminated imaging system can lead to a phenomenon called *speckle*, which gives a granular appearance to the image. Speckle occurs because coherent light, reflected from different points of an optically rough surface (rough on the scale of the wavelength of light), adds in or out of phase at the detector. These individual bright speckles are formed in regions where the amplitudes add and have dimensions of the order of the focal spot size d_r(3 dB). The speckly appearance of an image formed with laser illumination is unfamiliar and unpleasant, since we are used to viewing images in incoherent light. The actual amplitude and appearance of the speckles depend on how the sidelobe amplitudes from the randomly located reflectors add. In order to eliminate this phenomenon, incoherent illumination is generally used in standard microscopes. With incoherent illumination, for points which are more than an approximate distance d_r(3 dB) apart, the signal intensities rather than their amplitudes add. Since the intensity sidelobes are small and their phases vary randomly with time, speckle is not observed.

An example of how the coherence of the illumination affects the image of a knife edge or step is shown in Fig. 1.10. The figure illustrates the results of a theoretical calculation for the signal variations across a knife edge for coherent and incoherent illumination as a function of the normal-

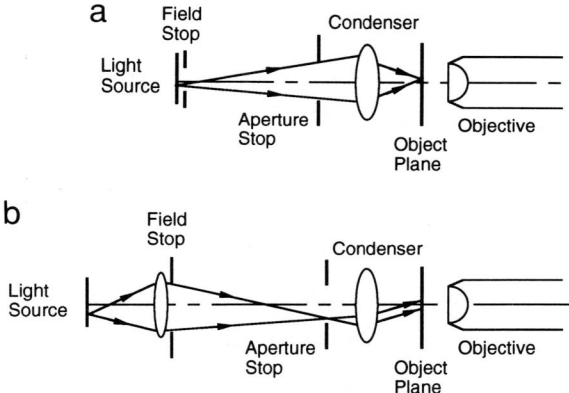

Figure 1.11 The illumination system of a standard microscope: (a) A critical illumination system; (b) a Köhler illumination system.

ized distance $X = 2\pi nx \sin \theta_0/\lambda$. As can be seen in the figure, there is quite severe "ringing" in the image when the edge is illuminated with coherent light from a focused cylindrical beam. On the other hand, if the illumination is incoherent, only the intensities add and the fringes virtually disappear. It will, however, be noted that there is more contrast in the coherent image. As is discussed in Section 1.3.4 and in Chapter 3, when an edge is imaged using a confocal scanning optical microscope, the amplitude sidelobes are equal to the intensity sidelobes for the incoherent case, the contrast is better because the sample is illuminated with coherent light, and the signal intensity falls off more rapidly across the edge.

Critical and Köhler Illumination Two different types of illumination systems are commonly employed in a standard microscope for illumination. With *critical illumination*, illustrated in Fig. 1.11(a), an incoherent source such as a filament lamp is used to illuminate the object through a *condenser* lens whose numerical aperture is controlled by the size of the *aperture stop*. In this system an image of the filament lamp, or other incoherent light source, is focused directly onto the sample.

A disadvantage of critical illumination is that if a source such as a filament lamp is imaged onto the object plane, the illumination may be highly non-uniform. Instead, as illustrated in Fig. 1.11(b), the most commonly used form of illumination is *Köhler illumination*. This type of illumination images the source onto the back focal plane of the condenser lens and provides more uniform illumination of the object while still acting as an incoherent source.[8]

1.2 The Standard Optical Microscope

The illumination of a standard optical microscope with either type of condenser is, in general, partially coherent. Diffraction causes the light from a point on the source to illuminate a region of finite extent determined by the Airy disk pattern of the condenser lens system. As a simple rule of thumb, the illumination of an object through a microscope condenser tends to be spatially coherent within a region the size of the main lobe of the PSF of the condenser; a region of diameter approximately $d_r(3\text{ dB})$. If the numerical apertures of the condenser and objective are comparable, then the spot size illuminated coherently is comparable to that of the detected spot. Regions outside this spot behave like separate sources and have random phase with respect to the spot. Thus for points farther apart than $d_r(3\text{ dB})$ the system behaves as if the source is incoherent, and the fields from points which can be observed separately by the objective are incoherent.

1.2.4 The Coherent Transfer Function, Line Spread Function, and Spatial Frequencies

The Coherent Transfer Function and Spatial Frequencies Abbé was the first to realize that a great deal of insight into the response of a lens system can be obtained by using Fourier theory.[11] Thus, we can use the PSF either to judge the transverse spatial variation of the fields at the focus or to determine the *spatial frequencies* to which the lens can respond. The highest spatial frequency transmitted by an optical system gives the period of the smallest diffraction grating which can be imaged. In this regard, it is a measure of the optical performance of the system in the same way that the highest temporal frequency transmitted by a circuit is a measure of its electrical performance. The finite size of the lens pupil limits the spatial frequency response of the system which causes the image of a perfect point to be smeared out into a spot occupying a finite area of the image plane. In the same way an impulse or delta function passing through a circuit is smeared out into a signal with a finite pulsewidth.

The spatial frequency response and the amplitude PSF of an optical system are related through the *coherent transfer function* (CTF) defined as

$$C(k_x, k_y) = \int_{-\infty}^{\infty} \int_{-\infty}^{\infty} h(x,y) e^{-j(k_x x + k_y y)} \, dx \, dy. \tag{1.8}$$

The inverse transform is given by

$$h(x,y) = \frac{1}{4\pi^2} \int_{-\infty}^{\infty} \int_{-\infty}^{\infty} C(k_x, k_y) e^{j(k_x x + k_y y)} \, dk_x \, dk_y. \tag{1.9}$$

By comparing Eq. (1.9) to Eq. (1.5) we see that $C(k_x, k_y) \propto P(x', y')$; the coherent transfer function is proportional to the pupil function.

The coherent transfer function, or for that matter any transfer function, simply specifies the complex weighting factor applied by the system to the frequency components at (k_x, k_y) relative to the weighting factor applied to the zero-frequency components. Comparing Eqs. (1.5) and (1.9), the spatial frequencies are defined by the relations

$$k_x = knx'/z_0; k_y = kny'/z_0 \qquad (1.10)$$

The spatial frequencies are the two-dimensional optical analog of the one-dimensional radian temporal frequency ω. In the paraxial approximation, the quantities defined by $k_x/kn = x'/z_0$ and $k_y/kn = y'/z_0$ are just the sines of the angles between the rays from x', y' to the focus and the axis.

The relationship between the CTF and the pupil function has some interesting implications. Since for a uniformly illuminated lens, the pupil function $P(r') = P(x', 0)$ is one over the area of the lens,

$$P(x') = 1 \qquad 0 < x' < a, \qquad (1.11)$$

It follows that the CTF takes the form

$$C(K, 0) = 1 \qquad 0 < K < 1, \qquad (1.12)$$

where $K = k_x/kn \sin \theta_0$. Thus there exists a finite passband in the spatial frequency domain within which a diffraction-limited system passes all frequency components without amplitude or phase distortion. At the boundary of this passband, the frequency response drops to zero, implying that the frequency components outside this passband are completely ignored. The CTF for a simple lens is illustrated in Fig. 1.12. This figure also plots the optical transfer function of the same lens along with experimental data confirming its shape. The optical transfer function will be discussed in Section 1.2.5.

The maximum values of the spatial frequencies are limited by the width or angular aperture of the pupil. In a cylindrical coordinate system the maximum angle is denoted θ_0, which is related to the numerical aperture of the lens, $N.A. = n \sin \theta_0$. The maximum spatial frequency imaged by an in-focus lens, illuminated with coherent illumination, is thus $k_x = kN.A.$ The maximum frequency response will be different for different optical systems and different illumination methods. For example, a standard optical microscope or a CSOM that uses light focused by a high-numerical-aperture condenser for illumination will have twice the maximum spatial frequency of the coherent system as discussed in Section 1.2.5.

1.2 The Standard Optical Microscope

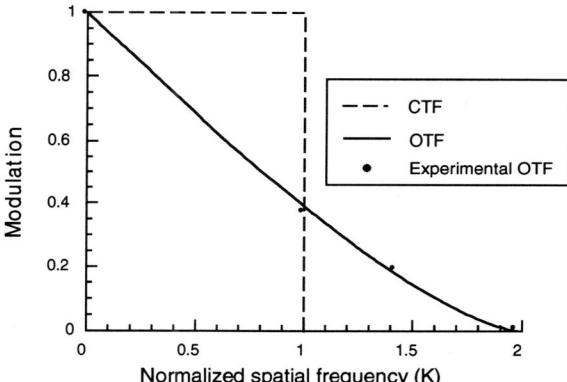

Figure 1.12 The coherence transfer function (dashed line) and the optical transfer function (full line) for a lens with a uniform pupil function. A comparison of the OTF with an experiment on an $N.A. = 0.55$ lens is also given (dots).

Following Abbé, we now consider the image of a grating with period ℓ as illustrated in Fig. 1.13. The grating will split a collimated input beam into several diffraction orders: a directly transmitted parallel beam and grating lobes or parallel beams propagating at angles $\theta_m \approx m\lambda/\ell$ to the axis. In this example, the $m = 0$ and $m = \pm 1$ beams pass through the lens, but the higher order grating lobes are diffracted at angles too large to be collected by the lens. The image of the grating that is formed contains spatial frequencies only up to values that can pass through the lens, and so the amplitude variation of the signal at the image plane is rounded off from the square wave form of the grating transmission. At the pupil plane or back focal plane of the lens, the grating lobes that pass through the lens form separate spots corresponding to each grating order. More generally, the Fourier transform of the transmission of an arbitrary object contains a continuum of spatial frequencies. Only those components corresponding to ray angles which pass through the lens form the image of the object.

If we consider the one-dimensional diffraction grating of Fig. 1.13, we can express the amplitude transmission $t(x)$ of this grating in terms of a Fourier expansion as

$$t(x) = \sum_{m=-\infty}^{n=\infty} A_m e^{-2jm\pi x/\ell}. \tag{1.13}$$

The lowest spatial frequency of this expansion, except for the constant term, is $1/\ell$. It is apparent that information on the period of the grating

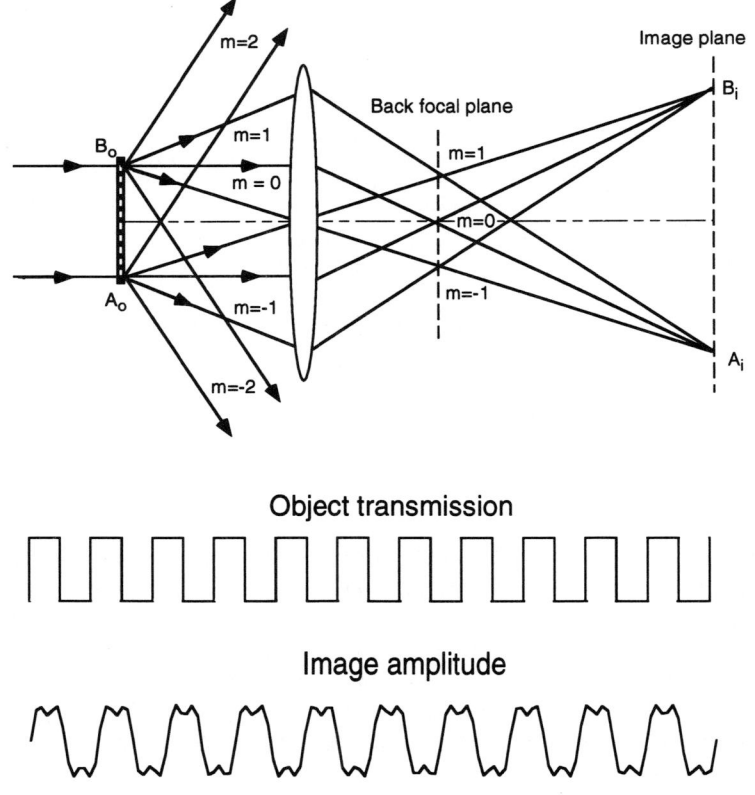

Figure 1.13 Abbé model for the image of a grating.

can be obtained with an optical system only if it can transmit a spatial frequency in the x direction greater than $1/\ell$.

The Line Spread Function Since we are dealing with a two-dimensional system, further definitions are needed for use with one-dimensional objects. For instance, the variation of the amplitude of the image received at the detector plane when the lens is illuminated by a perfect line source is called the *line spread function*, $l(x)$. This quantity is defined as

$$l(x) = \int_{y=-\infty}^{y=\infty} h(x,y)\, dy. \tag{1.14}$$

1.2 The Standard Optical Microscope

An example of the LSF for a coherent microscope is given in Section 3.4.3. The one-dimensional Fourier transform of the line spread function, $L(k_x)$, is

$$L(k_x) = \int_{x=-\infty}^{x=\infty} l(x) e^{-jk_x f_x x} \, dx. \tag{1.15}$$

Substituting Eq. (1.14) into (1.15), it will be observed that the Fourier transform of the line spread function is the coherence transfer function with zero spatial frequency in the y direction,

$$L(k_x) = C(k_x, 0) = \int_{-\infty}^{\infty} \int_{-\infty}^{\infty} h(x,y) e^{-jk_x x} \, dx \, dy. \tag{1.16}$$

If the system is cylindrically symmetric, then $L(k_x) = L(k_r) = C(k_x, 0)$, where k_r is the radial spatial frequency.

1.2.5 The Optical Transfer Function

In Section 1.2.4, we showed that useful information and the imaging characteristics of a lens system can be obtained by consideration of the amplitudes of the spatial frequency components. Since a detector responds to the intensity of the optical beam, intensity is the quantity most easily measured. It is, therefore, convenient to extend the spatial frequency formalism to deal with the intensity of the image and, in particular, the transverse intensity variation of a standard microscope with incoherent illumination.

Suppose that the intensity PSF of a lens is $I_h(x,y) = |h(x,y)|^2$. The Fourier transform of this response is called the *optical transfer function* (OTF) of the system and is given by

$$O(k_x, k_y) = \int_{-\infty}^{\infty} \int_{-\infty}^{\infty} |h(x,y)|^2 e^{-j(k_x x + k_y y)} \, dx \, dy. \tag{1.17}$$

The modulus of the OTF is known as the *modulation transfer function* (MTF). For a cylindrical system the OTF can also be obtained by integrating the incoherent LSF,

$$O(k_r) = \int_{-\infty}^{\infty} l'_S(x) \, dx, \tag{1.18}$$

where $l'_S(x)$ is the intensity line spread function for a one dimensional object defined in the same manner as the amplitude line spread function in Section 1.2.2. The form of the LSF for an incoherently illuminated standard microscope is given in Section 3.4.3 of Chapter 3.

The OTF is the Fourier transform of $|h(r)|^2$ and has the shape of the autoconvolution of $P(r')$. It is given by the expression[10]

$$O(K) = \frac{2}{\pi}[\cos^{-1}(K/2) - (K/2)\sqrt{1 - (K/2)^2}], \quad (1.19)$$

where, as above, $K = k_x/kn \sin\theta_0$. The OTF function has a roughly triangular shape and is illustrated in Fig. 1.12. This function has twice the spatial frequency bandwidth of the coherence function, with an amplitude which decreases with increase in spatial frequency and becomes zero at $K = 2$. This shape means that, unlike the CTF, higher frequencies are transmitted through the lens with some loss of fidelity.

The OTF and MTF are useful parameters because they are easily measured. Consider the image of a grating with a period ℓ. The power transmission of the grating is of the form

$$T_P(x) = B_0 + \sum_{m=1}^{\infty} B_m \cos(2\pi m x/\ell). \quad (1.20)$$

The strength of the spatial frequency component $k_x = 2\pi/\ell$ in the image is proportional to $B_1 \times O(K)$. By using a set of gratings with different periods but all with the same form factor, the quantity B_1 will remain constant. The image taken with each grating can be Fourier transformed and the strength of the sinusoidal component corresponding to the period ℓ can be determined. By this means, it is possible to measure the OTF of a lens system. An example of such a measurement made with an N.A. = 0.55 lens, normally used in a compact disk player, at $\lambda = 546$ nm, is shown in Fig. 1.12.[12]

In the preceding two sections we have seen that the transfer function approach represents a viable alternative to the PSF for describing the behavior of an optical system. Transfer function descriptions of the various imaging modes of the CSOM have been extensively covered in a book by Wilson and Sheppard.[13] In this text, we will generally use the PSF and point response to describe the resolution of a microscope.

1.2.6 The Rayleigh and Sparrow Two-Point Definitions

A detector, whether it is the human eye, a semiconductor diode, or a video camera, responds to the intensity of a signal rather than its amplitude. In a coherent imaging system, the amplitudes of signals from different parts of the image add, and the result is then squared to form the intensity image, while in an incoherent system the intensities add directly. Consequently, the use of the simple definition of resolution based on the half-

1.2 The Standard Optical Microscope

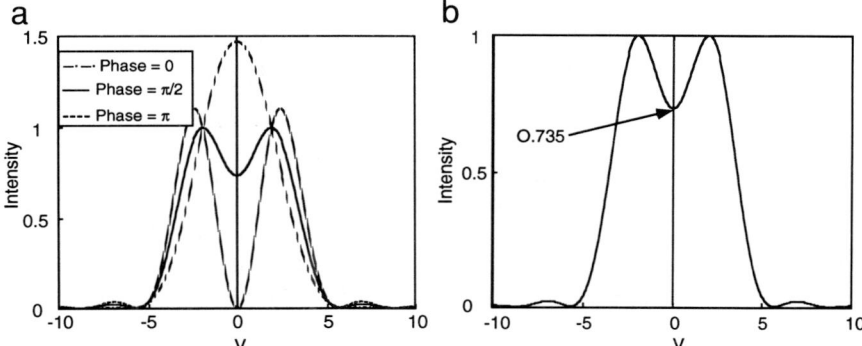

Figure 1.14 (a) Image intensity for two mutually coherent point sources separated by the Rayleigh distance. (b) Image intensity for two incoherent point sources separated by the Rayleigh distance.

power width is not always adequate. It is, therefore, common to employ the *Rayleigh criterion*, which states that two closely space illuminated points are distinguishable from each other if the maximum response to point A is located at the zero of the response to the point B.

Rayleigh Criterion When the illuminating signals have a phase difference ϕ, the intensity $I_c(v)$ at a point a normalized distance, $v = 2\pi r \sin\theta_0/\lambda$, from the midway point is

$$I_c(v) = \left| \frac{J_1(v - 1.91)}{v - 1.91} + e^{j\phi} \frac{J_1(v + 1.91)}{v + 1.91} \right|^2. \tag{1.21}$$

In Eq. (1.21), the subscript c denotes coherent single-frequency illumination, and it should be noted that $J_1(3.83) = 0$.[8,14] Equation (1.21) is plotted as a function of v in Fig. 1.14(a) for three different cases: $\phi = 0$, $\phi = \pi/2$, and $\phi = \pi$.

If the illumination of the two points is π out of phase, there must be a zero response at $v = 0$ midway between the two points, and the two points are easily distinguishable. If the illuminating fields are in phase, the intensity at $v = 0$ is 1.47 times the intensity at each of the points $v = \pm 1.91$, and the two points cannot be distinguished from each other. When the illuminating fields are $\pi/2$ out of phase, the intensity at $v = 0$ is 73.5% of the maximum intensity, and the points are distinguishable.

Normally, incoherent illumination is used with a standard microscope. In this case, the intensity $I_S(v)$ is given by the relation

$$I_S(v) = \left[\frac{J_1(v-1.91)}{v-1.91}\right]^2 + \left[\frac{J_1(v+1.91)}{v+1.91}\right]^2, \qquad (1.22)$$

where the subscript S denotes a standard microscope with incoherent illumination. This expression is plotted as a function of v in Fig. 1.14(b).

The intensity at $v = 0$ is 73.5% of its maximum value. The Rayleigh definition is therefore often stated in the form: *two points of equal brightness can be distinguished if there is a* 26.5% *dip in intensity between them*. This definition corresponds to the two incoherently illuminated points being separated by a distance d_S(Rayleigh), defined as

$$d_S(\text{Rayleigh}) = \frac{0.61\lambda}{N.A.}. \qquad (1.23)$$

Sparrow Criterion An alternative definition of two-point resolution, the *Sparrow criterion*, states that *two points of equal brightness can be distinguished if the intensity at the midway point is equal to that at the points*. Such a criterion is much more general in application than simply placing one point at the zero response to the other point.[7,14] It can apply equally well to coherent imaging or, for example, to a Gaussian beam which has no sharp spatial zero in response. The distance between two neighboring points which are just distinguishable using the Sparrow criterion in an incoherent imaging system is

$$d_S(\text{Sparrow}) = \frac{0.51\lambda}{N.A.}. \qquad (1.24)$$

We shall use this criterion as well as the Rayleigh criterion for determining the effective definition of the confocal microscope. It will be observed that, for incoherent imaging, the Sparrow criterion is identical to the single-point 3-dB definition. The reason is that it corresponds to the signal intensities of the two points adding to unity at the center point; thus, each contribution must be half its maximum value. For coherent imaging of two in-phase points, the Sparrow definition is approximately 1.5 times as large as for incoherent imaging.

1.2.7 Brightness of the Image

We have discussed various ways of defining the resolution of an imaging system and have been concerned with how the definition is af-

1.2 The Standard Optical Microscope

fected by the coherence of the light source and by the spatial frequency response of the lens system. However, good definition alone is not adequate to see the image; it also must be bright enough to observe, and we must choose the best mode, reflection or transmission imaging, fluorescence, phase contrast, and so on, with which to observe an object. In this section, we shall define brightness and show how it depends on the numerical aperture of the objective lens.

The *brightness* of an image is defined as the power per unit surface area emitted from a point on a light-emitting object which enters an aperture subtending a solid angle Ω at this point. Thus, if δP is the power emitted from an elemental area δS of the source into a solid angle $\delta \Omega$, the *photometric brightness* of an object is defined by the relation

$$\delta P = I \delta S = B \cos \theta \, \delta S \, \delta \Omega. \tag{1.25}$$

In Eq. (1.25), B is the photometric brightness of the object, θ is the angle between the ray and the normal to the surface, and the $\cos \theta$ term accounts for the projection of the elemental area δS in the θ direction. The intensity I is defined as $\delta P / \delta S$ and thus gives the apparent power per unit area emitted from the source. For diffuse reflectors or emitters, we can regard the brightness as being independent of angle θ so that the radiated power has a cosine variation with angle θ, a property known as *Lambert's law*.[8]

For simplicity, we will consider a critically illuminated reflecting microscope. In this case, the illumination source fills the back pupil plane of the objective with light. We will denote the coordinates between the source and the lens with a prime (′). Integrating Eq. (1.25) with respect to $\delta \Omega' = \sin \theta' \, \delta \theta' \delta \varphi'$ with the limits on θ' and ϕ' from 0 to θ_0' and 0 to 2π, respectively, the power incident on the sample from an element of area $\delta S'$ on the illuminating source is

$$\delta P = \pi B_S \sin^2 \theta_0' \delta S', \tag{1.26}$$

where B_S is the brightness of the source and θ_0' is the maximum ray angle subtended by the condenser lens at the source. This power is focused onto an area δS on the sample.

We can express Eq. (1.26) in sample coordinates by using the area relationship between the source and the sample, $M_r^2 \delta S = \delta S'$, and the sine condition, $n \sin \theta_0 = M \sin \theta'$. With these, substitutions, Eq. (1.26) becomes

$$\delta P = \pi B_S n^2 \sin^2 \theta_0 \, \delta S. \tag{1.27}$$

It then may be concluded that the power per unit area, or intensity of the reflected light, is

$$I = \pi B_S R_P (N.A.)^2, \tag{1.28}$$

where R_P is the power reflectivity of the sample. Thus the numerical aperture is a measure of both the light-gathering ability and the definition of a lens.

In the standard optical microscope, the reflected light travels back through the objective lens to the image plane. To calculate the power per unit area at the image plane, we follow the same arguments which led to Eq. (1.26). In this case, however, the brightness of the source in Eq. (1.26) is replaced by the reflected intensity of the sample, Eq. (1.28). With this substitution, and using the sine condition to express the result in terms of the numerical aperture of the objective, we obtain the result,[15]

$$\text{Image intensity} = \frac{\delta P}{\delta S'} = \frac{\pi B_S R_P (N.A.)^4}{M_r^2}. \tag{1.29}$$

A small increase in the numerical aperture of the objective will, thus, dramatically improve the image intensity, as well as improving the resolution. For this reason it is often preferable to work with as high a numerical aperture lens as possible at a given magnification. These requirements become critical for weakly reflecting and fluorescent samples. It should be noted that Eq. (1.29) is valid for Köhler illumination as well as critical illumination.

1.2.8 Imaging Techniques with the Standard Optical Microscope

Brightfield Imaging The range of applications of the standard optical microscope has led to the development of numerous imaging modes, most of which have been adapted to the CSOM and interference microscope. The most common is known as *brightfield* imaging, where images are produced by uniformly illuminating the entire sample so that the specimen appears as a dark image against a brightly lit background. Brightfield imaging is used as a general imaging technique for observation and inspection of samples. An example of a brightfield image of a box-in-box overlay structure used for measuring pattern registration on integrated circuits is shown in Fig. 1.15. This type of pattern and the metrology requirements for integrated circuit registration measurements are discussed in Chapter 5.

Darkfield Imaging An alternative technique, known as *darkfield* imaging, is a useful method of visualizing small particles and fine lines

1.2 The Standard Optical Microscope

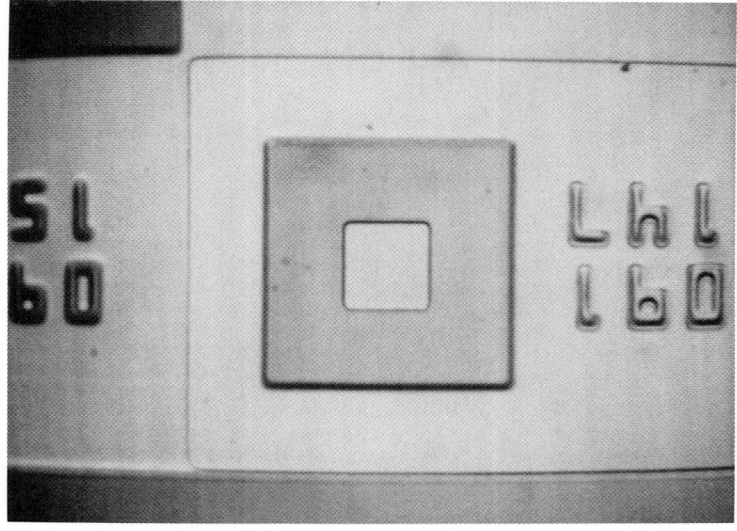

Figure 1.15 An example of a brightfield image of a box-in-box overlay structure.

in the microscope. In its most common implementation, the sample is illuminated with a hollow cone of light which is larger than the acceptance angle of the objective, as illustrated in Fig. 1.16. In this case specular reflectors do not reflect light into the objective and only the light which is scattered into the objective by particles or the edges of the sample is imaged. In a darkfield image, structures appear as bright lines against a dark background. An example of a darkfield image is shown in Fig. 1.17. This image is of a regular pattern of 2.4-μm lines on a silicon substrate. Only the edges of the lines and some small dust particles appear in the image.

Fluorescent Imaging A third technique, important for biological applications, is fluorescent imaging. In fluorescent imaging, the sample itself provides the light source used to form the image. Typically a specific area of a biological specimen is tagged with a molecule capable of producing fluorescent light in response to illumination by a shorter wavelength. The sample is then uniformly illuminated with blue or ultraviolet light to excite the fluorescent molecules. Radiation is emitted by these molecules at a longer wavelength than the illumination light, usually green or yellow,

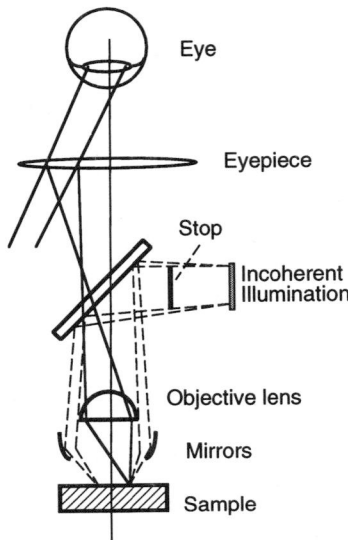

Figure 1.16 Schematic illustration showing the principle of darkfield imaging.

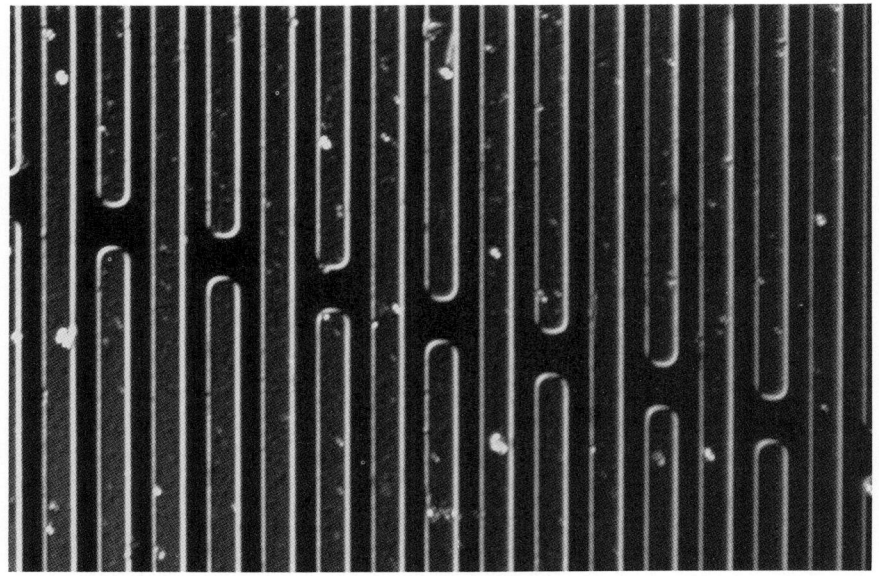

Figure 1.17 An example of a darkfield image of 2.4-μm lines on silicon.

1.2 The Standard Optical Microscope

and detected by the imaging system. A dichroic barrier filter is used to separate the illumination from the fluorescent signal.

Fluorescent molecules are available which adhere to, or tag, biological molecules with specific functions in the cell. Thus the location and material of the nucleus or other sites of specific biological activity can be observed. In fluorescent imaging, the objective lenses must have low chromatic aberration and high transmission at both the excitation and emission wavelengths. Furthermore, the fluorescent signals are often weak, which implies that the signal-to-noise ratio can be low. In this situation, sensitivity and glare from extraneous sources are major problems. In addition, if oil immersion lenses are used, the oil itself can fluoresce. Consequently, for fluorescent imaging with high-$N.A.$ lenses, laboratory-quality glycerin is used as the immersion fluid to minimize the auto-fluorescence. Fluorescent imaging will be discussed in more detail in Chapters 3 and 5. An example of a fluorescent image is given in Fig. 1.23.

Polarization Imaging Magnetic and birefringent films represent another class of samples commonly studied with an optical microscope.[16] With these samples, the plane of polarization of the incident light is rotated when it passes through or is reflected from the object. This phenomenon can be used to produce an image. In the most common configuration, a polarizer in a rotating mount is placed in the illumination path so that the incident polarization may be arbitrarily varied. A fixed or rotating analyzer is then used in front of the eyepiece to block light with the original polarization.

Phase Contrast Imaging The techniques described above rely on intensity changes caused by a change in transmission, reflectivity, fluorescence, or scattering to form an image. An alternative method is to image the phase rather than the amplitude of the transmitted or reflected light. Fritz Zernike received a Nobel Prize for his pioneering work on phase imaging with an optical microscope in 1935.[17] A simplified schematic of a Zernike phase microscope is shown in Fig. 1.18.[18] In Zernike phase contrast microscopy the object is illuminated by an annular source, and a phase plate is positioned in the back focal plane of the objective lens. The phase plate generally consists of a transparent disk in which the annulus has a smaller optical path length than the rest of the plate.

When light strikes the sample, a portion of the illumination is diffracted into higher order spectral components. The phase plate retards the phase of the undiffracted illumination by either $\pi/2$ or $3\pi/2$ radians relative to the diffracted light. The phase annulus is also usually darkened slightly in order to render the diffracted and undiffracted portions of the light

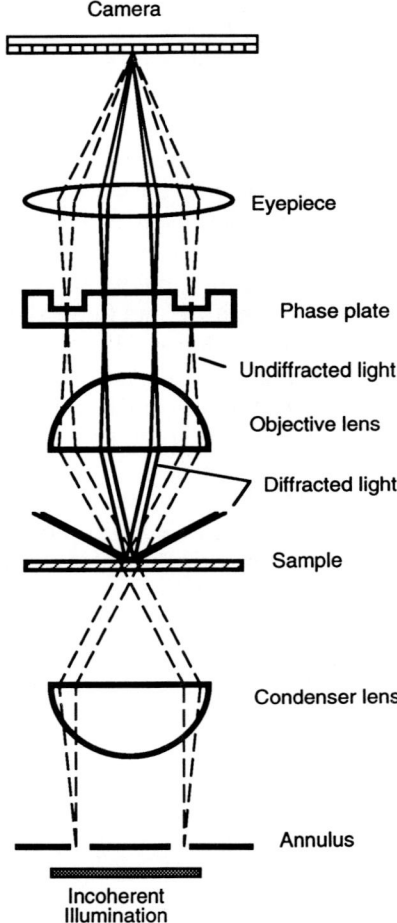

Figure 1.18 Simplified schematic of the principle underlying a Zernike phase contrast microscope.

energy roughly equal. When these two portions of the beam interfere at the detector, an image of the phase object is formed. The theory of phase imaging is discussed in more detail in Chapter 4.

Differential Interference Contrast Imaging Another important phase imaging technique called *differential interference contrast* (DIC) imaging was developed by G. Nomarski in 1955.[19] In its simplest form a Wollaston

1.3 The Confocal Microscope

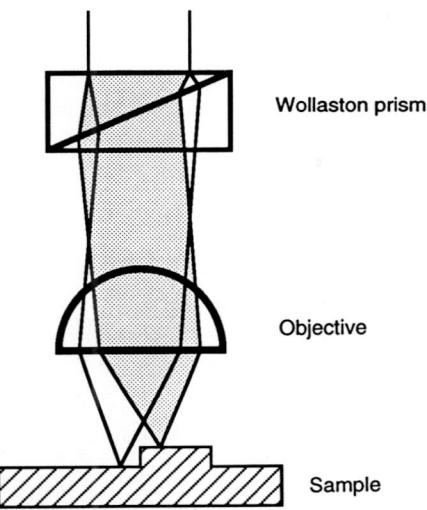

Figure 1.19 Simplified schematic of a differential interference contrast imaging system showing the Wollaston prism and objective lens.

prism, which is made from two bonded prisms of birefringent material, is used to produce two closely spaced spots on the sample, Fig. 1.19. When the two spots are reflected from different sides of a step on the sample, there will be a phase difference between them causing the step to appear brighter or darker than the background in the image. An example of a DIC image taken on a CSOM is shown in Fig. 1.20. This image was made on the same overlay structure as shown in Fig. 1.15. DIC imaging on both the standard and confocal scanning microscope will also be discussed in Chapter 4.

We will now turn our attention to the main subject of this book, the confocal scanning optical microscope.

1.3 The Confocal Microscope
1.3.1 Principle of Operation

The *confocal scanning optical microscope* (CSOM) differs from a conventional optical microscope in that it illuminates and images the object one point at a time through a pinhole as described in Section 1.1. One variation of the CSOM called the *confocal scanning laser microscope* (CSLM) is illustrated in Fig. 1.21. In this instrument, a collimated beam

Figure 1.20 An example of a DIC image of a box-in-box overlay structure.

from a laser is focused onto a pinhole which acts as a spatial filter. Light passes through the pinhole to a microscope objective and forms a diffraction-limited spot on the sample. A beamsplitter then deflects the reflected beam to a separate detector pinhole. The use of two pinholes eliminates the reflected light from the diaphragm surrounding the illumination pin-

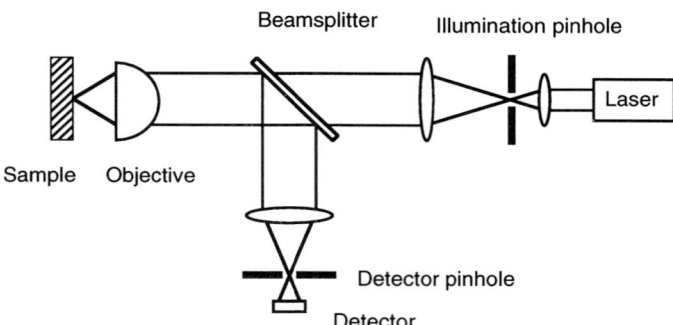

Figure 1.21 A CSLM employing two pinholes in the optical system.

1.3 The Confocal Microscope

hole. When the sample is moved out of the focal plane of the lens, the reflected light reaching the pinhole is defocused and does not pass through it, as shown in Section 1.1. Consequently, the light intensity received by a detector behind the pinhole decreases rapidly with the defocus distance, and the image disappears.

The microscope is *confocal* because the objective lens is used twice, both to illuminate and to image the sample. Since only one point is illuminated at a time, speckle is eliminated, but the sample or illumination beam must be raster scanned and the image must be built up pixel by pixel like a television picture. For these reasons the instrument is called a *scanning* optical microscope. In summary, the basic requirements for the CSOM are point illumination, point detection, a scanned image, and a confocal lens system.

Strictly speaking, the word confocal can apply to both a standard optical microscope and the CSOM. Both instruments can use a high-numerical-aperture lens to illuminate and image the sample; thus both microscopes fit the definition of confocal, two lenses sharing a common focus. The terminology confocal microscope as a synonym for the CSOM is, however, firmly fixed in the literature so we will use it throughout this book.

1.3.2 Scanning

In a CSOM either the sample or the illuminating beam must be scanned in order to build up the image point by point. Sample, objective, and beam scanning techniques all have been developed. Sample scanning typically takes of the order of 10 s per frame. It is a good choice for research on the instruments themselves, because it allows the optical system to be easily modified to produce different imaging configurations. Alternatively, the objective lens can be scanned. This technique is rarely used because it is difficult to maintain uniform illumination across the field of view.

The majority of commercial CSOMs employ some form of beam scanning, which is typically much faster than sample scanning. In addition, because the scan is demagnified by the objective lens, the mechanical tolerances on beam scanning systems are less critical than those for sample scanning. The simplest form of beam scanning is to raster scan the pinhole, or replace the pinhole with a single-mode optical fiber and scan the fiber.[20,21] A faster method is to scan the optical beam using a galvanometer mirror, as illustrated in a simplified form in Fig. 1.22, or to use an acousto-optic cell for a fast scan in at least one direction. Another useful technique is to replace the single pinhole with a Nipkow disk, which contains a large

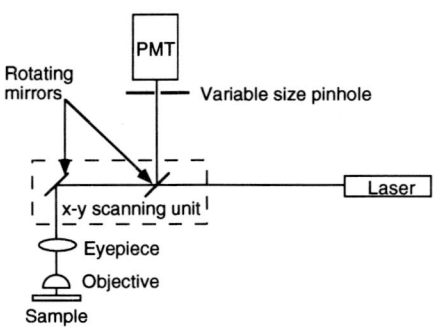

Figure 1.22 A simplified schematic of a beam-scanned CSLM.

number of pinholes. Many of these methods are capable of producing real-time images at video frame rates or faster. All these scanning techniques have been employed in various forms of commercial confocal microscopes. They will be discussed in more detail in Chapter 2.

1.3.3 Depth Response

The major advantage of the CSOM over a standard optical microscope is that a defocused image disappears in the CSOM, whereas it becomes blurred in a standard microscope. Before the development of the confocal microscope, to obtain good cross-sectional images of biological samples, it was necessary to slice the material into thin layers and mount each slice onto a microscope slide. A transmission microscope was usually employed to observe these thin samples.

The ability of the CSOM to use its depth resolution to eliminate reflections from glass slides and coverslips makes it possible to observe many types of samples with a reflection microscope without slicing them into thin sections. When examining relatively thick biological materials, a CSOM reflection microscope has the advantage that the details of an image are not obscured by glare from layers in front of or behind the region of interest.

With fluorescent imaging, the sample is illuminated at one wavelength which excites the fluorescent material, causing it to fluoresce at a longer wavelength than that of the incident light. The sample is observed through a filter which passes only the longer wavelength radiation. The ability to eliminate out-of-focus planes from the image removes the blurred fluorescent signal from regions of the sample where the beam is defocused.

1.3 The Confocal Microscope

Figure 1.23 illustrates this concept using two images of a human rib bone stained with brilliant sulphaflavine. Figure 1.23(a) is the image obtained with a standard optical microscope and Fig. 1.23(b) is the image obtained with a CSOM. Both images were made using an oil immersion lens on a microscope built by the Technical Instruments Company. Fine details of the bone structure can clearly be seen in the confocal image, whereas in the standard microscope image these details are obscured by fluorescent light originating from behind and in front of the layer of interest.

The range resolution of the CSOM makes it possible to measure quantitatively the profiles of features on the sample. The equation describing the form of the depth response can be simply understood. The receiving pinhole can be thought of as sampling a magnified reproduction of the fields on the axis of the objective lens. To measure the depth response, a mirror is moved axially through the focal plane of the lens. When the sample moves a distance z from the focal plane, the image of the illuminating pinhole moves a distance $2z$ away from the focal plane. It can be shown from Eq. (1.3) with $z \neq z_0$, $x = y = 0$, that the electric field amplitude varies along the axis of the lens approximately as $(\sin u/4)/(u/4)$,[8,22] where $u = 4nkz \sin^2(\theta_0/2) = 2nkz(1 - \cos\theta_0)$, and $k = 2\pi/\lambda$. Therefore, the amplitude variation, $V(z)$, of the light passing through an infinitesimal pinhole on the axis varies as

$$V(z) = \frac{\sin(u/2)}{u/2} = \frac{\sin[nkz(1 - \cos\theta_0)]}{nkz(1 - \cos\theta_0)}. \tag{1.30}$$

In Eq. (1.30), the normalized optical coordinate u is defined following the convention of Wilson and Sheppard.[13] This coordinate is slightly different from that used by Born and Wolf ($u = nkz \sin^2\theta_0$).[8] The two quantities, however, are equal in the paraxial approximation. Wilson's coordinate u was chosen because it is the most appropriate one for the CSOM and has the additional advantage that it is more accurate for high numerical apertures. The terminology $V(z)$, common throughout much of the literature, was first defined for the acoustic microscope, another type of confocal microscope.[23] In the acoustic microscope the voltage $V(z)$ is measured on an acoustic receiving transducer as a function of the defocus distance z.

The axial intensity $I(z)$ is proportional to $|V(z)|^2$. The measured depth response of a CSOM for an $80\times/0.90$ N.A. objective in air at $\lambda = 633$ nm is shown in Fig. 1.24. The depth resolution of the microscope is commonly defined as the distance between half-power points (3-dB points) of the intensity response given by the approximate formula,[24]

$$d_z(3 \text{ dB}) = \frac{0.45\lambda}{n(1 - \cos\theta_0)}. \tag{1.31}$$

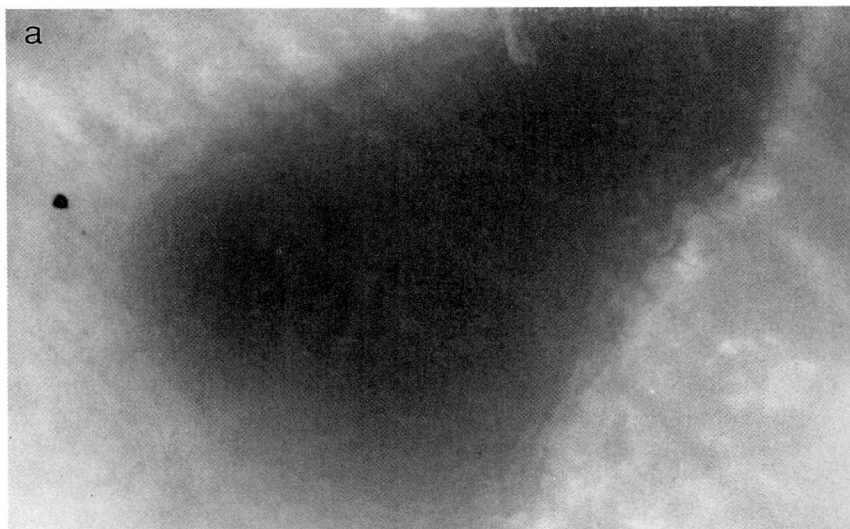

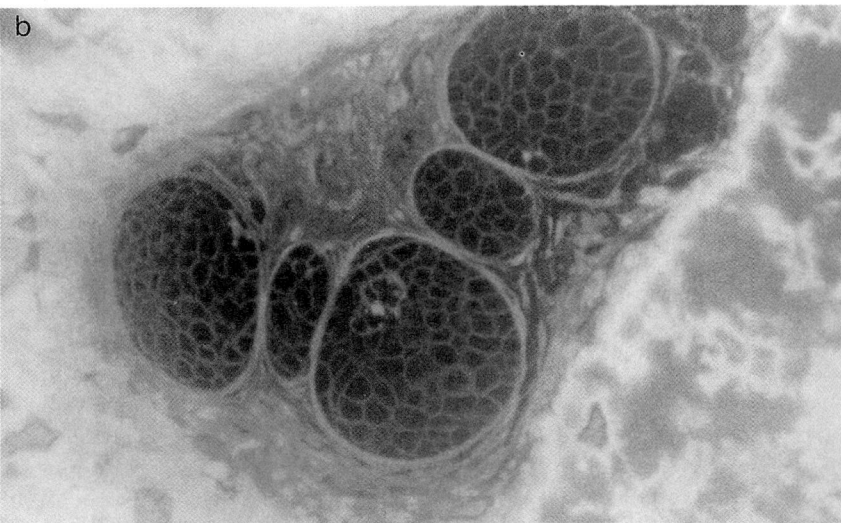

Figure 1.23 Fluorescent images of a human rib bone stained with brilliant sulphaflavine: (a) standard optical microscope image; (b) RSOM image. [Courtesy: G. Q. Xiao, "Confocal Optical Imaging Systems and their Applications in Microscopy and Range Sensing," Ph.D. Dissertation, Department of Physics, Stanford University, Stanford, California, USA (December 1989).]

1.3 The Confocal Microscope

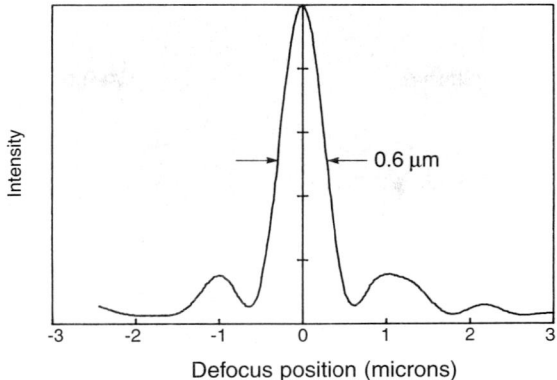

Figure 1.24 The measured depth response of a CSOM using a 80×/0.90 objective in air at λ = 633 nm.

In the paraxial approximation with $n = 1$, Eq. (1.31) reduces to the often quoted relationship

$$d_z(3 \text{ dB}) \approx \frac{0.89\lambda}{\sin^2 \theta_0} \approx \frac{0.89\lambda}{(N.A.)^2}. \tag{1.32}$$

Aberrations in the optical system have broadened the depth response shown in Fig. 1.24 and produced an asymmetrical sidelobe pattern. These results are typical of microscope objective lenses. The effect of lens aberrations on the depth response will be discussed in Chapter 3.

For a lens with an *N.A.* of less than ~0.4, the difference in the approximate width of the depth resolution calculated from Eqs. (1.31) and (1.32) is less than 5%. For larger numerical apertures, however, the difference can be greater than 30%. The 3-dB depth resolutions at various wavelengths for a lens with *N.A.* = 0.95 in air are calculated in Table 1.1, using Eq. (1.31).

Table 1.1 Depth Resolutions for *N.A.* = 0.95

Wavelength λ	Depth resolution $d_z(3 \text{ dB})$
633 nm	414 nm
546 nm	357 nm
436 nm	285 nm
365 nm	238 nm
248 nm	162 nm

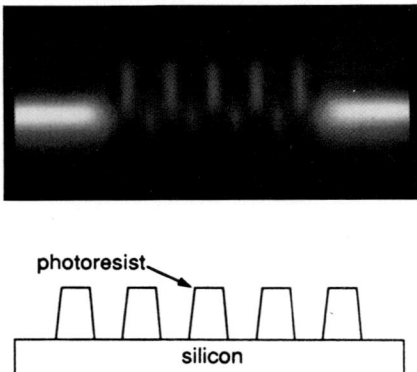

Figure 1.25 A cross-sectional image (cloud plot) of equal lines and spaces 0.6 μm wide in 1.0-μm-tall I-line photoresist on silicon. [Reprinted from Solid State Electronics **35**, T. R. Corle, "Submicron metrology in the semiconductor industry," 391–402. Copyright 1992 with kind permission from Elsevier Science Ltd., The Boulevard, Langford Lane, Kidlington OX5 1GB, UK.]

Cloud Plots The shallow depth response of the CSOM can be used to construct cross-sectional images optically, without physically cutting the sample. Due to their appearance, these optical cross sections are called cloud plots. To produce a cloud plot, the sample is scanned axially in the focus direction and the intensity variation at each pixel of a selected linescan is recorded. The resulting images record the intensity in the x–z plane. Figure 1.25 is a cloud plot image of equal lines and spaces 0.6 μm wide in 1.0 μm tall I-line (365 nm exposure wavelength) photoresist on silicon. Both the lines and spaces can be clearly seen in this image, the brightest regions corresponding to the positions of the reflecting surfaces. A detailed discussion of the important focus locations in this image and how they are used to generate measurements is given in Chapter 5.

1.3.4 The Point Spread Function and Two-Point Resolution

Let us now compare the imaging characteristics of a standard optical microscope with a CSOM. The main difference between these two instruments lies in the method of illumination and detection. The standard microscope uses a condenser lens to illuminate the entire sample uniformly from an incoherent light source. This type of illumination is spatially incoherent. A large-area detector, such as the human eye or a video camera made up of many pixels, detects the image. In the standard microscope, a

1.3 The Confocal Microscope

Table 1.2 Transverse Resolutions for $N.A. = 0.95$

Wavelength λ	Transverse resolution $d_{Cr}(3\ \text{dB})$
633 nm	248 nm
546 nm	214 nm
436 nm	171 nm
365 nm	143 nm
248 nm	97 nm

point on the sample is imaged as a diffraction-limited spot on the detector by the objective lens. The amplitude of the image field at the detector of a point object is given by the Airy function $h(r)$. Since the source is incoherent, in an ideal instrument the intensity PSF of the microscope is equal to the square of the amplitude response of the objective lens $I_s(r) = |h(r)|^2$.

In a CSOM, on the other hand, only one point on the sample is illuminated at a time through the objective. This point source illumination is spatially coherent. The amplitude of the illuminating field at the sample, rather than being uniform, is given by the PSF of the objective $h(r)$. In turn, the sample is imaged by the objective onto a point detector so that the amplitude PSF of this microscope is given by $A_C(r) = h^2(r)$, while the image intensity of a point is $I_C(r) = |h^2(r)|^2$, where the subscript C denotes a confocal microscope.

It is important to realize that for the CSOM, the amplitude PSF must be used in imaging calculations rather than the intensity PSF as in the standard microscope. For simple objects such as edges and points, the CSOM intensity image is the square of the intensity image produced by a standard microscope. The single-point resolution of the confocal microscope, defined as the width at the half-power points of the image of a point object, is

$$d_{Cr}(3\ \text{dB}) = \frac{0.37\lambda}{n \sin \theta_0} = \frac{0.37\lambda}{N.A.}. \tag{1.33}$$

This width is 73% of the single-point resolution of a standard optical microscope. The 3 dB transverse resolutions at various wavelengths for a lens with $N.A. = 0.95$ in air are calculated in Table 1.2, using Eq. (1.33).

Two-Point Resolution The exact formula for the two-point resolution of the CSOM is given in Chapter 3. Applying the criterion that two points can be distinguished when there is a 26.5% dip in amplitude between

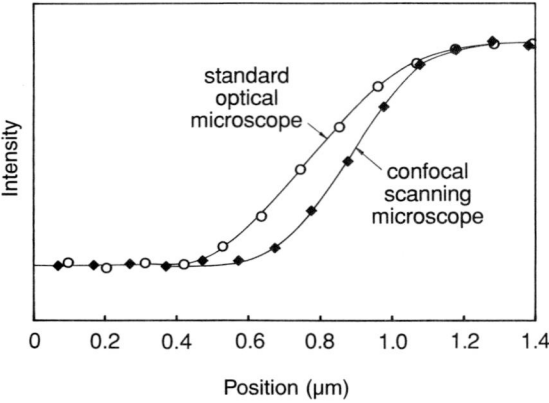

Figure 1.26 Comparison of the experimental edge response of a CSOM and a standard optical microscope.

them, the distance at which two points can be distinguished in the CSOM is only 8% less than that for the standard microscope. The Sparrow resolution is equal to that of the standard microscope as described in Chapter 3.

Edge Response Like the single-point response, the edge response of a CSOM is also the square of the response of a standard optical microscope. The improvement in resolution of the CSOM is illustrated in Fig. 1.26. In this figure, the experimental edge responses for both the RSOM and a standard microscope, using a 0.9 $N.A.$ lens at $\lambda = 550$ nm, are plotted. Because the PSF of the CSOM falls off more rapidly than that of the standard microscope, its edge response is also sharper. As expected, the 50% intensity point for the standard microscope corresponds to the 25% intensity point in a CSOM. The formula for the edge response is given in Chapter 3.

Transfer Function The amplitude response $A_C(r) = h^2(r)$ of the CSOM dictates the nature of the image. Since this amplitude response is identical to the intensity response of a standard microscope; it follows that the CTF of the CSOM is identical to the OTF of the standard microscope; that is, it has the same triangular shape as shown in Fig. 1.12. Thus, the spatial frequency response of the CSOM is the same as for the standard optical microscope and extends out to twice the frequency of a simple coherently illuminated lens.

1.3.5 History of the CSOM

Early History The ideas which led to the development of the modern CSOM first began to take shape in the early 1950s. In 1951 Young and Roberts suggested using the scanned spot of a cathode ray tube as the light source in order to build a scanning optical microscope.[25,26] The light was transmitted through conventional microscope optics to produce a single spot of light on the sample. Their microscope was designed for biological imaging but suffered from signal-to-noise problems due to the weak light source.

Another early contribution was made by Minsky, who was interested in artificial intelligence and wanted to obtain three-dimensional information about the brain. He realized that this information could not be obtained with a standard microscope because of its poor cross-sectioning ability. To get over this difficulty, he invented the confocal microscope in 1957. Minsky accurately described the salient features of this type of imaging in his patent and has recently written an extremely interesting article describing this early work and its problems.[27,28] Again, the lack of an adequate light source prevented the full development of his system. He also points out that the way in which images are displayed is of crucial importance. Although the definition was good, he made the mistake of using too much magnification by displaying the images on a large cathode ray tube, an unfamiliar format. Since his line scan was more closely spaced than the width of a single spot, the images looked "fuzzy" and thus did not impress his viewers with the sharpness of the results.

The development of the laser solved many of the problems encountered by the early researchers. Davidovits and Egger used a laser in 1969 to develop a working CSOM.[29] In 1979, Brakenhoff and his coworkers demonstrated a well-engineered CSLM, showing the improvement in definition and contrast that could be obtained with confocal imaging and demonstrating its use in biological applications.[30,31]

Much of the recent work in CSOMs was also stimulated by the development of the *scanning acoustic microscope* (SAM) by Lemons and Quate.[32] The SAM demonstrated that high-quality acoustic images free of speckle could be obtained with a coherent source, by scanning a single focused transducer to image only one point at a time. Kompfner, who was involved with the early acoustic work, suggested that this imaging technique might be excellent for observing internal features of translucent materials, such as body tissue, and started a group with Wilson and Sheppard at Oxford to try similar scanning microscope principles with optics.[13] Their experiments and theoretical analyses produced major improvements

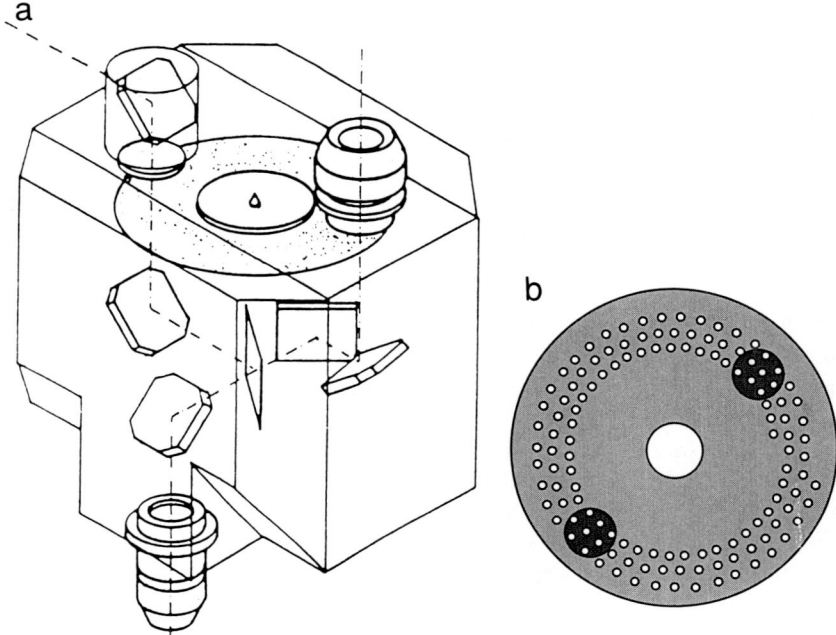

Figure 1.27 (a) The TSRLM; (b) the Nipkow disk showing the illuminating and received beams on the disk. [(a): From M. Petran, A. Boyde, and M. Hadravsky, "Direct view confocal microscopy," *Confocal Microscopy* T. Wilson, editor (Academic Press, 1990), with permission.]

in the understanding of CSOM design. For this reason much of the theory and terminology for CSOMs has its foundations in acoustic microscopy.

Nipkow Disk Microscopes At about the same time that Davidovits and Egger were building their microscope, Petran and Hadravsky, in Czechoslovakia, invented a different type of CSOM which operated in real time and produced an image which could be directly observed with the naked eye.[33,34] They suggested the use of a Nipkow disk to fulfill the requirements of point illumination and point detection. The Nipkow disk, in which a modulated light beam passed through holes drilled along a single spiral in a rotating disk, was invented in Germany in 1884 and was used in the first form of mechanically scanned television. This basic idea was adapted to provide the raster scan in a CSOM by using a disk consisting of an opaque disk into which many thousands of pinholes are drilled or etched in spiral patterns, as shown in Fig. 1.27.[35] Each illuminated

1.3 The Confocal Microscope

pinhole on the disk is imaged by the objective to a diffraction-limited spot on the sample. The light reflected from the sample can be seen in the eyepiece after it has passed back through a conjugate pinhole in the Nipkow disk. Several thousand points are simultaneously illuminated on the disk, achieving, in effect, several thousand confocal microscopes all running in parallel. Spinning the disk fills in the spaces between the holes and creates a real-time confocal image.

Petran and Hadravsky called their invention the *tandem scanning reflected light microscope* (TSRLM). In their original system the pinholes were etched into a thin sheet of copper foil. Several hundred pinholes were illuminated at one time. Typically, since only about 1% of the area of the disk is transparent, a relatively intense light source is required. Initially either the sun or a mercury vapor or other bright arc lamp was used as the light source. Later, as the microscope improved, a filament lamp could be used as the illumination source.

A major challenge encountered in these types of microscopes was to eliminate the reflected light from the top surface of the Nipkow disk. Since the transmission of the disk is only of the order of 1–2%, this light can easily cause enough glare to obscure the image of the object. The technique used by Petran and Hadravsky was to pass the light through one set of pinholes in the disk and return the reflected light through a conjugate set of pinholes on the opposite side, as shown in Fig 1.27(b).

The advantages of the TSRLM over a mechanically scanned CSOM for viewing samples are immediately apparent to the user. As the focus knob is turned, different features come into view in real time. The disadvantages, however, are poor light efficiency, mechanical and optical complexity, and the fact that the alignment between the input and output pinholes is highly critical. For these reasons, and the problems associated with developing this technology in Czechoslovakia and transferring it to the West during the late 1960s, in the 20 years since its invention very few of these microscopes were constructed. However, a commercial version is now available.

Stimulated by Petran's work, a simplified form of the tandem scanning microscope, called the *real-time scanning optical microscope* (RSOM) was developed by Xiao, Corle, and Kino. This microscope, illustrated in Fig. 1.28, uses the same set of pinholes for both illumination and imaging, reducing the number of optical components and simplifying the alignment.[36] To eliminate the light reflected from the top of the Nipkow disk, it is made of low-reflectivity black chrome deposited on glass. In addition, the highly polished disk is tilted so that the reflected light strikes a stop in the microscope. As a further step, the illumination

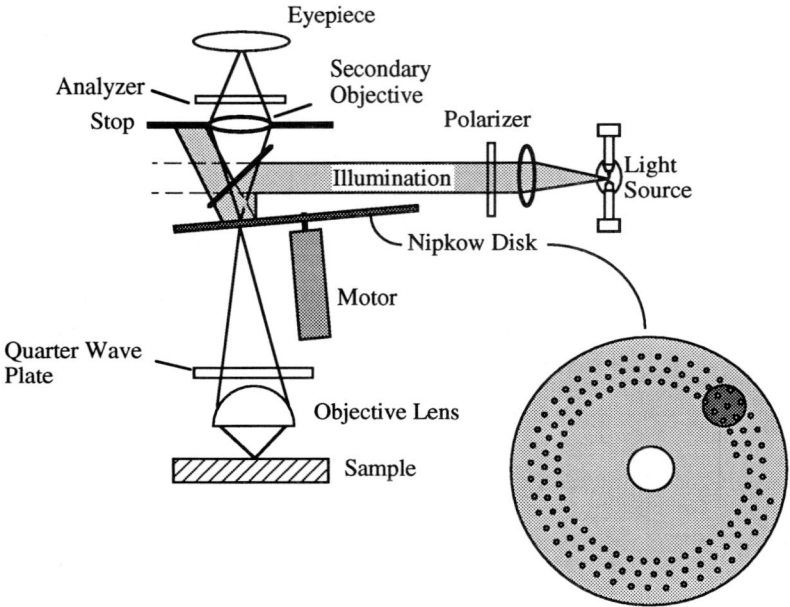

Figure 1.28 A simplified schematic of the RSOM.

is polarized, and an analyzer is placed in front of the eyepiece to eliminate any remaining light reflected from the disk. A quarter-wave plate placed near the objective rotates the polarization of the image to pass it through the analyzer.

1.4 Optical Interference Microscopes
1.4.1 Principle of Operation

We have described how the confocal microscope can be used to obtain good range resolution and cross-sectioning ability. Optical interference microscopes can also accomplish the same purpose by detecting the interference signal between beams reflected from the sample and a reference surface. Several types of interferometric microscope objectives have been developed. These lenses are based on the configurations of Michelson, Linnik, Mirau, and Tolanski interferometers. In this book we will primarily consider interference microscopes based on the Linnik and Mirau interferometers. We will describe applications of both these configurations for measuring surface roughness and height variations. In addition, two types

1.4 Optical Interference Microscopes

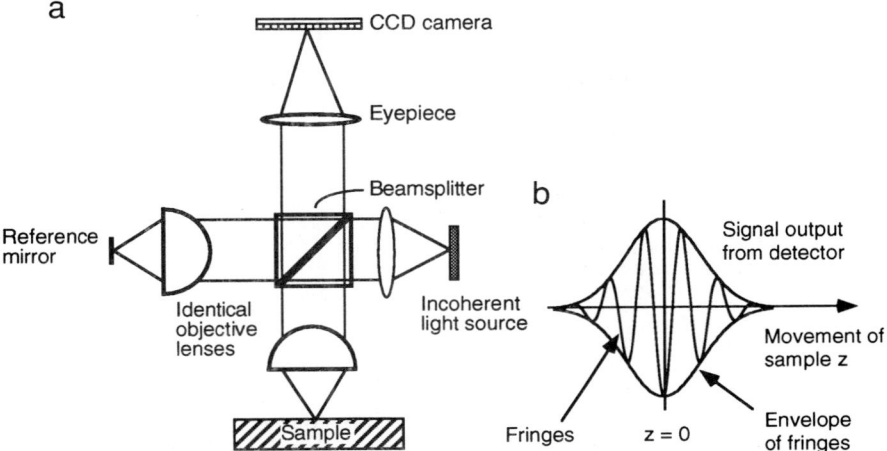

Figure 1.29 (a) A simplified schematic of the coherence probe interference microscope; (b) the detector signal output as a function of position z for an interference microscope. The envelope of this fringe signal is also shown.

of imaging microscopes developed for semiconductor metrology applications: the *coherence probe microscope* (CPM), based on the Linnik configuration, and the *mirau correlation microscope* (MCM) based on the Mirau interferometer will be discussed in detail.

The Linnik Microscope The Linnik interferometer-based microscope is in widespread commercial use. It is used by both the Wyko and Zygo corporations in versions of their microscopes, and in the CPM built by Davidson et al. at KLA Instruments. A simplified schematic of the Linnik microscope is shown in Fig. 1.29.[37,38] In this microscope, a spatially and temporally incoherent light source such as a filament or xenon arc lamp is split into two beams by a beamsplitter near the objective. Part of the light passes through an objective lens to the sample, while the remainder passes through an identical lens to the reference mirror. Light reflected from the object and the reference mirror travels back through the objectives, beamsplitter and interferes at a CCD detector array.

In this arrangement, one point on the spatially incoherent source is imaged to both a spot on a reference mirror and a spot on the sample; these spots in turn are imaged onto the detector plane. The detector gives an output signal which contains the product of the two images, so that

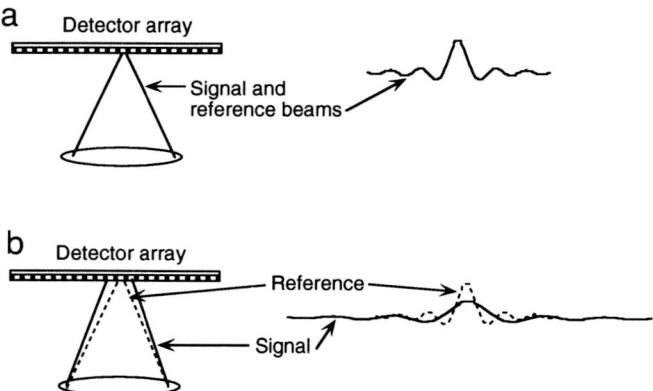

Figure 1.30 An illustration of the signal and reference beams in an interference microscope with plots of their field variations across the detector. (a) A focused signal beam; (b) a defocused signal beam.

the focused reference spot acts much like the pinhole in a confocal microscope by limiting the area on the detector over which the beam reflected from the sample produces a signal. As shown in Fig. 1.30(a), when the object is in focus, the two spots at the detector overlap perfectly and the maximum signal output is obtained from the detector. As the sample moves out of focus, Fig. 1.30(b), the spots no longer overlap each other perfectly and there is a phase difference between the two beams due to their different path lengths. Therefore, a product signal is obtained that takes the form of an interference or fringe pattern in the z direction whose envelope falls off in almost exactly the same way as the $V(z)$ term in the confocal microscope, Fig. 1.24.

An additional feature of great importance in the interference microscopes is that the fringes contain phase information which makes it possible to obtain depth measurements with accuracies of a small fraction of a wavelength. Therefore, these microscopes can also be used for measuring surface roughness and depth profiles with far more sensitivity than would be possible with a confocal microscope. The disadvantage of this type of microscope is that to obtain complete quantitative information, electronic processing is required for each pixel in the image.

As has been discussed above, the two signals, $S(x,y,z)$ from the sample and $R(x,y,z_0)$ from the reference mirror interfere at the detector only if their beam paths are almost identical. In this case the output current from one element of the CCD detector is of the form

$$I(x,y,z) = |S|^2 + |R|^2 + 2|SR|\, g(x,y,z - z_0) \cos \varphi(x,y,z - z_0). \tag{1.34}$$

1.4 Optical Interference Microscopes

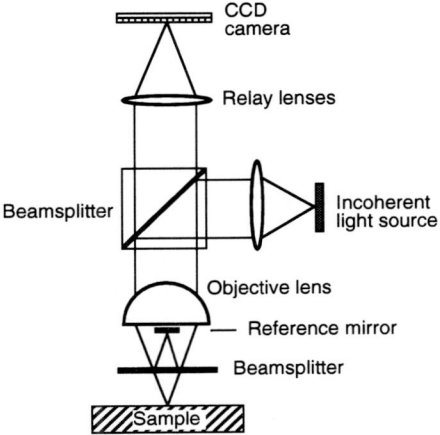

Figure 1.31 A simplified schematic of the Mirau correlation interference microscope.

In Eq. (1.34) z is the distance of the reflecting point from the focus, z_o is the distance of the reference mirror from the focus, $g(x,y,z - z_0)$ is the envelope of the correlation function for a spatially incoherent light source, and the term $\cos \varphi(x,y,z - z_0)$ represents the interference fringes. The signal is electronically processed to keep only the product term between the sample and reference signals. The signal processing techniques will be introduced in Section 1.4.2 and further discussed in Chapter 4.

The CPM is able to detect both amplitude and phase, and it is insensitive to range aberrations of the objective if identical lenses are used in the reference and beam paths. Its disadvantages are that electronic signal processing is required to obtain an image, and two identical objectives are required. Furthermore, because of the long beam paths involved the microscope must be built massively to avoid vibration problems.

The Mirau Correlation Microscope A related microscope, illustrated in Fig. 1.31, uses a Mirau interferometer consisting of a beamsplitter and a reference mirror positioned between the objective and the sample to generate the interference signals.[39] Because of the relatively short beam paths, the Mirau interferometer is less sensitive to vibrations than the Linnik interferometer. In addition, only one lens is required, although this lens must have a relatively long working distance in order to accommodate the beamsplitter and reference mirror. In addition, the beamsplitter used below the objective in the Mirau interference microscope can cause aberrations with a wide-aperture objective for which $N.A. > 0.5$, so that it is more difficult to work with.

Nikon instruments makes a Mirau objective which is used by Wyko and Zygo corporations in their microscopes. It employs a glass beamsplitter with an objective compensated for aberrations. A high numerical aperture Mirau microscope has been built by Chim et al. at Stanford University, which they call the *Mirau correlation microscope* (MCM). This device uses a beamsplitter only 800 Å thick made of a low-stress silicon nitride film grown on silicon. The reference mirror was built by depositing a platinum disk on a second silicon nitride film. These films introduced very little aberration into the 0.8 $N.A.$ beam which they used.[40] At the beamsplitter, approximately half the power of the converging beam is directed to the reference mirror; the other half is directed toward the sample. The reference mirror is positioned at the focal plane of the objective, while the sample is scanned along the vertical or z axis by a computer-controlled piezoelectric pusher. A CCD camera is used to obtain the interference signals, which are further processed to form the image. The design of the beamsplitter in the MCM is discussed in more detail in Chapter 2.

The reader will note that, for simplicity, the examples we have discussed use a critical illumination source with one point on the object giving rise to a focused spot on the detector which spatially correlates with a similar focused spot from the corresponding point on the reference mirror. In practice, a Köhler illumination system is employed to give uniform illumination and to make the device as similar as possible to a standard microscope. In this case, a point on the sample may be regarded as being spatially uncorrelated with neighboring points. The corresponding point on the reference mirror will be illuminated by the same light components from the source, which travel through the same path lengths to the corresponding point on the detector array. The spots formed at the detector array will have the same spatial and temporal variations and so will be perfectly correlated with each other. However, if the sample is out of focus, the situation shown in Fig. 1.30 will occur; the spatial and temporal correlation of the two spots will not be perfect and the output correlation signal will decrease just as for critical illumination.

1.4.2 Signal Processing Techniques

To measure the amplitude and phase of the reflected light in an optical interference microscope, either the sample can be tilted so that the interference fringes are visible or the reference or the sample can be scanned axially in the z direction. Scanning the sample produces an interference

1.4 Optical Interference Microscopes

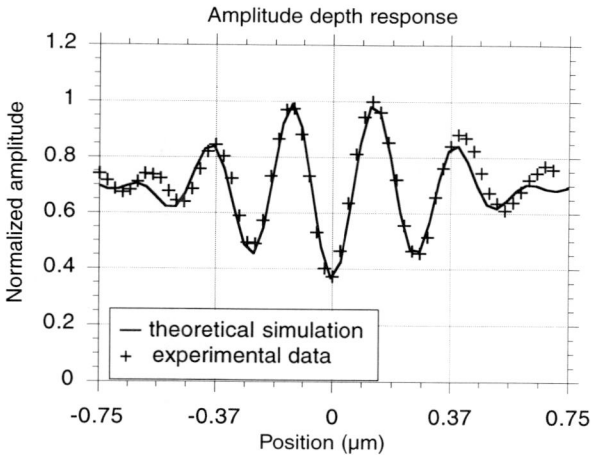

Figure 1.32 Comparison between the experimental and theoretical depth responses of the MCM using an 80×/0.8 N.A. objective lens at λ = 450 nm. [Courtesy: S. S. C. Chim, "The Mirau correlation microscope—a new tool for optical metrology," Ph.D. Dissertation, Department of Electrical Engineering, Stanford University, Stanford, California, USA, (June 1991).]

image that is similar to the depth response of a CSOM. The difference is that the signal has interference fringes superimposed on the structure, as illustrated in Fig. 1.29(b). The amplitude variations of the interference fringes reach a maximum when the reference and sample path lengths are identical, although this point is marked by an actual minimum in the signal. The minimum occurs because the beams which reflect off the two faces of the beamsplitter are π out of phase. An experimental depth response obtained with the MCM using an 80×/0.8 N.A. objective lens at λ = 450 nm is shown in Fig. 1.32 and compared with theory. The theory is calculated for N.A. = 0.75, in order to compensate for the attenuation of the large-angle rays in the particular objective lens used in the experiment. The agreement between experiment and theory is excellent.

Several different techniques have been used to calculate the envelope function of the depth response, $g(x,y,z - z_0)$. The technique invented by Davidson for the CPM assumes that the sample, which is typically a trench or line of photoresist on silicon, is uniform in one direction, i.e., the y direction.[41] If the sample is tilted by a small angle ψ, then a scan of the CCD data in that direction automatically produces a scaled version of the fringes. The change in y is equivalent to a change in z. By measuring the maximum peak-to-peak excursion of this scan, the function $g(x,y,z - z_0)$ can be calculated. The width of the envelope is defined as the depth

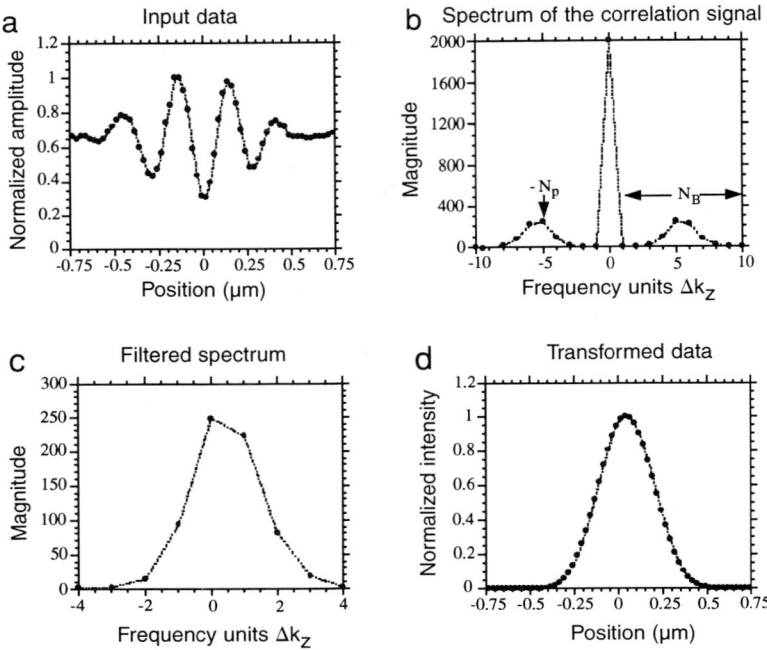

Figure 1.33 A flowchart summarizing the Fourier transform data analysis technique of Chim et al. (a) Input data from the microscope, (b) spectrum of the input data, (c) the filtered spectrum, and (d) the amplitude of the filtered data in the space domain. [Courtesy: S. S. C. Chim, "The Mirau correlation microscope—a new tool for optical metrology," Ph.D. Dissertation, Department of Electrical Engineering, Stanford University, Stanford, California, USA (June 1991).]

of focus of the microscope. This technique has the advantage that no mechanical scanning in the z direction is required. Its limitation is that the sample must be uniform in one direction.

An alternative technique is used in the MCM. The product term in Eq. (1.34) is a modulated cosine which has an envelope with a maximum amplitude located at $z = z_0$. Chim and Kino regard the carrier wave (the cosine term) as the sum of two exponentials, $\exp(j\phi)$ and $\exp(-j\phi)$, and use signal processing to eliminate one of these terms. The magnitude of the signal $g(x,y,z - z_0) \exp(j\phi)$ then depends only on the magnitude of the envelope $g(x,y,z - z_0)$, and the phase can be determined accurately from the exponential term.

For the MCM, intensity measurements are made at each pixel in the image at 64 different focus positions, as shown in Fig. 1.33(a), and then

1.4 Optical Interference Microscopes

Fourier transformed to yield the frequency spectrum of the z-scan, as in Fig. 1.33(b). The negative spatial frequency and the near-zero frequency terms are eliminated, and the packet of positive spatial frequency components is centered around zero, as in Fig. 1.33(c). After this process is carried out, the inverse Fourier transform is taken to yield both the amplitude and phase of the signal in a form equivalent to $g(x,y,z - z_0) \exp j\phi$. The amplitude response corresponding to the envelope of the detected signal, $g(x,y,z - z_0)$, is shown in Fig. 1.33(d).

To measure surface roughness or surface profiles, phase information is required. One technique which is employed in the Wyko and Zygo profiling microscopes is to use narrowband filtered light or a laser source and move the reference mirror rather than the sample, to at least four different positions each one eighth of a wavelength apart.[42,43] Assuming that the amplitude $g(z - z_0)$ is constant over this range, it is then possible to eliminate amplitude information and determine from these four measurements the position z of the sample surface. Typically five measurements are made to give some redundancy. These techniques are discussed in more detail in Chapter 4.

1.4.3 Depth and Transverse Resolution

Images obtained with an interference microscope have definitions in the transverse and range directions comparable to or better than those of a confocal microscope. Two different phenomena are responsible for determining the range resolution or width of the correlation envelope $g(x,y,z - z_0)$. With low numerical aperture lenses, the bandwidth of the illumination is the dominant factor. In this case, when the source has a uniform illumination over a wavelength band from λ_1 to λ_2, the range resolution between the half-power points of the depth response is determined by the correlation length and given approximately by

$$d_z(3 \text{ dB}) = \frac{1.78\pi}{\Delta k(1 + \cos \theta_0)}, \tag{1.35}$$

where $\Delta k = 2\pi/\lambda_1 - 2\pi/\lambda_2$. In an optical interference microscope using a low numerical aperture objective, the axial spread of the depth response fringes can be limited to ~ 1 μm.[44] This depth response is considerably narrower than could be obtained with a CSOM using the same lens.

If a high numerical aperture lens is used, a narrowband system has the same range and transverse resolution as a confocal microscope with the same objective lens, due to the fact that the reference spot at the detector acts much like a pinhole for the signal beam, as already discussed.

The range resolution for the narrowband high-numerical-aperture interference microscope is given by the same formula as for the CSOM

$$d_z(3 \text{ dB}) = \frac{0.45\lambda}{n(1 - \cos\theta_0)}. \tag{1.36}$$

There is, however, an important difference between the two microscopes. Axial chromatic and spherical aberrations do not affect the depth response in the interference microscope. The reason is that both the sample and reference beams travel through the same objective lens. Phase changes introduced by lens aberrations canceled when the product of one signal, which varies as $\exp -j\phi$, is taken at the CCD array with the complex conjugate of the other, which varies as $\exp j\phi$. It will be seen that a plot of the depth response taken with the MCM, Fig. 1.32, is almost ideally symmetric and smooth. This result should be compared with that for the CSOM in Fig. 1.24 and those of Chapter 3, where due to aberrations the curve is not as symmetric.

Cloud Plots The shallow depth response of the MCM can be used to construct cross-sectional images, or cloud plots like the CSOM. An example of a cross-sectional image of a trench in photoresist laid down on silicon, illustrated schematically in Fig. 1.34(a), is given in Fig. 1.34(b).[44] To make a cloud plot in the MCM, the reference plane is kept fixed, usually at the focal plane of the objective, and the intensity as a function of the transverse position x is displayed. The sample is then moved a small fraction of a wavelength in the z direction and a further intensity scan made in the x direction. By repeating this process for different z positions, an x–z intensity plot, or cloud plot is generated. The raw input data in Fig. 1.34(b) show fringes falling off in intensity with distance from the focus. By processing the signal to obtain the envelope of the fringes, amplitude and intensity cloud plots are obtained. It will be observed that the intensity is maximum when the beam is focused on the bottom of the trench and passes through another maximum when it is focused on the bottom of the resist. The data can also be processed to give phase information, and a cloud plot made with an intensity proportional to the phase change is shown in Fig. 1.34(b). Finally, an image is shown of a scan in the x–y direction with the beam focused on the bottom of the sample.

For a large numerical aperture lens in an interference microscope, the range resolution will be narrowed as the bandwidth is increased, particularly if the center frequency of the illumination remains the same. Theoretical and experimental results taken with a 0.8 $N.A.$ lens by Chim

1.4 Optical Interference Microscopes

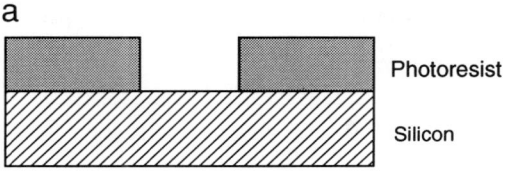

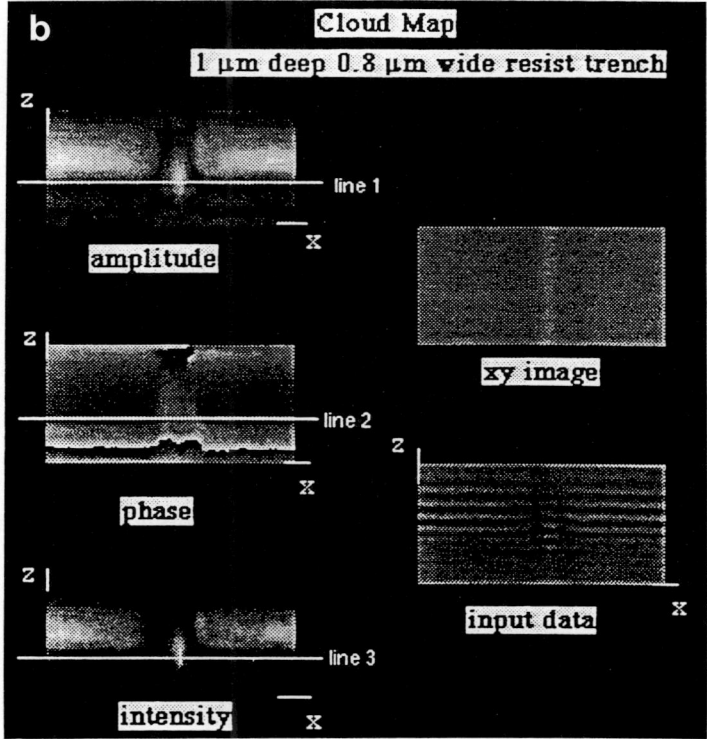

Figure 1.34 (a) Schematic cross section of a trench 0.8 μm wide in 1-μm-thick photoresist, (b) Cloud plots of this sample taken with an $N.A.$ = 0.8 objective in the Mirau correlation microscope. [Courtesy (b): S. S. C. Chim, "The Mirau correlation microscope—a new tool for optical metrology," Ph.D. Dissertation, Department of Electrical Engineering, Stanford University, Stanford, California, USA (June 1991).]

are shown in Table 1.3. It will be seen from the table that the range resolution improves with increasing bandwidth.

In addition to cross-sectional images, the MCM interferogram can be processed to view the sample from the top. Sixteen images of an integrated

Table 1.3 Measured and Theoretical 3-dB Depth Resolutions for the MCM

Spectrum	d_z (experimental)	d_z (theoretical)
400–500 nm	0.55 μm	0.55 μm
400–550 nm	0.54 μm	0.53 μm
400–800 nm	0.47 μm	0.45 μm

circuit taken with the microscope focused on different layers of an integrated circuit 0.132 μm apart are shown in Fig. 1.35. These results are similar to those obtained with the CSOM. In addition, this microscope is capable of quantitative measurements of phase. On the other hand, optical interference microscopes of this type are not suitable for fluorescence imaging because the depth resolution of the device depends on correlating a reference signal with the signal from the sample, but no reference is available for the fluorescent image signal.

Measurement of Transparent Films Another application of interference microscopy is to measure the thickness of a transparent thin film, using the phase difference between the reflections from the top and bottom surfaces of the film. Since the spacing between these two surfaces cannot be varied, fringe information can be obtained by varying the optical wavelength or by Fourier transforming the image to obtain the interference

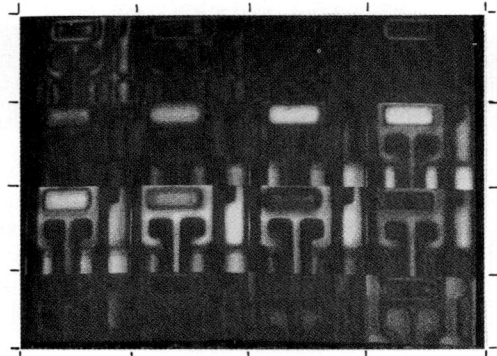

Figure 1.35 Sixteen images of an integrated circuit taken with the MCM microscope focused on layers 0.132 μm apart. [Courtesy: S. S. C. Chim, "The Mirau correlation microscope—a new tool for optical metrology," Ph.D. Dissertation, Department of Electrical Engineering, Stanford University, Stanford, California, USA (June 1991).]

Table 1.4 Comparison of Different Microscopes

	CSOM	CPM/MCM	NSOM	SAM	SEM	STM	AFM
Approximate transverse resolution	0.2–10 μm	0.2–10 μm	1–200 nm	10 nm–2 mm	3–100 nm	0.1–1 nm	0.1 nm–1 μm
Frame size, 2000 spots/frame	400 μm–2 cm		1–200 μm	10 μm–2 m	6–200 μm	0.2–1 μm	0.2 μm–2 mm
Approximate best 3-dB depth resolution	0.3 μm	400 μm–2 cm 0.1 nm (phase)	10 nm	10 nm–5 mm	1 μm	0.1 nm	0.5 nm
Advantages	Familiarity, no vacuum, cross-sectioning, RSOM real-time, direct view in color. Fluorescence, polarization and phase contrast modes.	Good resolution, highly accurate phase measurements, cross-sectioning.	Much better resolution than RSOM, no vacuum, fluorescence.	Measures material properties. Suitable for wide range of samples. Particularly effective on large samples.	Superb 3D-like images, excellent resolution, large depth of view. The standard of the semiconductor industry.	Atomic resolution. Gives information obtainable no other way. Measures chemical properties.	Atomic resolution. Measures magnetic, electric, van der Waals forces, and friction. Can profile relatively large structures.
Disadvantages	Resolution limited by diffraction. Cannot measure electrical properties.	Transverse resolution limited by diffraction. Cannot measure electrical properties.	Spacing to sample very small. Except for SIM efficiency and sensitivity poor.	Needs liquid medium for operation. Slow because of mechanical scan.	Needs vacuum. Hence not suitable for many biological samples. Can damage semiconductor samples.	Must be used on conducting structures. Relatively slow scan. Spacing of probe to sample very small. Probe easily damaged.	Relatively slow scan. Spacing of probe to sample very small. Difficult to use in deep holes. Probe easily damaged.

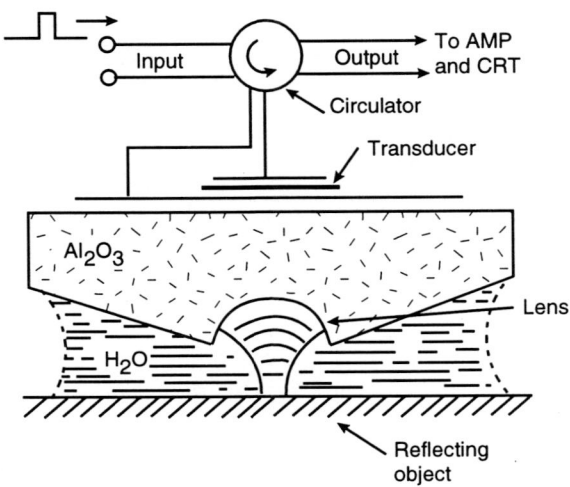

Figure 1.36 A schematic of the lens system of the scanning acoustic microscope (SAM). [From G. S. Kino, *Acoustic Waves: Devices, Imaging, and Analog Signal Processing*, Copyright © 1987. Reprinted by permission of Prentice-Hall, Inc., Upper Saddle River, NJ.]

pattern as a function of ray angle. Such techniques, described in Chapter 5, make it possible to measure the thickness of a film down to approximately 20 nm with good accuracy.

1.5 Comparison of Scanning Optical Microscopes with Other Types of Scanning Microscopes

Scanning optical microscopes represent one of a large variety of scanning microscopes. The best known of these is the *scanning electron microscope* (SEM).[45] Other variations include the *scanning acoustic microscope* (SAM), the *scanning tunneling microscope* (STM), and the *scanning force microscope* (SFM). Table 1.4 compares the approximate resolutions in the transverse and range directions and the advantages and disadvantages of some of these microscopes.

The Scanning Acoustic Microscope This microscope, illustrated in Fig. 1.36, uses an acoustic beam excited by an acoustoelectric transducer and focused through an acoustic lens to scan the object.[46] The object is usually immersed in water, since it is a better propagating medium than air for acoustic waves. The lens is concave shaped because the acoustic velocity in the sapphire lens material is seven times the velocity in water.

1.5 Comparison of Scanning Optical Microscopes with Other Types

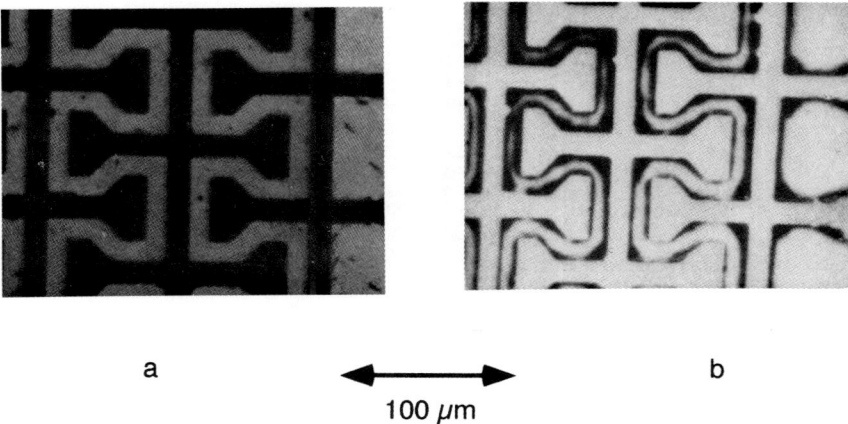

a   ←——→   b
   100 μm

Figure 1.37 Optical and acoustic images of a 100-nm layer of chrome on glass. The acoustic image clearly shows poor adhesion of the metal film: (a) optical image; (b) image taken with a SAM operating at 2.6 GHz. [From G. S. Kino, *Acoustic Waves: Devices, Imaging, and Analog Signal Processing*, Copyright © 1987. Reprinted by permission of Prentice-Hall, Inc., Upper Saddle River, NJ.]

Waves reflected from the object yield an electrical output at the transducer. Like the scanning optical microscope, either the transducer is scanned over the surface of the object or the object is scanned and a raster display of the output is shown on a video screen. The resolution of the SAM is somewhat less than an acoustic wavelength at $N.A. \approx 0.75$ in both the transverse and range directions. Thus, for acoustic waves of frequency 3 GHz ($\lambda = 0.5$ μm) propagating in water, the FWHM transverse resolution is approximately 0.35 μm. Better resolution and lower attenuation in the operating medium can be obtained by using other coupling media, such as liquid helium. The major advantage of the acoustic microscope is its ability to examine material properties and the interior of optically opaque materials. An example is shown in Fig. 1.37(b) of an acoustic image of a 100 nm thick layer of chrome on glass taken at a frequency of 2.6 MHz. Figure 1.37(a) is an optical image of the same sample. The poor adhesion of the chromium layer can be clearly seen in the acoustic image. One of the most important applications of acoustic microscopy uses low-frequency acoustic waves, on the order of 20 MHz, to image the internal structure of circuit boards and composite materials with resolutions of the order of 100 μm.

The Scanning Electron Microscope This microscope uses a focused electron beam to scan an object in a vacuum. The image is obtained by

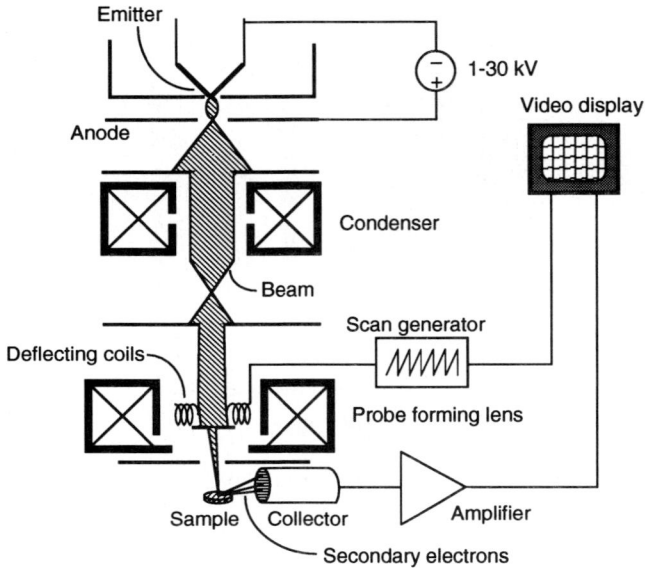

Figure 1.38 Simplified schematic of a scanning electron microscope.

collecting secondary or backscattered electrons from the source on which the focused beam impinges, as illustrated in the simplified schematic of Fig. 1.38. Since the number of electrons collected depends on the nature of the material and its geometry, excellent images of geometrical features are obtained. An example of a sectioned sample of photoresist trenches coated with gold and imaged with a 30-keV beam is shown in Fig. 1.39. In the low submicron range, the SEM has better resolution than a CSOM because the diffraction wavelength of an electron beam is so small; for 2-keV electrons, typical wavelengths are on the order of 0.04 nm, compared with 300–500 nm for an optical microscope. The resolution of an SEM in many applications is, however, worse than the theoretical limits given by diffraction due to aberrations caused by electron scattering as the beam penetrates into the sample. Typical definitions are on the order of 10 nm for 2-keV electrons and 2.5 nm for 10-keV electrons.

Another difference between the SEM and CSOM is that, although the wavelength is extremely short, the numerical aperture is very small. So SEM images remain in focus over a relatively large range and have a three-dimensional appearance, as can be seen in Fig. 1.39. On the other hand, not much depth information is provided. To get depth information, the sample must be cross sectioned, destroying it for further use.

1.5 Comparison of Scanning Optical Microscopes with Other Types 59

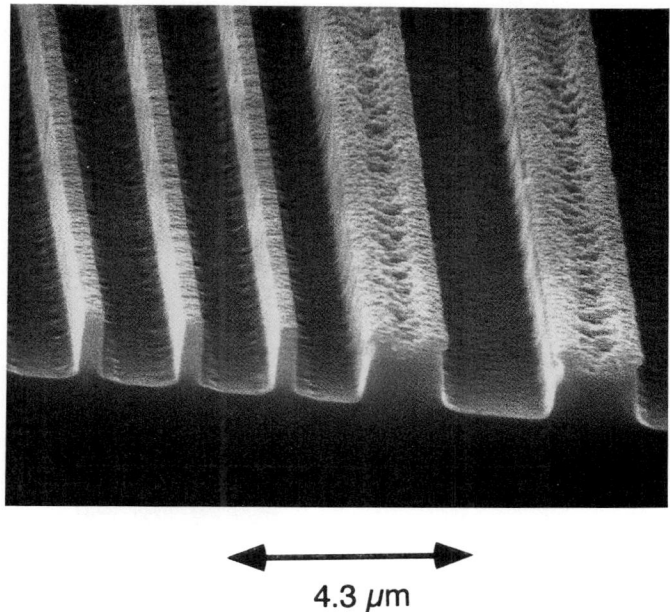

4.3 μm

Figure 1.39 Image of photoresist taken with a scanning electron microscope with a beam energy of 30-keV.

The fact that the sample must be placed in a partial vacuum is the major disadvantage of the SEM, compared with an optical microscope. This limitation renders the SEM unsuitable for use with live materials. In addition, the possibility of damage to biological or semiconductor samples by the high-energy electrons and the need to coat surfaces with a metal so that they do not become negatively charged present problems. In recent years, some of these difficulties have been obviated by working with beam acceleration potentials below 1-keV. In a low-voltage SEM, the secondary emission coefficient of insulators is greater than unity, so charging does not take place.

The Scanning Tunneling Microscope The resolution of the scanning optical, scanning acoustic, and scanning electron microscopes are determined by diffraction, and hence by the wavelength of the source. As has been discussed in Section 1.1, another class of microscopes, the near-field microscopes, have a resolution which depends on the way quasistatic fields fall off from a pinhole or probe. The most familiar of these is the *scanning tunneling microscope* (STM).

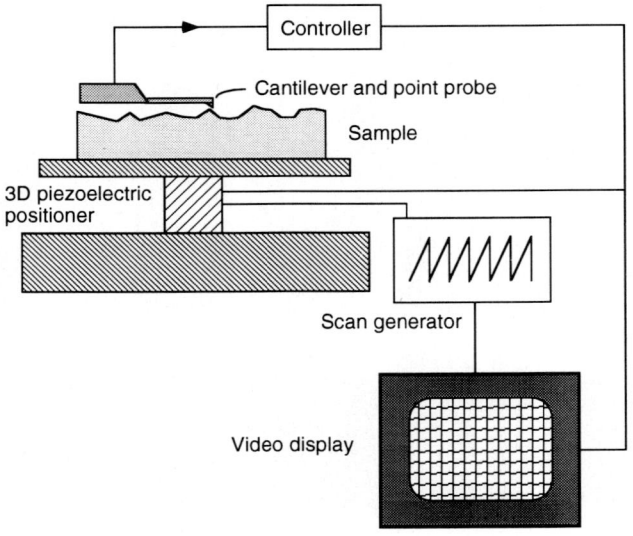

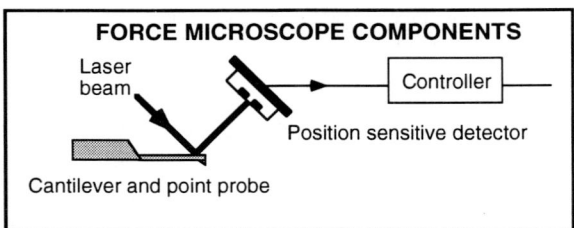

Figure 1.40 Schematic of an STM with an inset showing how the system is adapted for use as an atomic force microscope (AFM).

The STM, illustrated schematically in Fig. 1.40, works on the principle that when a conducting point with a tip of radius of the order of 10 nm is placed within a few tenths of a nanometer from a sample, a potential applied between the conducting point and the object causes a tunneling current to pass between them. This current increases exponentially with applied voltage and decreases exponentially with the spacing between the point and the object source.[47] Generally, an image is obtained by scanning the point over the surface of the object and measuring the spacing required to maintain a constant current. The definition is typically on the order of

1.5 Comparison of Scanning Optical Microscopes with Other Types 61

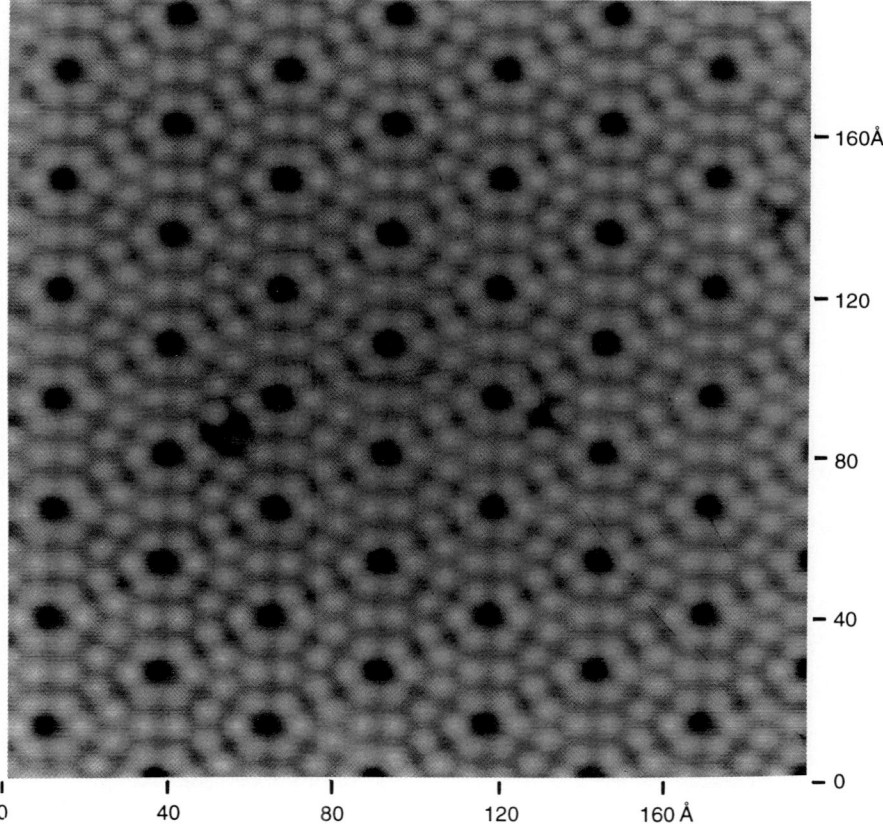

Figure 1.41 An STM image of the (111) surface of silicon showing the 7 × 7 reconstruction of the surface. [Courtesy: C. F. Quate, Stanford University, Stanford, California, USA and Park Scientific Instruments.]

one or two tenths of a nanometer. The STM is used for measurements at the atomic level, producing superb images of atomic structure at the surface of an object. A picture taken with an STM of the atomic structure of the surface of silicon is shown in Fig. 1.41.

The scanning tunneling microscopes are not suitable for measuring the internal properties of materials because the definition decreases so rapidly away from the probe and their fields do not penetrate into the interior of a solid. They are, however, ideal for measuring surface potentials or profiles. Since current flows between the probe and the sample, the sample must be a conductor.

RIE lines: height = 1μm; bottom width = 0.45μm

Figure 1.42 An AFM image of 1-μm-tall photoresist lines with a 0.45-μm spacing at the bottom of the trench. [Courtesy: Y. Martin and K. Wickramasinghe, IBM Yorktown Heights, New York, USA.]

Scanning Force Microscopy This microscope is a derivative of the STM which measures the force between the tip and the sample and is suitable for imaging both conductors and nonconductors.[48,49] As will be seen from the inset in Fig. 1.40, with a cantilever and point probe, the deflection of the cantilever is used to determine the force between the probe and the sample. In the example shown, the deflection of a light beam reflected from the surface of the cantilever and impinging on a position-sensitive detector determines the position of the cantilever. Another technique is to use a simple fiber-optic interferometer which measures the interference between light beams reflected from the end of a fiber and from the cantilever. Yet another approach is to use a piezoresistive silicon cantilever with a resistance that changes as the cantilever is deflected.

A large number of force microscopes have been developed. One type is capable of measuring van der Waals forces between the tip and a dielectric; the resolution is on the order of 10 nm. Atomic resolution can be obtained by measuring the repulsive force when the tip touches the surface of the sample. Another type of force microscope is based on using a magnetic wire sharpened to a fine point and measuring the force between it and a magnetic material, such as a magnetic disk in a disk memory. When the point probe is scanned over the disk, this technique provides superb profiles of the domain fields with definitions in the range of a few hundred angstroms.

A scanning microscope can also be constructed utilizing a miniature thermocouple.[50] A current is passed through this thermocouple and heats it slightly. When the thermocouple is placed near the object to be imaged, there is slight cooling due to heat transfer between the thermocouple and the sample. Again, by scanning to keep the temperature constant, it is possible to trace the profile of the object. The transverse definition of this technique is approximately 50 nm, while the range resolution is on the order of 1 nm.

A major advantage of this type of microscope is that it can be used for surface profiling. However, there may be difficulties in measuring, for instance, the profile of a deep hole because the probe cannot be pushed too far into the hole and there may be a force between the sides of the probe and the top corner of the hole. An AFM image of photoresist lines 1 μm tall with a spacing at the bottom of the trench of 0.45 μm is shown in Fig. 1.42. This image illustrates the excellent resolution which can be achieved when using a scanning probe system in real-world imaging applications.

Despite the many recent advances in microscopy, the optical microscope remains at the forefront of modern science. Its major strengths are its versatility, familiarity, and ease of use. As measurement challenges increase, new ideas have enabled the optical microscope to keep pace with developments in technology and to be increasingly useful for research and for routine measurements in biological and industrial applications.

References

1. G. S. Kino and T. R. Corle, "Confocal scanning optical microscopy," Physics Today 55–62 (September 1989).

2. K. Carlsson and N. Åslund, "Confocal imaging for 3-D digital microscopy," Appl. Opt. **26**, 3232–3243 (1987).
3. D. W. Pohl, W. Denk, and M. Lanz, "Optical stethoscopy image recording with resolution λ/20," Appl. Phys. Lett. **44**, 651–653 (1984).
4. E. Betzig, A. Lewis, A. Harootunian, M. Isaacson, and E. Kratschmer, "Near-field scanning optical microscopy (NSOM)," Biophys. J. **49**, 269–279 (1986).
5. E. Betzig, J. K. Trautman, T. D. Harris, J. S. Weiner, and R. L. Kostelak, "Breaking the diffraction barrier: optical microscopy on a nanometric scale," Science **251**, 1468–1470 (1991).
6. S. M. Mansfield and G. S. Kino, "Solid Immersion Microscope," Appl. Phys. Lett. **57**, 2615–2616 (1990).
7. E. Hecht and A. Zajac, *Optics* (Addison Wesley, 1979).
8. M. Born and E. Wolf, *Principles of Optics*, 6th edition (Pergamon Press, 1980).
9. J. W. Goodman, *Introduction to Fourier Optics* (McGraw-Hill, 1968).
10. R. N. Bracewell, *The Fourier Transform and Its Applications*, 2nd Edition (McGraw-Hill, 1986).
11. H. Volkman, "Ernst Abbé and his work," Appl. Opt. **5**, 1720 (1966).
12. S. Hayashi and G. S. Kino, "Solid immersion lens for optical storage," Three-Dimensional Microscopy: Image Acquisition and Processing II, T. Wilson and C.J. Cogswell, editors, SPIE **2412**, 80–87 (1995).
13. T. Wilson and C. J. R. Sheppard, *Scanning Optical Microscopy* (Academic Press, 1984).
14. G. S. Kino, "Fundamentals of Scanning Systems," *Scanned Image Microscopy*, E. A. Ash, editor (Academic Press, 1980).
15. D. L. Taylor and E. D. Salmon, "Basic fluorescence microscopy," Fluorescence Microscopy of Living Cells in Culture Part A Y. L. Wang and D. L. Taylor, editors, Methods in Cell Biology **29**, 207–237 (1989).
16. T. Wilson and C. J. R. Sheppard, "Imaging of birefringent objects in scanning microscopes," Appl. Opt. **24**, 2081–2084 (1985).
17. F. Zernike, "Das Phasenkontrastverfaren bei der mikroskopischen Beobachtung," Z. Tech. Phys. **16**, 454–457 (1935).
18. M. Spencer, *Fundamentals of Light Microscopy* (Cambridge University Press, 1982).
19. G. Nomarski, "Microinterferometre differentiel a ondes polarisés," J. Phys. Radium **16**, 9S–13S (1955).
20. T. Dabbs and M. Glass, "Fiber-optic confocal microscope: FOCON," App. Opt. **31**, 3030–3035 (1992).
21. S. Kimura and T. Wilson, "Confocal scanning optical microscope using a single-mode fiber for signal detection," Appl. Opt. **30**, 21432150 (1991).
22. G. S. Kino, *Acoustic Waves: Devices, Imaging, and Analog Signal Processing* (Prentice-Hall Inc., 1987).
23. C. J. R. Sheppard and T. Wilson, "Effects of high angles of convergence on V(z) in the scanning acoustic microscope," Appl. Phys. Lett. **38**, 858–859 (1981).
24. T. R. Corle, C-H Chou, and G. S. Kino, "Depth response of confocal optical microscopes," Opt. Lett. **11**, 770–772 (1986).
25. J. Z. Young and F. Roberts, "A flying-spot microscope," Nature **167**, 231 (1951).

26. F. Roberts and J. Z. Young, "The flying-spot microscope," Proc. IEE **99**, Pt. III A, 747–757 (1952).
27. M. Minsky, "Microscopy apparatus," U.S. Patent 3,013,467, December 19, 1961.
28. M. Minsky, "Memoir on inventing the confocal scanning microscope," Scanning **10**, 128–138 (1988).
29. P. Davidovits and M. D. Egger, "Scanning laser microscope," Nature **223**, 831 (1969).
30. G. J. Brakenhoff, P. Blom, and P. Barends, "Confocal scanning light microscopy with high aperture immersion lenses," Jorn. Micro. **117**, Pt. 2, 219–232 (1979).
31. G. J. Brakenhoff, "Imaging modes in confocal scanning light microscopy (CSLM)," Jorn. Micro. **117**, Pt. 2, 233–242 (1979).
32. R. A. Lemons and C. F. Quate, "Acoustic Mcroscopy," Chapter I in *Physical Acoustics: Principles and Methods, XIV*, W. P. Mason and R. N. Thurston, editors, 1–92 (Academic Press, 1979).
33. M. Petran, M. Hadravsky, M. D. Egger, and R. Galambos, "Tandem-scanning reflected light microscope," J. Opt. Soc. Am., **58**, 661–664 (1968).
34. M. Petran, M. Hadravsky, and A. Boyde, "The tandem scanning reflected light microscope," Scanning **7**, 97–108 (1985).
35. P. Nipkow, German Patent 30,105, January 15, 1884.
36. G. Q. Xiao, T. R. Corle, and G. S. Kino, "Real-time confocal scanning optical microscope," Appl. Phys. Lett. **53**, 716–718 (1988).
37. M. Davidson, K. Kaufman, I. Mazor, and F. Cohen, "An application of interference microscopy to integrated circuit inspection and metrology," *Integrated Circuit Metrology, Inspection, and Process Control*, Kevin M. Monahan, editor, SPIE **775**, 233–247 (1987).
38. M. Davidson, K. Kaufman, I. Mazor, and F. Cohen, "The coherence probe microscope," Solid State Technology, 57–59 (September 1987).
39. S. S. C. Chim and G. S. Kino, "Correlation microscope," Opt. Lett. **15**, 579–581 (1990).
40. S. S. C. Chim, P. A. Beck, and G. S. Kino, "A novel thin film interferometer," Rev. of Sci. Inst. **61**, 980–983 (1990).
41. M. Davidson, K. Kaufman, I. Mazor, and F. Cohen, "An application of interference microscopy to integrated circuit inspection and metrology," *Integrated Circuit Metrology, Inspection, and Process Control*, Kevin M. Monahan, editor, SPIE **775**, 233-247 (1987).
42. J. C. Wyant, C. L. Koliopoulos, B. Bushan, and D. Basila, "Development of a three-dimensional noncontact digital optical profiler," Trans. ASME, J. Tribol. **108**, 1–8 (1986).
43. J. F. Beigen and R. Smythe, "High resolution phase measuring laser interferometric microscope for engineering surface metrology," in Proc. Fourth International Conference on Metrology and Properties of Engineering Surfaces (National Bureau of Standards, Washington DC, 13–15 Apr. 1988).
44. S. S. C. Chim, "The Mirau correlation microscope–a new tool for optical metrology," Ph.D. Dissertation, Electrical Engineering Department, Stanford University, Stanford, California (June 1991).

45. D. McMullan, "Scanning Electron Microscopy 1928–1965," Scanning **17**, 175–185 (1995).
46. C. F. Quate, "Acoustic microscopy," Physics Today **38**, 34–42 (August 1985).
47. P. K. Hansma and J. Tersoff, "Scanning tunneling microscopy," J. Appl. Phys. **61**(2), R1–R24 (1987).
48. P. K. Hansma, V. B. Elings, O. Marti, and C. E. Bracker, "Scanning tunneling microscopy and atomic force microscopy: Applications to biology and technology," Science **242**, 209–216 (1988).
49. W. Blaine Stine, "High resolution analysis of biological samples by scanning probe microscopy," USA Microscopy and Analysis 19–21 (November 1995).
50. C. C. Williams and H. K. Wickramasinghe, "Scanning thermal profiler," Appl. Phys. Lett. **49**, 1587–1589 (1986).

CHAPTER 2

Instruments

2.1 Introduction

The basic requirements for a *confocal scanning optical microscope* (CSOM) are point illumination, point detection, a confocal lens system, and a method of scanning the image. A number of different optical systems have been used to fulfill these requirements. They range in complexity from simple microscopes, using a laser, pinhole, and a scanning stage, to more complicated instruments employing white light sources, galvanometer mirrors, acousto-optic cells, or Nipkow disks for scanning. In this chapter we will discuss a few variations of the CSOM. The different approaches will be compared, pointing out the most common applications for each instrument.

The chapter begins with Section 2.2, The Confocal Scanning Laser Microscope, which gives an overview of a simple stage scanned *confocal scanning laser microscope* (CSLM). In this overview each of the components will be examined and some of the principles used for their selection will be explored. We will then show how these principles have been applied in several representative commercial CSLMs. Section 2.3, Nipkow Disk Scanning Microscopes, will follow essentially the same pattern. A discussion of the basic components is followed by illustrations using commercial examples. The chapter continues with Section 2.4, Slit Microscopes, Section 2.5, Alternative Imaging Configurations, and Section 2.6, Interference Optical Microscopes.

Interference microscopes are a second class of depth-sensitive optical microscopes that have been widely developed. Interference CSOMs which can measure phase were an early innnovation of several of the university research groups. Although they represent an important extension of the CSOM's capabilities, they have tended to remain confined to university laboratories. Interference microscopes based on Michelson, Linnik, or

Mirau interferometers are a widely used range of metrology tools with applications in everything from surface roughness measurements to semiconductor characterization.

Recently, another type of scanning microscope has excited great interest, then *near-field scanning optical microscope* (NSOM). This device passes light through a small pinhole placed close to the sample to obtain an image with a resolution comparable to the pinhole size. Images with definitions in the 20–30 nm range, and in one case as low as 1 nm, have been obtained. Another near-field device, the *solid immersion microscope* (SIM), uses high-refractive-index materials to decrease the effective wavelength and is based on the Nipkow disk microscope. Because of the great importance of these new concepts, we will discuss them in Section 2.7, along with a more detailed theory of their operation in Chapter 3.

2.2 The Confocal Scanning Laser Microscope

The simplicity and versatility of the CSLM have made it the workhorse of research with scanning optical microscopes. Stage scanning CSLMs have been widely developed in a variety of research laboratories. They are often the first instrument built by new researchers and the first instrument on which new ideas are tried. One of the early commercial versions, built by SiScan Instruments, was targeted to semiconductor metrology. It was able to scan a single site on a silicon wafer with a frame time of the order of 10 s. Beam scanning CSLMs, however, have had by far the most extensive development as commercial products. Today, they are accepted as indispensable research tools for a wide range of imaging applications.

In this section, the basic components used to construct a CSLM will be described. We will begin with the components used to build a simple stage scanning microscope and then move on to discuss enhancements such as beam scanning. The discussion will be followed by Section 2.2.8, which contains descriptions of different commercial CSLMs. These examples are meant to illustrate the design tradeoffs made by different manufacturers in their products.

A simplified schematic of a stage scanning CSLM is shown in Fig. 2.1. In this instrument the laser light is passed through a pinhole and beam expander. It is then focused by the objective lens to a single spot on the sample, satisfying the point illumination criterion. The objective lens, working in a reflection mode, acts as a confocal lens system satisfying the second requirement. Light reflected from the sample passes back through the objective and is deflected by the beamsplitter into the perpendicular arm of the microscope. There, a lens focuses the light onto a

2.2 The Confocal Scanning Laser Microscope

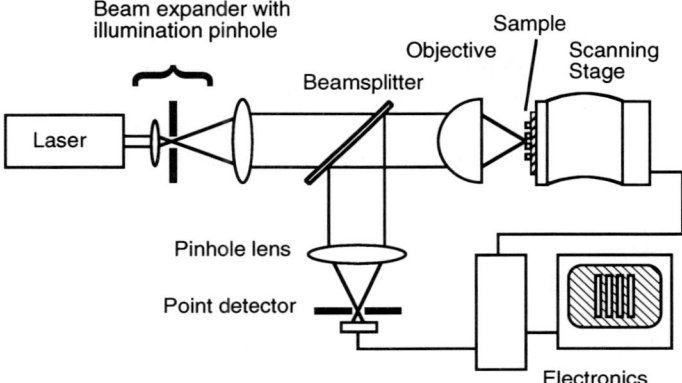

Figure 2.1 Simplified schematic of a stage scanning confocal scanning laser microscope (CSLM).

pinhole and a photodiode that acts as the point detector. In order to form an image, the stage is scanned in an x,y raster pattern and the image is built up point by point. The system in this configuration offers maximum flexibility and ease of alignment. We will now examine each of the components in greater detail.

2.2.1 The Illumination Source

The role of the illuminator is to provide a stable source of spatially coherent light for the microscope. The CSLM shown in Fig. 2.1 uses a laser, pinhole, and beam expander to accomplish this goal. Lasers are common sources for CSOMs because they provide an inexpensive bright single-frequency light source. A single-frequency source simplifies the design of the barrier filters and dichroic mirrors used to separate the excitation and fluorescent beams used in many biological applications. In addition, many fluorescent dyes require significant pump energy to generate adequate light for imaging, so that a relatively bright source is required.

When selecting a laser for use in a CSLM, important performance specifications that must be evaluated include its mode structure, intensity, pointing, and wavelength stability. The laser should have its output in a TEM_{00} mode. A pure TEM_{00} output allows a beam expander to illuminate the objective lens uniformly without adding a pinhole to the illumination path to form a spatial filter. In such a system the point source criterion can be fulfilled without the use of a pinhole because the light appears to originate from a point source at infinity. If the laser beam is imperfect, however, it can be "cleaned up" by focusing it onto a pinhole somewhat smaller than the focused beam spot size.

The intensity stability of the laser is the most important specification because a change in the source intensity can be interpreted as a change in reflected intensity of the sample. In addition to intensity stability, lasers are specified by their pointing stability. The pointing stability is a critical factor when a spatial filter, or pinhole, is used in the beam expander. In this case instability in the pointing direction of the laser will be converted into intensity changes of the light at the sample. When no pinhole is used in the illumination system, on the other hand, pointing instability can change the apparent position of the spot on the sample. In a properly designed instrument the change of position is generally small compared with the spot size of the beam on the sample.

When choosing a laser for a CSLM, wavelength stability is less important than intensity or pointing stability. Most applications are not wavelength sensitive, so that a small change in the source wavelength will have little affect on the image.

Many types of lasers have been used in CSLMs. Helium-neon gas lasers are a good choice because they are inexpensive, reliable, can be amplitude stabilized for critical measurement applications, and are available in a variety of wavelengths. Diode lasers are another stable and inexpensive alternative provided care is taken to eliminate feedback of the reflected beam into the laser, which can cause intensity fluctuations. Argon lasers that provide green or blue light are particularly useful sources for fluorescence imaging. HeCd lasers are generally not used because of problems with intensity and pointing stability.[1]

In microscopes designed for reflection mode imaging in materials science applications, a broadband mercury or xenon arc lamp, rather than a laser, is the preferred illumination source.[2] Broadband illumination reduces the coherent interference of light reflected from different layers of the sample, which can significantly alter the image. This feature is especially important when imaging integrated circuits that are composed of multiple dielectric layers on a reflecting substrate.[3]

Arc lamps have several advantages compared to other types of broadband light sources such as filament lamps. They are compact, efficient, and have high point brightness so that they are easily focused onto the illumination pinhole. In addition, arc lamps have a much larger bandwidth than other light sources. Light from a xenon lamp, for example, spans a region from about 230 nm to greater than 1 μm.

A potential disadvantage of imaging with a broadband light source is that axial chromatic aberration in the optical system will cause different

2.2 The Confocal Scanning Laser Microscope

focal planes on the sample to have different colors. This aberration can be a benefit in inspection applications where areas of the sample at different focus positions are easily distinguished by their different colors.[4] However, in most precision metrology applications, axial chromatic aberration will increase the width of the depth response and thus degrade the resolution of the microscope.

In addition to housing the light source, the illumination system should contain a set of lenses to focus the light onto a pinhole and then relay the light from the pinhole to the rest of the optical system. A beam expander with a built-in pinhole fulfills this requirement for the simple CSLM shown in Fig. 2.1. The beam expander should also have sufficient magnifications to fill the pupil of the objective lens uniformly with light. If the objective pupil is not uniformly filled, the decrease in light intensity near the edges of the lens will lower the effective numerical aperture of the lens and broaden its depth response as shown in Chapter 3. A good rule of thumb is that the size of the expanded beam at the objective pupil should be 5–10 × larger than the objective pupil.

2.2.2 The Objective Lens

Undoubtedly, the most important component of a CSOM is the objective lens. Recently, at a conference on scanning optical microscopes a panel of experts was asked to name the greatest challenges in scanning microscopy today. Eight of the twelve people on the panel named the availability of consistent, high-quality objective lenses as a key factor.[5] The problem lies with the complexity of the objective. High-quality microscope objectives consist of many lenses that are assembled and aligned by hand. Consequently, even the best objectives from the same manufacturer can exhibit large performance differences. Typically less than 25% of the objectives tested have adequate imaging performance for use in a high-definition confocal microscope. Rigorous lens selection is thus essential to achieve optimum performance from a CSOM.

The most common aberrations that affect the resolution of a CSOM are spherical aberration and axial chromatic aberration. Since the objective is used for both illumination and receiving the reflected light, the effect of lens aberrations is increased in a confocal imaging system. Aberrations in the lens system must therefore be kept to an absolute minimum. Spherical aberration can be controlled by selecting and testing high-quality lenses. Chromatic aberration is often inherent in the design of the objective lens. Since conventional optical microscopes are not as sensitive as the CSOM to axial color variations in the focused beam, conventional objective lenses are not always corrected for axial color variation. Care should be taken to select apochromatically corrected lenses when building a CSOM using a broadband light source. These lenses are color corrected at three wavelengths rather than two as with most other lenses.

The objective lens used in the CSLM, shown in Fig. 2.1 is infinity corrected, meaning that the lens aberrations are minimized for an infinite tube length, i.e., when plane waves from a point source located at infinity are focused onto the sample. By contrast, finite tube length objectives require the point source to be located a fixed distance behind the objective, typically 150–210 mm behind the lens, depending on the lens design.[6] Historically, most finite tube length objectives were designed for biological applications, so they are oil immersion, water immersion, or coverslip corrected. On the other hand, most infinite tube length lenses are so-called dry lenses designed for observing samples without the use of immersion fluids, although there are exceptions to this rule.

Adequate testing of an objective lens is imperative before using it in a CSOM. At a minimum, the testing procedure should involve measuring both the depth response and transverse resolution of the lens. The depth response can be determined using a simple direct measurement. A depth response curve is made by scanning a mirror sample axially through the focal plane of the lens and measuring the output of the detector as a function of defocus distance. Ideally the depth response curve should be symmetric with a full width at half maximum that approaches the theoretical limits. If broadband illumination is used in the microscope, the lens should display these properties over the complete wavelength range; often, the ideal depth response is not obtained. Variations in the FWHM greater than 50% are observed in different lenses of the same type from one manufacturer.

The transverse resolution of an objective cannot be directly measured as easily as the depth response. It can, however, be inferred by measuring several different submicron structures of known dimensions such as diatom scales or the patterns on a state-of-the-art integrated circuit and plotting the results against the known dimensions of the structure. A nonlinear relationship between the measured and nominal widths indicates aberration in the image caused by the objective or scanning mechanism. Errors in the scanning system can be identified because they appear as a systematic departure from linearity independent of the objective lens. Once these are eliminated, the residual aberrations can be assigned to the optical system. Alternatively, if a well-designed set of samples is used with a constant reflectivity difference between the lines and spaces of the grating, the depth of the modulation can be measured to obtain the transfer function of the system. Another useful technique is to measure the 20–80% edge response of a cleaved silicon sample, or of a metal film on glass.

2.2.3 The Scanning Stage

After passing through the objective lens the light strikes the sample. In the microscope of Fig. 2.1, the stage is scanned in an x,y raster pattern to build up an image of the sample point by point. Stage scanning makes the introduction of phase plates or other modifications relatively simple because the illumination beam remains stationary. Since in a stage scanning microscope the objective is used on axis, aberration correction is also much easier than with a beam scanning microscope which uses the entire field of view of the lens to form an image.

High accuracy and vibration stability are required for scanning stages in a CSOM. Typically, the stage resolution should be 0.1 μm or better. The scanning stage must meet these precision specifications while simultaneously providing a sufficient scanning range for the size and weight of the sample. Early scanning stages were based on copper leaf springs, wire frame oscillators, and even stereo speaker systems.[7] Today, however, it is possible to buy high-quality, flexure scanning stages commercially.

In addition to the stage being stable, it must be possible to accurately measure and report the stage position to the scan electronics. There are many ways to generate the (x,y) position signals from the stage. These include using linear variable differential transformers, laser interferometers, Doppler velocity meters, encoders, and even the driving signal for the stage. For precision measurements it is better to measure the stage position and feed this signal into a scan converter than it is to infer the stage position from the driving signal.

Once all these scanning requirements are met, the stage must still be mounted into the system in a way that precisely controls the objective-sample separation. Maintaining this distance to within the depth of focus of the microscope is critical to successful operation of the CSOM. Most CSOMs use a type of *piezoelectric transducer* (PZT) based pusher to focus the microscope and to help maintain the objective-sample separation. PZT pushers have the advantage of being fast and extremely rigid, so they keep the sample in a fixed position. Their disadvantages are that they have a limited range of travel and suffer from hysteresis, so they are best used in a closed-loop feedback system.

2.2.4 The Intermediate Optics

After reflecting from the sample, the light travels back through the objective lens and reflects off the beamsplitter to the pinhole lens and point detector.[8] The beamsplitter and pinhole lens constitute the intermediate optical system of the CSLM shown in Fig. 2.1. In this microscope, the use of both collimated laser light and stage scanning simplifies the intermediate optical system.

The optical quality of the beamsplitter and the pinhole lens is not as critical as the objective lens. For the beamsplitter the transmitted wavefront distortion is important only if it broadens the size of the focused beam on the sample. A 50/50 beamsplitter with a transmitted wavefront error of $\lambda/4$ single pass is sufficient for most imaging applications. Ghost reflections from the second surface of the beamsplitter are blocked by the point detector provided the substrate of the beamsplitter is thick enough. An alternative to the standard thin-film beamsplitter is a polarizing beamsplitter which also increases the efficiency of the microscope as discussed in Section 2.3.

The pinhole lens should have a moderate to long focal length. A longer focal length lens enables the detector pinhole to be larger and thus easier to align. In practice, we have found that a camera lens with a focal length of about 50 mm works well.[9] The aberrations in this lens also affect the imaging properties of the microscope; however, since the light passes through the lens only once, the imaging system is less sensitive to pinhole relay lens aberrations than to objective aberrations.

2.2.5 The Pinhole

The pinhole is an important component for determining both the axial and transverse resolutions of the microscope.[10] The theoretical effect of pinhole size on resolution is discussed in Chapter 3; here we will give a more pragmatic method for selecting the appropriate pinhole. To select a pinhole, both the reflectivity of the sample and the required resolution should be considered. The tradeoffs are simple; a larger pinhole transmits more light to the detector, generating a larger signal but with less resolution. A smaller pinhole has theoretically better resolution, but transmits less light to the sample, so the signal-to-noise ratio decreases. In practice, the pinhole should have approximately the same diameter as the half-power width of the Airy pattern produced by the pinhole lens. The relevant formulas are given in Chapter 3. For example, if the entrance pupil diameter of the pinhole lens is 5 mm and its focal length is 50 mm, then the pinhole diameter should be about 10 μm.

2.2.6 The Detector and Electronics

After passing through the pinhole the light impinges on the photodiode generating an electrical signal. In low-light applications a photomultiplier tube (PMT) may be substituted for the photodiode. Additional components may include a relay lens to image the detector pinhole onto the photodiode and a narrowband laser line filter in front of the detector to prevent stray room light from contributing a spurious background signal to the image.

The electrical signal produced by the photodiode is sent, after amplification, to a scan converter, where it is combined with the x and y position signals to produce an image for display. Many types of scan converters are commercially available with various features for image digitization and analysis.

The simple CSOM we have described above is both functional and versatile. Its design forms the basis for many of the more complicated microscopes discussed below.

2.2.7 Beam Scanning Techniques

Because of their inherent speed advantage, beam scanning microscopes have been the preferred approach for commercial CSLMs. Beam scanning CSLMs commonly use galvanometer mirrors and in a few cases *acousto-optic* (AO) deflectors as the beam steering devices.[11,12,13,14,15] A rotating polygon scanner is rarely used because it is difficult to make the device accurate and vibration free enough for microscopy applications. Galvanometer mirrors have the advantage that the scan is independent of wavelength allowing them to be used for white light or fluorescent imaging. In the AO cell, on the other hand, the beam deflection is proportional to the optical wavelength and the frequency of the driving signal. Since the driving frequency is electronically controlled, these devices give a fast scan.[16,17] The acoustic wave in the AO cell behaves like a moving diffraction grating which splits the input light beam into undiffracted and diffracted components. The diffracted beam is deflected by an angle $\theta = (\omega_A/\omega_O)(c/v_A)$ where ω_O is the optical frequency, ω_A is the acoustic frequency, c is the velocity of light, and v_A the acoustic wave velocity in the cell. Typically this angle is small, of the order of 1–2°. In addition, the diffracted light beam is either upshifted or downshifted to a frequency $\omega_O \pm \omega_A$ by the acoustic cell, depending on alignment of the cell to the optical beam.[18]

In addition to the scanning mechanism, beam scanning microscopes must employ additional transfer lenses to image the center of the scanner (the point about which the beam rotates) into the back focal or telecentric plane located near the pupil of the objective lens as illustrated in Fig 2.2(a).[19] Microscope objectives are designed to be telecentric, meaning that the magnification is indepedent of the focus position. As illustrated in Fig. 2.2(b), light reflected from any point on the sample will pass through the telecentric plane as a rectilinear beam. The beam angle to the axis varies with the position of the light spot

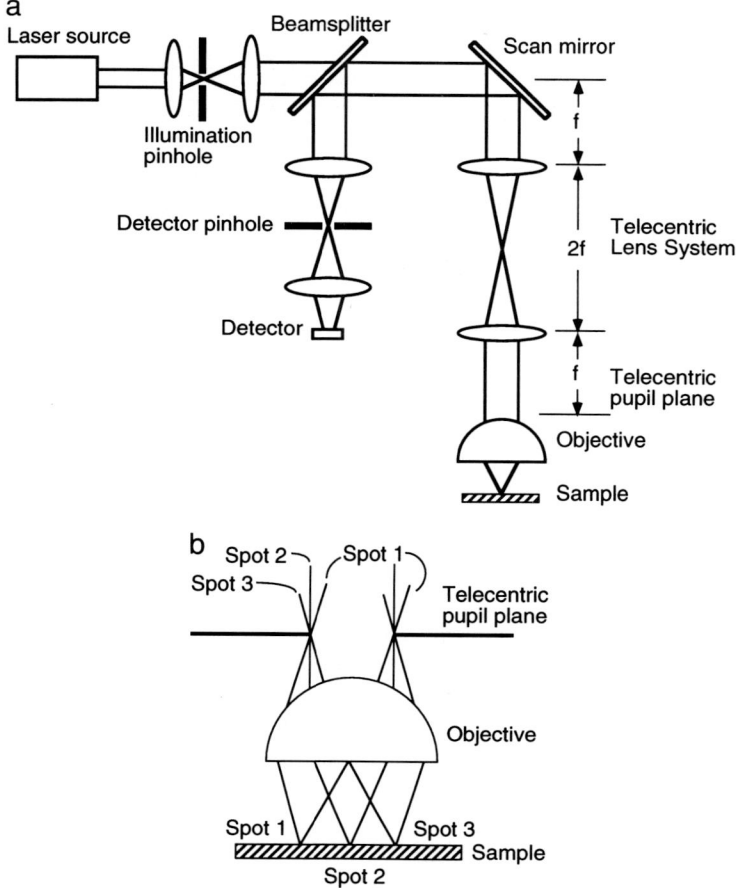

Figure 2.2 (a) A schematic of a CSLM with a telecentric lens system. (b) Illustration of beam rotation at the telecentric pupil plane.

in the sample plane, and the beam will appear to be rotating around a point in the pupil plane.[19] It is this plane to which an image of the scanning mechanism must be projected.

The transfer lens also prevents vignetting of the beam as it scans over the sample by ensuring that the scanned beam rotates around the telecentric plane of the objective rather than scanning across it. If more than one scanner is used, then both scanning mechanisms must be placed in conjugate telecentric planes. However, any number of telecentric planes may be generated by adding additional transfer lenses to the system, so

2.2 The Confocal Scanning Laser Microscope

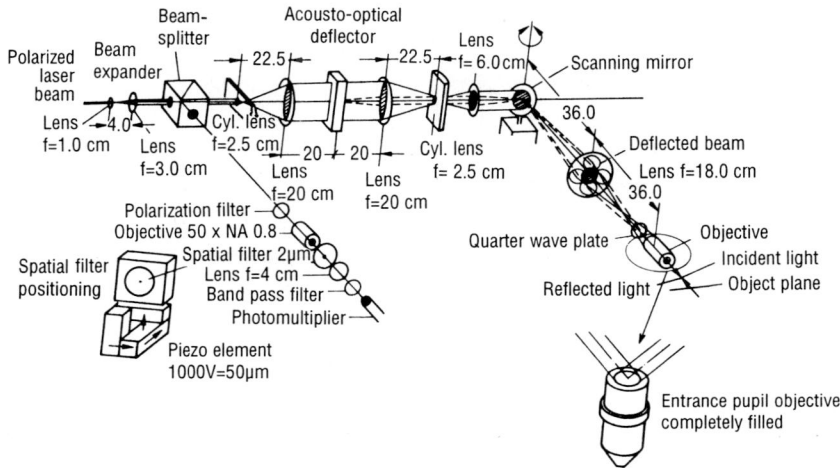

Figure 2.3 Illustration of a beam scanning CSLM which uses both an AO cell and a galvanometer mirror to scan the beam. [From A. Draaijer and P. M. Houpt, "A real-time confocal laser scanning microscope (CSLM)" SPIE **809**, 85–88 (1987), with permission.]

as to accommodate two galvanometer mirrors for x and y scans. An alternative is to place the x and y scanners in proximity so that they are within the depth of focus of the transfer lens system.

A more complete schematic of a CSLM that employs mixed scanning methods and a transfer lens system is shown in Fig. 2.3.[20] The acousto-optic cell scans the beam in the y direction, while a mirror is used for the fast or x scan. Relay lenses are located between the scanners to image the scanning mirror and AO cell to the telecentric plane of the objective. The use of infinity corrected microscope objectives simplifies the design of the transfer optics. In this case the transfer system consists of a series of lenses optimized for a 1:1 conjugate ratio placed in the parallel light space between the scanners. If finite conjugate objectives are used the transfer lens system must be modified to account for the source location.

In the microscope shown in Fig. 2.3 the light reflected from the sample is descanned by the galvanometer mirror and AO cell before it is focused on the stationary point detector. The position of the spot on the sample is thus determined by measuring the angle of the galvanometer mirror and the frequency of the AO deflector. Typical frame rates for galvanometer mirror scanning CSLMs are 1–15 frames per second.

A dramatically simplified slit scanning CSLM that uses the back side of the scanning mirror to relay the light to a charge-coupled device

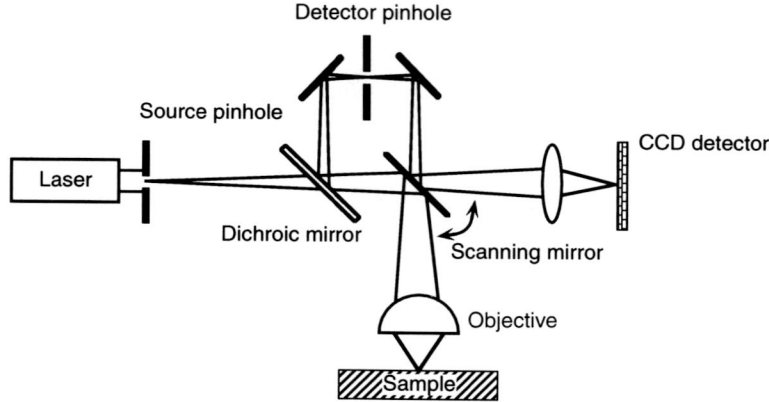

Figure 2.4 Schematic of a CSLM using bilateral scanning and a CCD array at the detector.

(CCD) array or other pixillated detector was invented by Svishchev in 1969 (Fig. 2.4).[21] Other versions of this microscope were subsequently built by Brakenhoff and Visscher[22] and Masters and Thaer.[23,24] This microscope eliminates the need for precision control of the galvanometer mirrors in order to determine the spot location and thus is able to generate a faster and more accurate scan than the traditional beam scanning microscope.

In the bilateral CSOM shown in Fig. 2.4, the light from a laser or other spatially coherent source is scanned over the specimen by the front surface of the scanning mirror. The reflected light is descanned by the same mirror and passes through the detector pinhole placed at the image plane of the objective. Light that passes through the detector pinhole reflects off the back side of the scanning mirror before striking a CCD array detector. Reflection off the back of the scanning mirror effectively scans an image of the detector pinhole over the CCD array. Since the mirror is a rigid structure, this scanning occurs in exact synchronization with the spot on the object. The technology has been commercialized by Meridian Instruments, who have used it to produce a fluorescent microscope which images at 120 frames per second, albeit using slits rather than pinholes as described in Section 2.4.2.

The Macroscope   The scan range in a beam scanned CSLM is controlled by both the scanners themselves and the focal length of the objective lens. The scan area can be dramatically increased in a CSLM by replacing the telecentric lens system shown in Fig. 2.2(a) with an $f*\theta$ scan lens as demonstrated by Dixon et al.[25,26] In their instrument they put the scan lens on a slider so that it could be easily moved in and out of the optical path. In this instrument the field of view size could be varied from 50 to 75,000 μm.

2.2.8 Commercial Examples

The ideas discussed above form the basis for a variety of successful commercial microscopes. In this section we show how the basic principles have been applied in two representative microscopes: the Biorad MCR-1000 and the Technical Instruments Brite*i confocal scanning laser microscope. Both these microscopes were designed to mount on top of the trinocular observation tube of a standard optical microscope. The instruments project a scanned laser beam onto the sample through the objectives of the standard microscope. The reflected light returns to the confocal scanning head, where it is descanned before being relayed to a PMT.

The Biorad MRC-1000 The Biorad MRC series pioneered the application of confocal microscopy to biological applications and has been the most successful group of commercial CSLMs yet introduced. A schematic of the MRC-1000, the latest instrument in the series, is shown in Fig. 2.5. Light from a laser is coupled into the microscope through a single-mode optical fiber. To support multiple wavelength or color imaging two or more lasers can be combined using a dichroic beamsplitter. The fiber serves as the point illumination source for the microscope. The light exiting the fiber then passes through a dichroic beamsplitter, used for fluorescent imaging, before being relayed to the scanners. Galvanometer mirrors are employed for both x- and y-axis scans. A pair of concave relay mirrors is placed between the scanners. The mirrors image the y-axis scanner onto the x-axis scanner, which is located in a telecentric image plane of the microscope, as shown in Fig. 2.2. The use of all reflective components in the relay system gives the microscope high-UV efficiency and makes the scan achromatic.

Light reflected from the sample travels back through the conventional microscope optics to the scanning head, where the galvanometer mirrors descan the image before it is transmitted through a series of dichroic and conventional beamsplitters to the PMTs. The scanning head can contain up to three PMTs for multiple-wavelength fluorescent imaging or simultaneous reflection mode, polarization, and fluorescent imaging. The PMTs

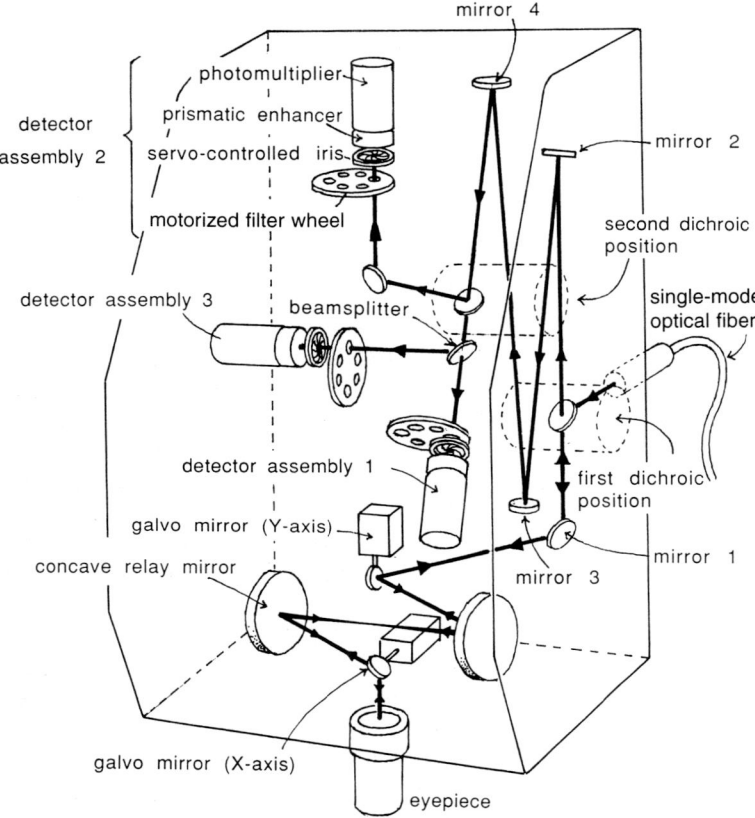

Figure 2.5 Diagram of the light path of a Biorad MRC-1000 confocal laser scanning microscope. [Courtesy: W. B. Amos, Laboratory of Molecular Biology, Cambridge, UK.]

are able to work in photon counting mode, which is preferred for low-light-level imaging since it allows the correct black level to be subtracted from the image and it is largely immune to nonlinearity in the analog-to-digital (A/D) conversion. In addition, the PMTs contain a proprietary prismatic enhancer that increases their quantum efficiency by two to four times.[27]

A unique feature of the Biorad MRC-1000 is that the specimen is imaged at very high magnification onto the point detectors. This characteristic allows an iris diaphragm rather than a pinhole to be used as the point detector aperture. The diameter of the iris diaphragm can be varied to

2.2 The Confocal Scanning Laser Microscope

find the best compromise between signal intensity and confocal optical sectioning ability.

Technical Instruments Brite*i Microscope The Technical Instruments Brite*i confocal microscope is a second example of a commercial confocal laser scanning microscope. The purpose of this design is to simplify the system and lower the cost of the device without making too many compromises on image quality. An optical schematic of the instrument is shown in Fig. 2.6. It uses a single-mode, non-polarization-preserving fiber to bring the light into the scanning head assembly. This fiber also acts as a spatial filter for the optical system. The fiber coupling optics uses an achromatic lens with a useful spectral range of 450–650 nm. This wavelength range is suitable for most common gas lasers. An adjustment screw has been added to control the centering of the fiber at its input end, thus making it possible to adjust the illumination intensity continuously. In addition, a set of neutral density filters is provided at the output of the fiber to adjust the illumination intensity in coarser increments. Laser line filters in the illumination path isolate the system from light caused by secondary lines and fluorescent emission from the laser tube.

After passing through the laser line and neutral density filters, a relay lens focuses the end of the fiber onto the pinholes. Two different pinhole sizes are provided, 20 μm and 50 μm, along with a through beam path that does not contain a pinhole. In this instrument, the same pinhole is used for both illumination and imaging. After the light passes through the pinhole, a second lens collimates the light before it is reflected from the scanners.

Galvanometer mirrors are used for both x and y scans in the Brite*i CSLM. The scan rate of these mirrors is 1 frame/second for a 640 × 480 scan. The scan rate can be slowed by the user to allow longer integration on the detectors if desired. Unlike the Biorad instrument, no relay lenses are used between the scanners. Instead, the Technical Instruments microscope relies on placing the two scanners in proximity to preserve telecentric operation. While this approach is not ideal, if the focal length of the scan lens is long enough such that both mirrors are within its nominal depth of field, it represents a practical, more compact and lower cost alternative to the use of relay lenses.

After passing through the scanners, the light is relayed to the microscope by the scan lens, where it is focused onto the sample by the objective lens. The reflected or fluorescent light from the sample retraces its path through the optical system and is descanned by the galvanometer mirrors and focused onto the pinhole plane. From the pinholes the light is transmit-

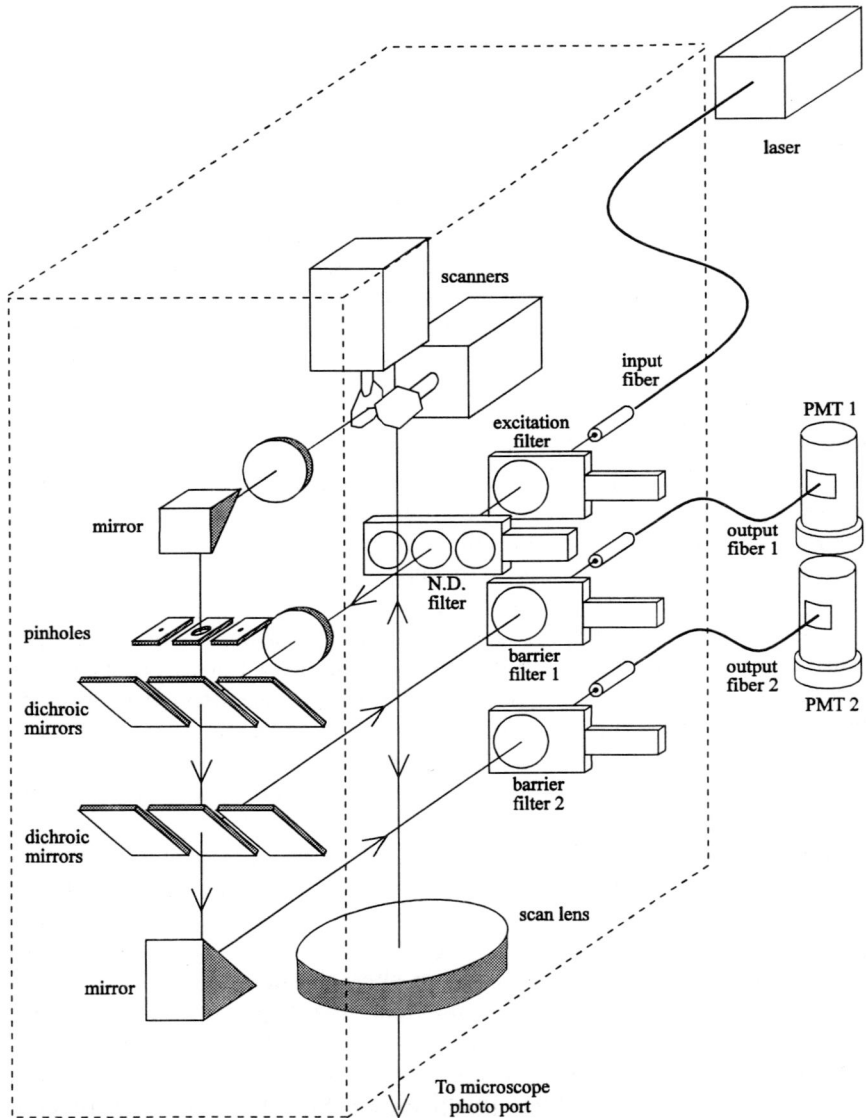

Figure 2.6 Diagram of the Technical Instruments Brite*i scanning optical microscope. [Courtesy: V. Cejna and G. Q. Xiao, Technical Instruments Corporation, San Jose, California, USA.]

2.2 The Confocal Scanning Laser Microscope

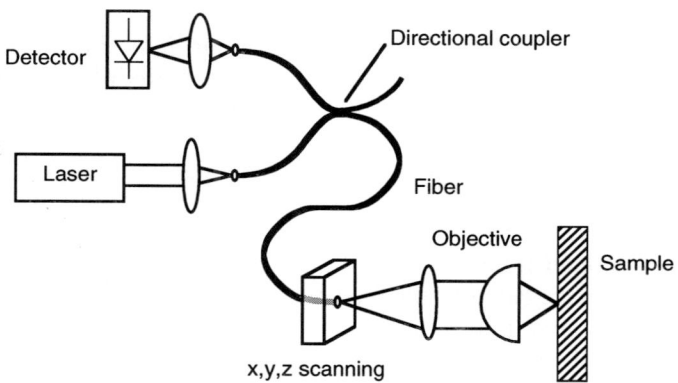

Figure 2.7 Simplified schematic of a fiber optic scanning microscope.

ted by a set of dichroic mirrors to the output fibers and hence to the PMT detectors.

Removing the PMT detectors from the scan head has two advantages. First, it places the PMTs close to the primary electronics, thus reducing electronic noise that may accumulate if the signal is transmitted over a long wire. Second, it reduces the size and weight of the scan head so that it can be more readily supported by a standard microscope frame. The Technical Instruments Brite*i scanning head measures only 3.5" × 6.5" × 8.5" and weighs less than 8 lb, comparable in size and weight to many intensified CCD cameras. The disadvantage of this arrangement, compared with using on-board PMTs, is that the light efficiency is lower due to light losses when coupling into and out of the second set of fibers.

2.2.9 Fiber-Optic Scanning Microscopes

Scanned Fiber Systems A variation of the beam scanning microscope uses an optical fiber and directional coupler as the point source, point detector, and beamsplitter (Fig. 2.7).[28,29,30,31,32,33,34] The end of the fiber closest to the objective is moved in the x, y, and z directions to scan the beam over the sample. This configuration has the advantage of great simplicity. Other advantages are that, since the image of the fiber is demagnified at the sample, precision scanning of the fiber is not required, and the fiber itself can be used for simple image processing. A single-mode optical fiber can be used to produce a reflection image; if a two-mode fiber with a split detector is used, a differential amplitude or differential phase image can be obtained, as described in Chapter 4.[35,36] In addition, a relatively rapid scan is possible due to the low mass of the fiber.[37] The disadvantages of this approach are its poor light efficiency and the restriction to a relatively narrow range of wavelengths with a given fiber.

Hybrid Systems There are numerous variations on the basic fiber-optic microscope design that combine elements of fiber microscopes and conventional scanning optical microscopes. In a system designed by Harris and Delaney, the fiber replaces the relay optics and pinhole with the directional coupler functioning as the beamsplitter.[38] Galvanometer mirrors scan the image between the fixed fiber and the objective. A combined beam scanning, fiber optic confocal microscope has been built by Gmitro and Aziz.[39] In their microscope a fiber bundle transfers the image of the sample from a potentially inaccessible place, such as inside the human body, to an external location where it can be probed by a standard confocal microscope. Since the individual fibers act like point sources and detectors, the fiber bundle preserves the shallow depth response of the microscope. The disadvantage, however, of using multiple fibers is that the cladding on the fibers will cause discontinuities to appear in the image. In addition the number of fibers in the fiber bundle will limit the number of pixels in the image. Typical fiber bundles will have far fewer fibers than are needed to generate a good-quality TV image.

2.3 Nipkow Disk Scanning Microscopes

Nipkow disk scanning or direct view microscopes are another class of widely used scanning optical microscopes. The technology was pioneered by Petran and Hadravsky in Czechoslovakia in the late 1960s with the introduction of the *tandem scanning reflected light microscope* (TSRLM).[40] Since then, their ideas have been widely adapted to a variety of scanning microscopes.

2.3.1 One-Sided and Two-Sided Designs

The Two-Sided System The original Petran microscope used a two-sided Nipkow disk design as shown in Fig. 2.8. The microscope illuminated one side of the Nipkow disk and the image was observed through a conjugate set of pinholes on the opposite side of the disk as discussed in Chapter 1. The advantage of this arrangement is that light reflected from the top of the disk is returned to the illumination source and so does not cause glare. The design is optimum for a brightfield microscope observing weakly reflecting objects. In these applications the complete rejection of stray light from the source is important for successful imaging. The

2.3 Nipkow Disk Scanning Microscopes

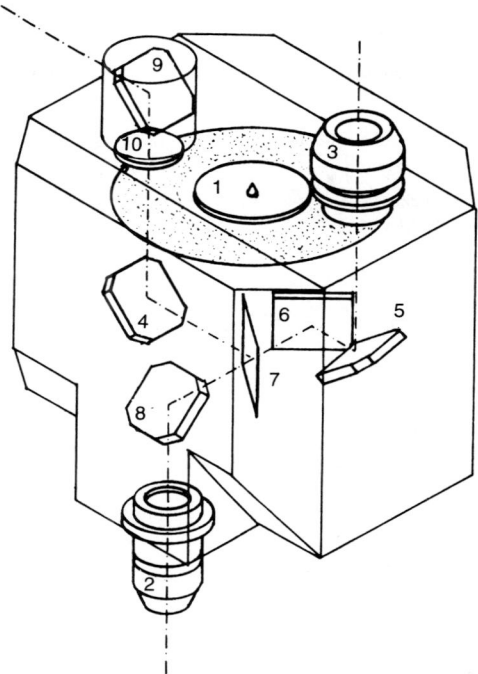

Figure 2.8 Schematic of the tandem scanning reflected light microscope (TSRLM). [From M. Petran, A. Boyde, and M. Hadravsky, "Direct View Confocal Microscopy," *Confocal Microscopy* T. Wilson, editor (Academic Press, 1990), with permission.]

disadvantage of the two-sided design is that adjusting and maintaining the alignment of the microscope is difficult. Alignment between the conjugate set of pinholes on the rotating disk is complicated by the presence of several internal mirrors and the need for the Nipkow disk to be accurately spun about its central axis.

The One-Sided System An alternative to the two-sided TSRLM design is a one-sided disk microscope. In the one-sided design the same set of pinholes is used for both illumination and imaging. This arrangement removes the alignment problem. The idea of using only one set of pinholes was initially suggested by Egger and Petran; however, it was rejected because of the perceived difficulty of eliminating the light reflected from the top of the disk.[40,41,42] In 1975 Frosh and Korth also filed a patent in which they outlined a one-sided design. Their system used reflections

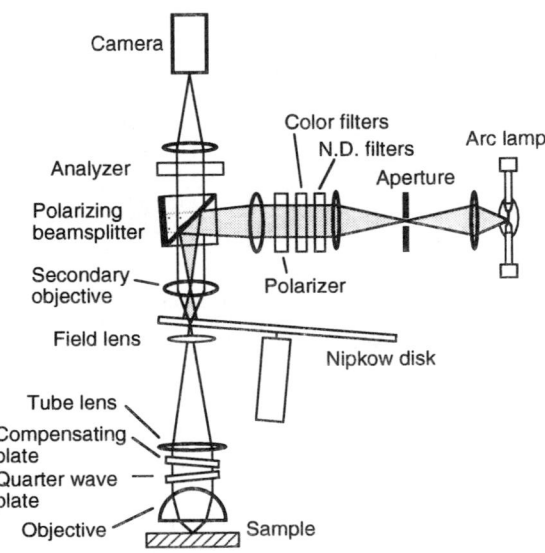

Figure 2.9 Detailed schematic of the real-time scanning optical microscope (RSOM) developed by Xiao, Corle, and Kino.

from a complex multielement disk to eliminate light from the illumination source.[43]

The first widely successful one-sided Nipkow disk microscope was developed by Xiao, Corle, and Kino at Stanford University (Fig. 2.9). They called their instrument the *real-time scanning optical microscope* (RSOM).[44,45] Two different techniques are used in the RSOM to eliminate light reflected from the top of the disk. First, the disk is tilted by a small angle, approximately 5°. Tilting the disk deflects the light away from the eyepieces of the microscope to an internal stop. In addition, an optical isolator is used to deflect light reflected from the top of the disk back to the light source while passing light reflected from the sample to the eyepieces.

We will now go on to examine the components of the RSOM in greater detail, much as we have done for the CSLM.

2.3.2 The Nipkow Disk

The rotating Nipkow disk is the key technology that makes the direct view microscopes possible. Although the disk is conceptually based on Nipkow's original design for a mechanically scanned television, it was

2.3 Nipkow Disk Scanning Microscopes

modified by Petran and Hadravsky to improve the light efficiency and increase the number of lines in the image. A single modulated light source was used in the original application of the Nipkow disk, so that only one pinhole of a set arranged along a rotating single spiral was illuminated at a time. For the microscopy application, because they had an extended light source and a CCD camera or the human eye as the receiver, Petran and Hadravsky could use parallel processing. Their disk contained many thousands of pinholes arranged in multiple interleaved spirals. Several hundred pinholes at a time were simultaneously illuminated. By illuminating many pinholes at once, Petran and Hadravsky were able to increase both the light efficiency and scanning speed of the microscope. Devices based on this principle operate at frame rates of several hundred per second, with several thousand lines in the image.

Pinhole Size and Layout The optimum size and spacing of the pinholes depend on the focal length of the tube lens and the diameter of the objective pupil.[7,46,47,48] The dimensions required are discussed in detail in Chapter 3. As a rule, the diameter of the pinholes should be roughly equal to the 3-dB width of the Airy pattern projected onto the disk by the objective lens and the center-to-center spacing should be great enough to eliminate the light reflected from out-of-focus planes. A typical microscope objective with a pupil diameter of 1 cm operating at a tube length of 180 mm will have a 3-dB spot size at the disk of 20 μm. If the desired background intensity of an out-of-focus sample is 2% of its average peak intensity, then the center-to-center spacing should be 134 μm.

The layout of the pinholes on the disk is driven by the requirement that the scanned field of the object be uniformly illuminated. If several interleaved spirals are used, either the center-to-center hole spacing or the hole diameter must change corresponding to the radial location of the pinholes on the disk. Given this choice, it is better to vary the center-to-center hole spacing rather than the pinhole diameter. If the pinhole diameter varies, the instrument will have different depth and transverse resolutions at different locations in the field of view.

The arrangement of the pinholes can be either a quasirectangular or a hexagonal grid. A hexagonal grid has a higher hole density than the rectangular array for a given center-to-center spacing, thus maximizing the overall transmission of the disk. Additionally, the holes can be made in a variety of shapes. If they are smaller than the central lobe of the objective diffraction pattern at the disk, then the exact shape is not important. A commercial instrument designed by Tracor Northern successfully used square holes in the Nipkow disk, although in general hexagonal or octagonal holes are recommended.[46]

Both the resolution and light efficiency of a Nipkow disk microscope are influenced by the spacing of the pinholes in the disk. When the spacing is large, or when the sample is near the focal plane, the form of the depth response is exactly the same as for a single-pinhole CSOM using the same size pinhole. As the spacing between the pinholes decreases, a portion of the defocused light reflected from the object can return through alternative pinholes to the eyepiece. In the geometric optics approximation, the defocus distance at which light begins to pass through neighboring pinholes is given approximately by

$$z = \frac{(c-c)}{2M(N.A.)},\qquad(2.1)$$

where $(c - c)$ is the center-to-center spacing of the pinholes, M is the magnification of the objective, and $N.A.$ is its numerical aperture.

Beyond this point, the depth response is different from the value for a single-pinhole CSOM. As the defocus becomes more extreme, as discussed in Chapter 3, the proportion of reflected light passing through the disk approaches the density of pinholes on the disk,

$$\kappa = \frac{\pi a^2}{A_d},\qquad(2.2)$$

where a is the radius of the pinholes and A_d is the disk area per pinhole. For a disk with a hexagonal array of pinholes, $A_d = \sqrt{3}(c-c)^2/2$.

Speckle Due to Close Pinhole Spacing A second problem that can arise if the pinholes are too closely spaced is interference between the light signals from different pinholes causing speckle. Speckle effects are apparent only when a narrowband light source such as a laser is employed. The use of a rotating diffuser or phase randomizer placed in front of the laser source will reduce the effect. With a broadband arc source, the bandwidth of the illumination is usually adequate to eliminate it entirely.

It is common for the disks used in direct view scanning microscopes to have multiple bands of holes with different pinhole sizes and spacing on the same disk. The different bands have different transmissivities and are designed to accommodate samples with different reflectivities and, sometimes, different objectives. For low-reflectivity samples, slits rather than pinholes can be used on the disk. The light transmission of a slit pattern is about 10%, compared with approximately 2% for pinholes, but as discussed in Chapter 3, the use of slits leads to higher sidelobe levels.

Disk Fabrication There are many different materials and methods of fabricating Nipkow disks for use in a scanning optical microscope. The Petran microscope, for example, used a thin (10–20 μm) copper or nickel foil in which the pinholes were etched using standard photolithographic procedures to determine their location.[42] These disks were supported by a frame around their periphery. A second technique, developed by Tracor Northern, is to fabricate the disk out of a silicon wafer. The hole pattern is again created using a photomask. An anisotropic etch is used to make holes in the wafer.[46] The holes in this disk are square and they have the same orientation.

A third technique, first used by Xiao and Kino[49] and now employed by most manufacturers, is to fabricate the disk directly out of a standard chrome glass photomask of the type used in the semiconductor industry. These disks are self-supporting and the precise hole pattern can be generated in any desired shape. The glass disks are coated with either photographic emulsion or antireflective (AR) chromium oxide. Although photographic emulsions have a lower intrinsic reflectivity over the visible range, the granular constituents of the emulsion can scatter and depolarize the incident light. Since the RSOM uses tilting and/or polarization isolation to eliminate the light reflected from the top of the disk, it is important that the incident light on the disk be specularly reflected, have its polarization preserved, and the reflected light level be as low as possible. A standard industrial AR chrome glass plate reflects about 18% of the light at 546 nm. However, through the use of additional coatings on these plates, the reflectivity can be reduced to 1 or 2%. Standard chrome-glass photomasks are also slightly transparent, about 0.1%. This transmission is significant when compared with the transmission of the hole pattern, ~2%. If it is not eliminated it will contribute to unwanted background light in the final image. The transmission of light through the chrome areas of the plate can be reduced to acceptable levels by using plates coated with a double thickness of chrome.

In addition to the Nipkow disk, other multiple-pinhole CSOMs have been proposed.[50] Wayland has used a plate containing an array of pinholes that was vibrated back and forth. This system was used for microvascular and neurophysiological studies.[51] It has also been suggested that a disk with a random pattern of holes might serve as a scanning mechanism, or a vibrating array of diode lasers as point sources, although to the best of our knowledge no one has built such an instrument.[52,53]

2.3.3 Illumination of the Disk

The low light efficiency of Nipkow disk-based scanning optical microscopes requires a bright light source in order to see the image. Both lasers and mercury arc lamps have been used as sources; however, arc lamps are preferred for two reasons. First, coherent interference between the light reflected through adjacent pinholes in the disk can cause lines to appear in the image. Second, most direct view microscopes were designed for reflection mode imaging of solid objects such as semiconductors, bone, and teeth. In these applications broadband illumination reduces the coherent interference effects between different layers of the sample, resulting in an image that is easier to interpret and one in which quantitative measurements of dimensions are more accurate. Another feature of broadband illumination is that it provides direct color images that can show different depths in different colors.

In the RSOM shown in Fig. 2.9 a xenon arc lamp is used as the illumination source. A condenser lens system images the light from the lamp onto a small aperture that filters the light and acts as a small uniform source for the illumination system. From this aperture the light passes through a set of neutral density and wavelength filters, a polarizer, polarizing beamsplitting cube, and low-power objective lens, called the secondary objective, before striking the Nipkow disk.

The neutral density filters adjust the intensity of the light striking the sample so that the camera signal does not saturate when high-reflectivity samples are observed. These filters limit the intensity of the light passed to the rest of the optical system and, correspondingly, the amount of stray light that must be removed by the polarizers. The wavelength filters usually contain a variety of narrow and broadband filters. For semiconductor measurements two filters are especially useful: a broadband blue (400–480 nm) filter for critical dimension measurements on photoresist lines and a long-wavelength filter, >525 nm, which does not expose photoresist. It is also advisable to have a narrowband mercury line filter at 546.1 nm. This mercury line is a common design wavelength for many microscope objectives, and therefore illumination with this wavelength will produce the narrowest depth response in the microscope.

As shown in Fig. 2.9, a Köhler illumination system images the light source aperture onto the pupil plane of the secondary objective lens located just above the disk. If the disk were not present, light passing through a point on the axis of the pupil plane would be focused by the field lens to a small spot on the axis of the pupil of the main objective.

When the Nipkow disk is present, several hundred holes in the disk are simultaneously illuminated. Each of these pinholes diffracts the light passing through it into an Airy pattern which is observable at the tube lens or objective lens pupil plane. The central rays passing through each pinhole will be focused by the field lens and hence each diffraction pattern

2.3 Nipkow Disk Scanning Microscopes

will be centered on the axis of the pupil of the objective lens. Consequently, the objective lens will be uniformly illuminated by all the pinholes in the field of view of the microscope. This overlapping of the diffraction patterns ensures that diffraction-limited images will be produced over the entire field of view. Furthermore, if the pinhole size is chosen correctly, the pupil illumination from each pinhole will be almost uniform.

With a finite-size light source, this arrangement ensures uniform illumination of the pupil of the secondary objective lens and, in turn, the pupil of the primary objective lens. If an extended source aperture rather than a point aperture is used to illuminate the disk, the total amount of light reaching the sample will increase. However, the diffraction patterns will be washed out; furthermore, they will not be centered on the axis or completely overlap on the pupil of the objective. The resulting nonuniform illumination of the pupil is equivalent to apodization, which will adversely affect the resolution of the microscope;[7] this image degradation is more pronounced at the edges of the field of view.

One method of correctly setting the size of the source aperture is to observe the pinhole diffraction pattern on a piece of paper placed over the objective pupil, then increase the size of the source aperture until the sidelobes begin to wash out. As long as the sidelobes in the pinhole diffraction patterns are visible, the image degradation will be negligible. Typical source aperture sizes will be approximately 0.5 mm depending on the design of the illumination system.

2.3.4 The Tilted Disk and Optical Isolator

The RSOM designed by Xiao et al. uses both a tilted disk and an optical isolator to eliminate the light reflected from the top of the disk. Either of these methods is sufficient to reduce the light to acceptable levels. However, for the best performance it is desirable to combine both techniques.

Tilting the disk has the advantage that conventional beamsplitters can be used in the optical system so that the light is unpolarized when it strikes the sample. This method is best suited for polarization and differential interference contrast imaging, where it is advantageous to be able to select any desired input polarization. The disadvantage of tilting the disk is that it causes the focal plane at both the sample and the camera or eyepieces to be tilted. The effect at the sample can be neglected because any tilt or wobble in the disk is demagnified by the square of the objective lens magnification. For a 50× lens and a Nipkow disk tilt of 5°, the focal plane tilt is only 0.002°. The small tilt of the Nipkow disk has a much greater effect on the image at the camera.

Typical magnifications for the Nipkow disk to camera relay optics are 2–5× so that any tilt in the disk is magnified by a factor of 4–25× at the camera. Unless a custom optical system is designed to compensate for the tilt, it should be kept to a minimum.

The second method employed to eliminating light from the top of the disk in the RSOM uses a polarization-based optical isolator. To construct an optical isolator the illuminating beam is polarized and an analyzer positioned in front of the eyepieces with its polarization direction at a right angle to the polarizer. A quarter-wave plate above the objective lens rotates the polarization direction of the light reflected from the sample so that it is transmitted by the analyzer. This combination of polarization components effectively eliminates reflected light from the disk while passing light that has been reflected from the sample.

In the RSOM shown in Fig. 2.9, the light passes through both a prepolarizer and a polarizing beamsplitter before it strikes the top of the disk. These two polarization components are necessary in order to obtain sufficient performance from the optical isolator. For example, if the disk transmission is 1% (10^{-2}) and we want a signal-to-noise ratio of 1000 in the image (noise power less than one thousandth of the signal power), the optical isolator must reduce the intensity of the light reflected from the top of the disk by at least a factor of 10^{-5}. Typical polarization rejection ratios for thin-film polarizing cubes are on the order of 10^{-3}. As a result, sufficient isolation cannot be obtained from a single cube. If two cubes are used in series, however, sufficient isolation can be obtained. An alternative is to use calcite polarizers that give sufficient polarization rejection ratios in a single optical element.

In addition to eliminating the light reflected from the top of the disk, polarization isolation improves the light efficiency of the microscope. In a system using a conventional beamsplitter with a reflection efficiency of $\Gamma_{bs} \approx 50\%$ and a Nipkow disk efficiency of $\Gamma_{ND} \approx 1\%$, the overall efficiency is approximately $\Gamma \approx \Gamma_{bs} \times \Gamma_{ND} \times \Gamma_{bs} \approx 0.25\%$. If a combination of polarizer and polarizing beamsplitter is used, there will be an initial 50% reduction in intensity at the polarizer $\Gamma_p \approx 50\%$. However, the polarizing beamsplitter will direct virtually all of the light to the sample, thus $\Gamma_{pbs} \approx 100\%$. The light efficiency of an instrument using polarization components throughout is $\Gamma \approx \Gamma_p \times \Gamma_{pbs} \times \Gamma_{ND} \times \Gamma_{pbs} \approx 0.5\%$, twice as great as that of the system with conventional beamsplitters. Another advantage of polarizing beamsplitters is that they are usually made of thin films and so do not produce ghost images. Even with the use of polarization components, however, the light efficiency of a Nipkow disk scanning microscope is poor!

2.3.5 The Field Lens, Tube Lens, and Objective Lens

After passing through the Nipkow disk, as discussed above, the diffraction patterns produced by the pinholes are centered on the pupil of the objective lens by the field lens. Since the field lens relays only pupil planes, aberrations in this component are less critical than in the tube lens and objective lens. Usually a moderate-power single lens element is sufficient.

The tube lens, shown in Fig. 2.9, collimates the diffracted light produced by the pinholes. The addition of this lens is necessary only if infinity-corrected objectives are used. Generally a good-quality triplet is needed for broadband imaging applications. For finite tube length lenses, the pinholes on the disk will be imaged directly to the sample by the objective. Since the light passes through this lens twice, the tube lens aberrations are also significant. The aberrations in the tube lens are usually combined mathematically with the objective aberrations into one aberration term. If care is not taken, the tube lens can be the dominant contributor to this combined aberration term.

As in the CSLM, the objective lens is one of the most critical components in the RSOM. When using broadband illumination, the best results are obtained with apochromatically-corrected lenses. These lenses are color corrected at three wavelengths rather than just two as with conventional achromatic objectives. The objectives should also be corrected for axial chromatic aberration. Most objectives today are highly corrected for lateral color; however, due to the poor z-resolution of the standard brightfield microscope, they are not as highly corrected for axial color. An example is shown in Fig. 2.10, where the experimental depth response for an objective illuminated by two different bandpass filters, 505–545 nm (dashed line), and 400–480 nm (solid line) are shown. As the bandwidth of the illumination increases so does the width of the depth response, despite the fact that the central wavelength decreases. This behavior is typical of most commercial objectives, which have worse performance in the blue end of the spectrum. Broadening of the depth response is a disadvantage in metrology where it is often desirable to know the exact z-location of a particular structure. In an inspection system, however, axial color can be used to color code images so that image planes at different focal positions appear with different colors.

Lastly, it is important to realize that the field lens, tube lens, and objective lens should form a fully corrected optical system. Most microscope manufacturers use the entire optical train from objective to eyepiece

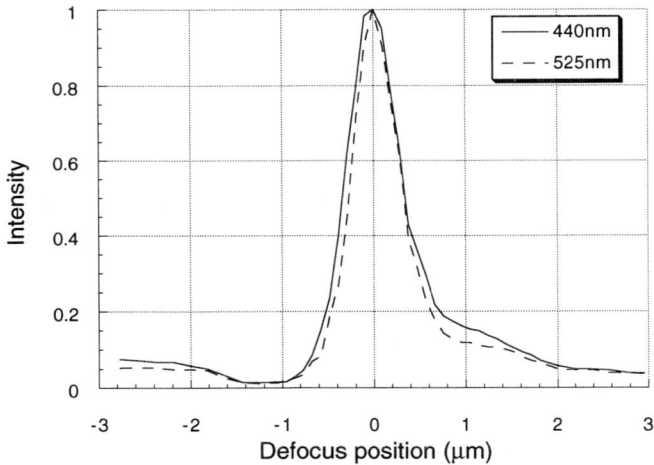

Figure 2.10 Experimental depth response for an objective illuminated by different wavelength bands: 505–545 nm (dashed line), 400–480 nm (solid line).

to correct aberrations. As a result, most objective lenses have some residual aberrations that are corrected in the eyepiece. Two exceptions are Nikon and Zeiss. Nikon fully corrects the objective lenses themselves, while Zeiss uses the objective lens and tube lens combination for aberration correction. The difference in aberration correction strategy does not indicate that these manufacturers' lenses are preferred for confocal scanning microscopy. Indeed, we have built systems using Leitz, Olympus, Nikon, and Zeiss objectives and cannot recommend one unequivocally over another. The best advice we can offer is to try the lens in your system.

2.3.6 The Imaging Path

In the RSOM design, light reflected off the sample travels back through the objective lens and the tube lens to the same pinhole in the Nipkow disk. A magnified, sampled image of the sample is formed on the top surface of the disk. The lens system above the disk is, in fact, a low-magnification microscope with an objective lens, tube lens, and eyepieces or camera relay optics, with the disk located in the object plane. Since it is essentially a geometric image of the disk that is relayed to the eyepieces, highly corrected optical components are not required in this part of the imaging system. In the configuration shown in Fig. 2.9, an infinite tube length low-power secondary objective lens is placed below a polarizing

2.3 Nipkow Disk Scanning Microscopes

beamsplitter cube. However, the illumination light that is reflected from the beamsplitter and lens elements of the secondary objective must be eliminated. The light reflected from the secondary objective is usually not a major problem because the lens surfaces are curved and hence tend to direct the reflected light away from the eyepieces. The part of the illuminating beam reflected from the lower face of the beamsplitter cube can be eliminated by rotating the cube slightly about an axis normal to the beam-splitting surface.[54] An alternative design is to place the upper objective above the main beamsplitter. In this case a pellicle type of beamsplitter should be used to minimize the aberrations introduced into the focused beam by the glass of the beamsplitter cube.

2.3.7 Commercial Examples

A simplified diagram of the K2-IND confocal scanning optical microscope module from Technical Instruments integrated with a Reichert standard optical microscope frame is shown in Fig. 2.11(a), along with a more detailed diagram in Fig. 2.11(b) showing the very similar add-on module for use with a wide variety of microscopes. The integrated system has all lenses designed to fully compensate for aberrations introduced by the beamsplitter and other RSOM components.

This microscope incorporates many of the design features discussed earlier in this chapter. The Nipkow disk is illuminated by an arc lamp and condenser system that contains a small source aperture that ensures uniform illumination of both the disk and the objective pupil. The disk itself contains two different aperture patterns in concentric rings. One ring contains 45 μm-pinholes with 2% light transmission; the other has 25-μm pinholes with 0.6% light transmission. A third possibility is to remove the disk from the beam path entirely, thus allowing the use of the system as a standard microscope. The desired hole pattern is selected by translating the disk in the optical beam, a feature not possible with the TSRLM.

Light that passes through the disk is focused onto the sample by the objective lens. The field lens below the disk matches the pupil of the objective lens into the illumination system as discussed in Section 2.3.3. Reflected or fluorescent light from the sample retraces its path through the optics to the Nipkow disk. The image formed at the top surface of the disk is relayed to the eyepieces by the lens assembly located just above the beamsplitter. In the Reichert form of the microscope, this lens assembly has been specially corrected so that the beamsplitter does not introduce additional aberrations into the system.

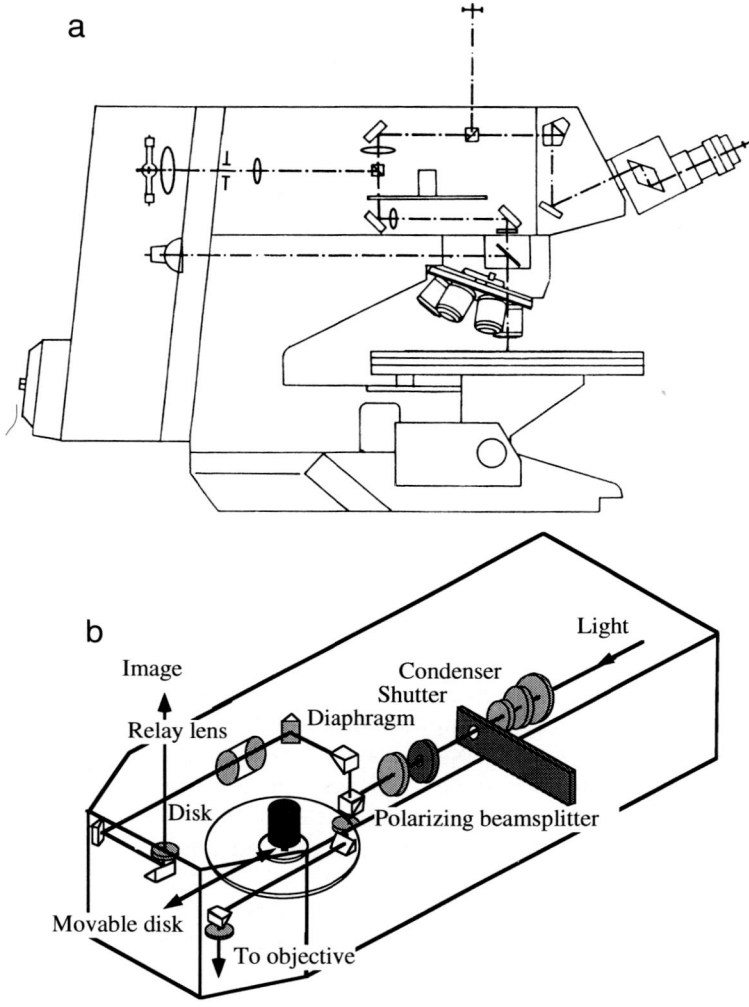

Figure 2.11 (a) A simplified diagram of the Technical Instruments CSOM integrated into a Reichert microscope. (b) A diagram of the K2-IND CSOM module. [Courtesy: V. Cejna and G. Q. Xiao, Technical Instruments Corporation, San Jose, California, USA.]

A polarization-based optical isolator and a tilted disk are used to eliminate the light reflected from the top of the disk. The main beamsplitter located just above the disk is a polarizing beamsplitter. It is the only polarizing element in the optical system, although provision has been made to insert additional polarization components into the illuminator if it is necessary. The λ/4 waveplate is located just after the second fold mirror. In the Reichert microscope, there is the capability of differential interference contrast imaging by replacing the λ/4 plate with a Wollaston prism, as discussed in Chapter 4. In the biological form of this microscope, the polarizing beamsplitter is replaced by a movable dichroic beamsplitter which allows different choices of fluorescence lines.

2.4 Slit Microscopes

The light efficiency of a scanning optical microscope can be improved by using slits rather than pinholes as the source.[55,56] This option is particularly attractive in the Nipkow disk microscopes, where the light efficiency is low. Slits allow more light through the limiting apertures resulting in a higher intensity at the image. In addition, in a beam scanning CSLM, the beam need only be scanned in one direction, simplifying the optical system. The disadvantage of the slit microscope is that the background level when the system is defocused is high and the transverse resolution along the direction of the slit is the same as for a standard microscope. Other disadvantages are that the use of coherent illumination, such as that from a laser, can cause speckle to appear in the nonscanned direction of the image.

In this section the designs of several different scanning slit microscopes will be reviewed. Two of these instruments are representative of the types of microscopes used in ophthalmology.[57] The third uses bilateral scanning and the fourth is a hybrid slit/pinhole microscope. The latter two instruments are commercial microscopes that were designed to compete with the CSLMs described in Sections 2.2.8 and 2.3.7.

2.4.1 Ophthalmologic Slit Microscopes

An early form of confocal slit microscope was built by Maurice, who used it for observation of the cornea of the eye.[58] The basic configuration is shown in Fig. 2.12. In this microscope, one side of the objective is illuminated through a slit. Light reflected from the object passes back through the other side of the objective, then through a second slit located at the intermediate image plane of the microscope, before striking a photographic film plate. The film and the object are scanned synchronously to form an image. With this method Maurice was able to obtain high-quality images of the eye that rival those obtained by the best scanning microscopes today.

The idea was later improved by Koester, who used the scanning mirror system, shown in Fig. 2.13, to form real-time images.[59] In Koester's

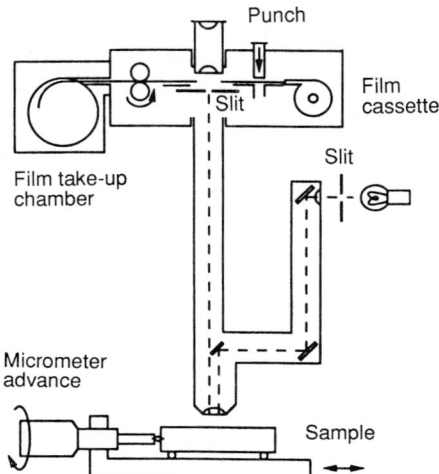

Figure 2.12 Schematic of an early confocal slit microscope. [From C. J. Koester, "A comparison of various optical sectioning methods: The scanning slit confocal microscope," *Handbook of Biological Confocal Microscopy,* J. B. Pawley, editor, 189–194 (Plenum Publishing, 1990), with permission.]

microscope the illuminated image of a slit is scanned across the focal plane by one face of an oscillating triangular mirror. The illumination is limited by an aperture to one side of the objective pupil so that light which is reflected from the sample can be spatially separated from the illumination. This light is then reflected from the second face of the triangular mirror before being focused onto a second slit. Because the beam reflected from the sample is reflected, in turn, from the second facet of the oscillating mirror, the image is descanned and thus is stationary at the second slit.

After passing through the detector slit, the light beam is directed to the third face of the triangular mirror that scans the image over the detector, allowing a stationary multielement detector to be used. In this microscope the oscillating mirror was vibrated at a frequency of 1 kHz, to produce a real-time image. The principle of operation is very similar to that of the bilateral scanning microscope described below. The disadvantage of this arrangement compared with a bilateral scanning microscope is that since only one side of the objective is used, the point spread function is not symmetrical and hence the resolution is somewhat degraded.

2.4 Slit Microscopes

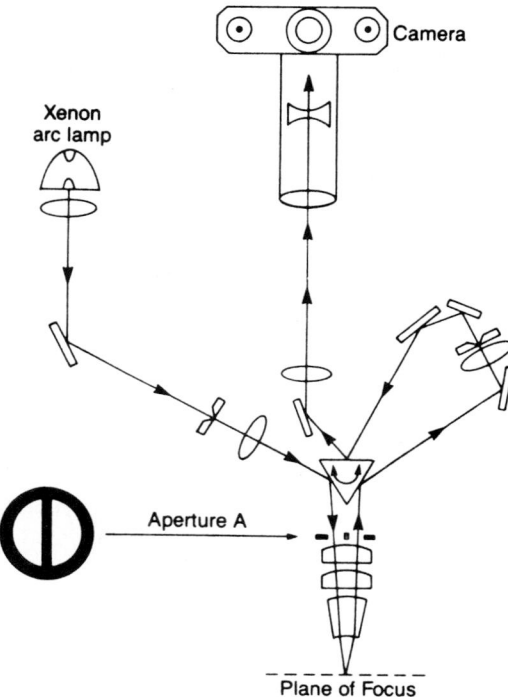

Figure 2.13 Schematic of an improved scanning mirror slit microscope. [From C. J. Koester, "A comparison of various optical sectioning methods: the scanning slit confocal microscope," *Handbook of Biological Confocal Microscopy,* J. B. Pawley, editor, 189–194 (Plenum Publishing, 1990), with permission.]

2.4.2 Bilateral Scanning Slit Microscopes

A commercial implementation of the bilateral microscope described in Section 2.2.7 is shown in Fig. 2.14. The INSIGHT PLUS Bilateral Scanning optical microscope was developed by Meridian Instruments, Inc. to compete with other fluorescence imaging laser scanning microscopes. The aim of the design is to obtain real-time imaging of fluorescent samples with good sensitivity. Like the commercial CSLMs described in Section 2.2.8, the INSIGHT PLUS is a scanning head designed to mount onto an upright or inverted standard optical microscope.

In this instrument, the light is generated by an external laser source. Laser options include single-line argon (488 nm), multiline argon

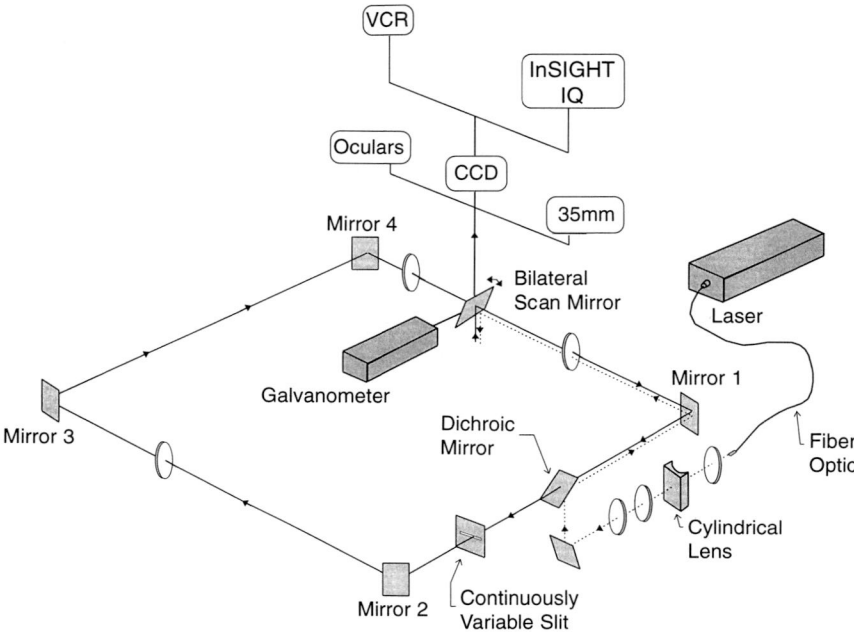

Figure 2.14 Simplified diagram of a commercial bilateral scanning slit microscope. [Courtesy: A. W. de Feijter, Meridian Instruments Inc., Okemos, Michigan, USA.]

(simultaneous 488 nm and 514 nm), or krypton (simultaneous 568 nm and 647 nm). The laser light is coupled into a single-mode optical fiber, which transmits the light to the scanning head. For dual laser operation, two lasers, typically krypton and argon, are each coupled into a separate optical fiber. The two fibers are brought into proximity at the input to the scanning head.

The output of the fiber passes through a cylindrical lens, laser line filter, and an optional neutral density filter before being reflected through a dichroic beamsplitter to the patented bilateral scanning mirror; the beamsplitter behaves like a perfect mirror to the input beam. The cylindrical lens spreads the image of the fiber in one direction so that a line of light rather than a point of light is imaged onto the sample. For fluorescence imaging, the dichroic mirror is typically a long-pass dichroic filter optimized for the wavelength of the laser. When multiple lasers are used for illumination, a multiband polychromatic mirror may be substituted for the dichroic mirror shown in the figure. For reflection mode imaging, the dichroic mirror is replaced by a 20%/80% (reflection/transmission) polarizing beamsplitting cube. In this case a quarter-wave plate placed immediately prior to the objective lens completes the optical isolator. All these different beamsplitter assemblies can be prealigned to permit rapid exchange without the need to realign the entire optical path.

The dichroic mirror reflects the laser light via mirror 1 to the galvanometer-driven bilateral scanning mirror. The scan lens between mirror 1 and the bilateral scanning mirror ensures that the line illumination is focused onto the sample. The bilateral scanning mirror oscillates at 60 Hz, thus scanning the image at 120 frames per second. Reflected or fluorescent light returns along the same path and is descanned by the bilateral scanning mirror before being focused on the slit aperture. The continuously variable slit aperture blocks the out-of-focus signal and allows for the optimum tradeoff between depth resolution and light efficiency. After passing through the slit aperture the signal is directed by mirrors 2, 3, and 4 to the back side of the bilateral scanning mirror, which scans the line signal over the detector. The lenses located near mirrors 3 and 4 focus the slit image onto the detector plane.

The detector may be a human eye observing the sample through a set of eyepieces or any of a variety of cameras. Intensified or cooled CCD cameras can be used for detection of low-light-level fluorescence. For multiple-wavelength fluorescence imaging, either a color CCD alone or a monochrome CCD camera in combination with a computer-controlled filter wheel is used. From the camera the image can be sent to a VCR or a computer frame grabber for further processing.

2.4.3 Hybrid Slit Microscopes

Another type of scanning slit microscope that has been successfully produced for the commercial market by Lasertec in Japan is shown in Fig. 2.15. The microscope differs from the conventional slit microscope because it uses a point source to illuminate the sample and a linear CCD array to form a slit detector. This design produces images that are free from both speckle and geometric distortions caused by a nonlinear scan. In the Lasertec microscope, polarization is used to separate the illuminating from the reflected beam, and scanning is carried out in the x and y directions with an AO scanner and a scanning mirror, respectively. Relay lenses are used to image the scanners into the telecentric pupil of the objective lens.

The Lasertec microscope represents one of a class of hybrid instruments. These are scanning optical microscopes that combine pinholes and slits in order to increase the light efficiency and scanning speed of the instruments. There are three common arrangements for hybrid confocal

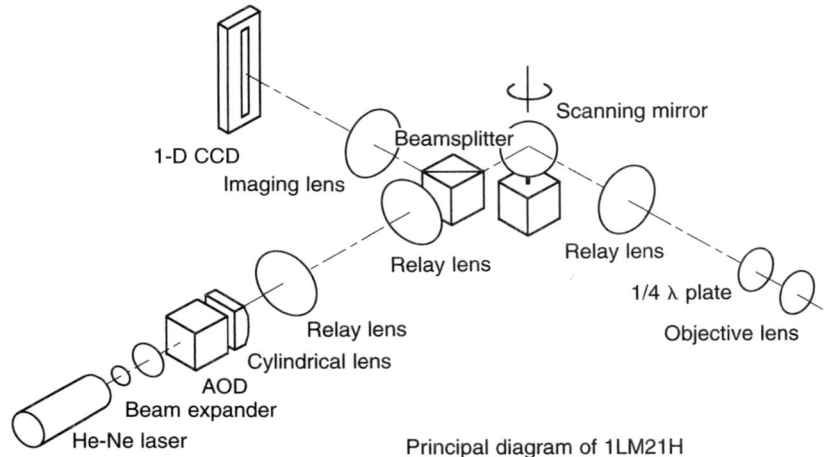

Figure 2.15 Schematic of a confocal scanning slit microscope developed by Lasertec Corporation [Courtesy: T. Ode, Lasertec Corporation, Yokohama, Japan.]

microscopes:[60] (1) a slit for illumination and a point detector, as shown in Fig. 2.16(a); (2) point illumination and a slit detector as shown in Fig. 2.16(b); (3) point illumination with a one-dimensional CCD array detector, as shown in Fig. 2.16(c).

The slit illumination method shown in Fig. 2.16(a) has the highest efficiency because more of the source power is used to illuminate the sample. In addition, with slit illumination the beam need only be scanned in one direction, simplifying the optical system. The disadvantages of this arrangement are that the use of coherent laser illumination can cause speckle to appear in the nonscanned direction. In addition, the use of cylindrical optics for beam steering can result in different numerical apertures for the illumination in the scanned and nonscanned directions.

Some of these issues are addressed by the use of point illumination and a slit detector as shown in Fig. 2.16(b). Point illumination eliminates the speckle in the image and allows equal numerical apertures to be used in both directions. Since the light is now scanned in both directions, an extra set of scanners is required. The resolution of this arrangement is set, in the direction parallel to the slit, by the spot size of the illumination. Because the entire focused beam, including its sidelobes, is detected in this direction, the sidelobe level is higher than for a CSOM using point illumination and point detection. In the direction perpendicular to the slit, the resolution is comparable to that of a CSOM. In addition, a complication

2.4 Slit Microscopes

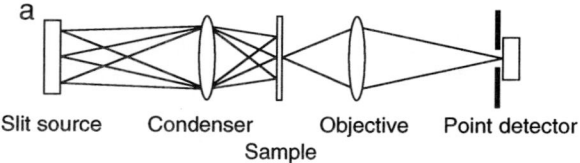

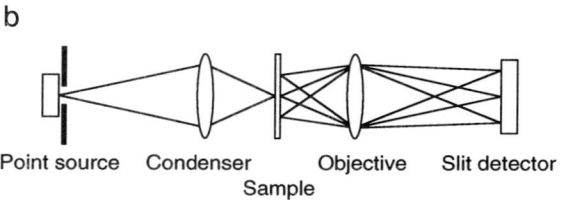

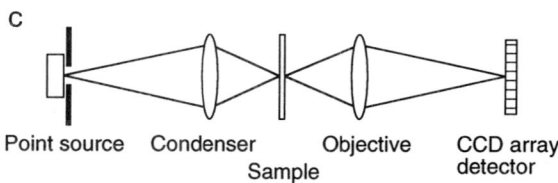

Figure 2.16 Three common arrangements for hybrid confocal microscopes: (a) slit illumination and a point detector, (b) point illumination and a slit detector, and (c) point illumination with a one-dimensional CCD array detector.

of this design is that geometric distortion of the scan will show up as a geometric distortion of the image.

The microscope shown schematically in Fig. 2.16(c) is closest to the classical CSOM and so has the best imaging performance. In this arrangement the resolution is determined by a combination of the spot size of the illumination and the size of the detector elements of the CCD array. The use of a CCD array eliminates geometric distortion in the direction parallel to the array because each pixel in the detector can be assigned to a unique location on the sample. Geometric distortions in the direction perpendicular to the detector array will still, however, pose a problem.

These instruments illustrate the variety of slit microscopes that have been devised. As with more conventional confocal systems, slit microscopes trade off light efficiency and resolution. The success of the slit approach depends on the type of sample being observed and the required image resolution.

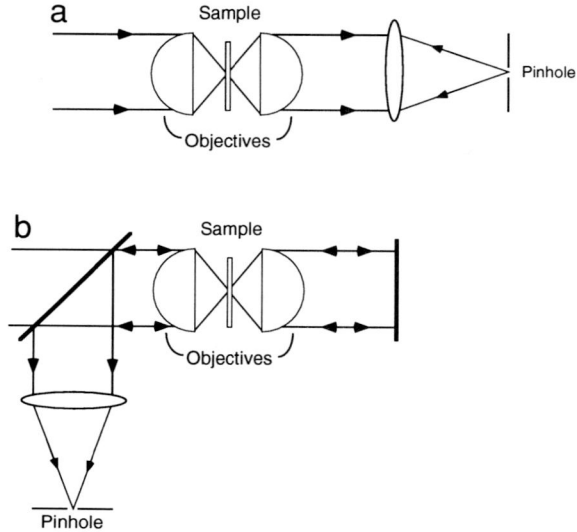

Figure 2.17 Simplified schematics of CSTMs: (a) a single-pass transmission microscope; (b) a two-pass transmission microscope.

2.5 Confocal Transmission Microscopes

The CSOM has been used mainly in the reflection mode. However, it is also possible to make a *confocal scanning transmission microscope* (CSTM). Brakenhoff constructed the first microscope of this type by placing two objectives on each side of a transparent sample with the second lens followed by a pinhole, as illustrated schematically in Fig. 2.17(a).[61] A CSTM will have the range resolution and remove glare from out-of-focus layers in a transparent material like a reflection confocal microscope. These instruments are particularly suitable for inspection of weakly-reflecting biological materials and for samples such as photomasks, which are designed for operation in transmitted light. The microscope is difficult to use, however, because two objective lenses are required, and the optimum distance and alignment between the two objectives are determined by refraction in the transparent medium through which the focused beam passes. In addition, many versions of this system can be used only with stage scanning.

Several attempts have been made to make the CSTM more flexible and compatible with the reflecting CSOM. Sheppard and Wilson have developed one approach to this problem as shown in Fig. 2.17(b).[62] In

2.5 Confocal Transmission Microscopes

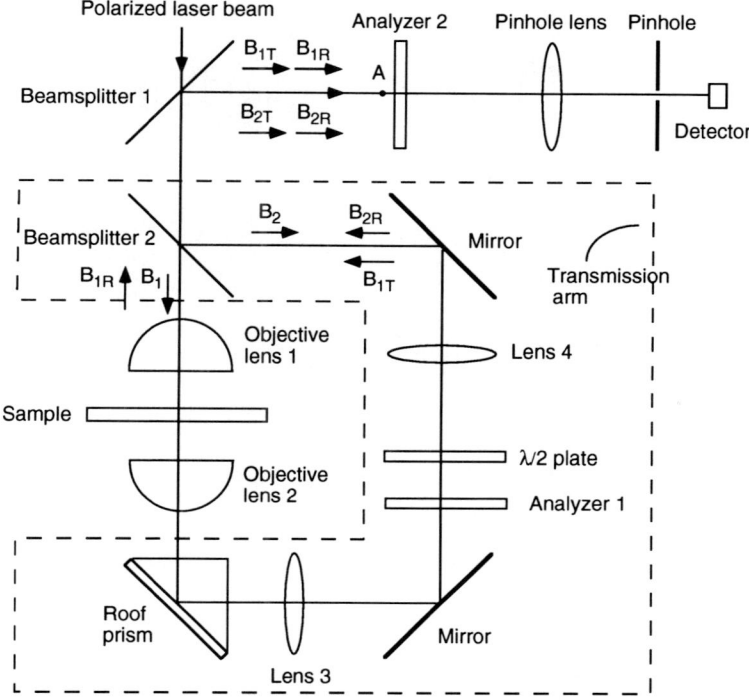

Figure 2.18 A flexible transmission or reflection confocal microscope.

their instrument the light transmitted by the sample is collected by a second objective lens and then reflected by a mirror, so that it passes back through the sample. Since the beam passes through the sample twice, aberrations caused by the absorbing and refracting characteristics of the sample will distort the beam so that the image quality may be worse than for the single-pass system shown in Fig. 2.17(a). On the other hand, the amplitude PSF of the transmission-reflection system should be $h^4(r)$, which should be narrower than the $h^2(r)$ response of a CSOM. The distortion and alignment problems, however, tend to outweigh this advantage.

A powerful approach devised by Dixon et al. is shown in Fig. 2.18.[63,64] The idea is to make a one-pass transmission system that, with small changes, can become a reflection microscope with reflection from either the back or front of the sample. The microscope can also be used in a combined mode in which the sum of the reflections from both the top and bottom of the sample is observed. This combined mode is not the same

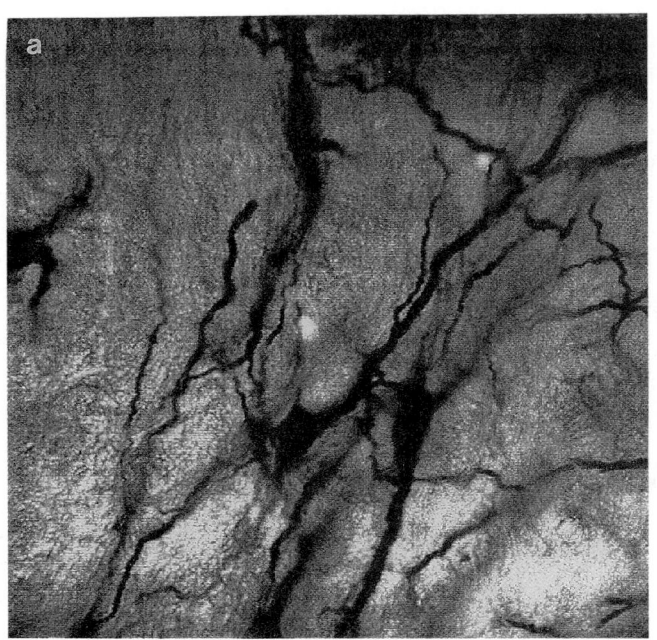

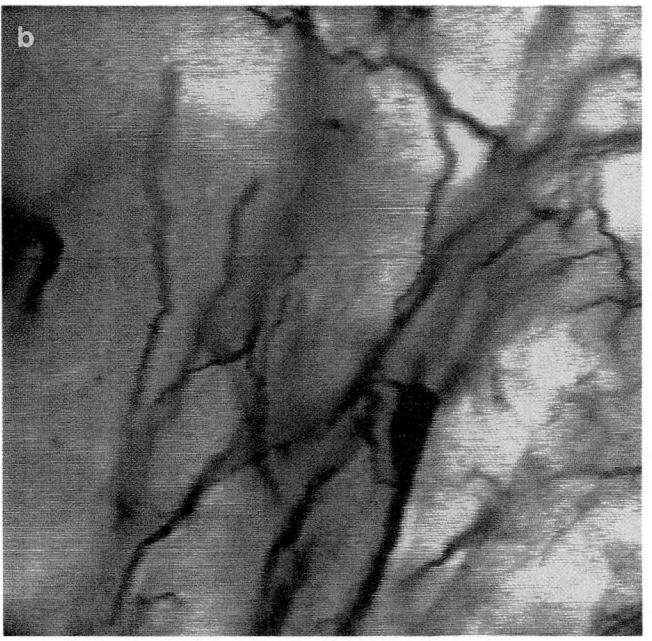

2.5 Confocal Transmission Microscopes

as that of the 4Pi microscope discussed below and in Chapter 4, since the path lengths from the two reflections are different. The microscope uses a polarized laser source. By suitable rotation of analyzers and a half-wave plate, the beam can be directed along different paths. In the system shown in Fig. 2.18, beamsplitter 2 splits the input beam into two parts B_1 and B_2, which propagate in an anticlockwise or clockwise direction, respectively, around the instrument. In transmission, analyzer 1 is aligned with the polarization of beam B_1. The transmitted light beam B_{1T} passes through the half-wave plate, where its polarization is rotated by 90°. It is then reflected by beamsplitters 2 and 1 through analyzer 2 to the pinhole and detector. The detector can receive the transmitted beam only if analyzer 2 is rotated by 90° from analyzer 1. On the other hand, if analyzer 2 is kept at the 0° position, the transmitted signal will not be received and only a signal reflected from the top side of the sample, B_{1R}, in the normal reflection confocal microscope mode will be imaged. Other modes of operation are possible. For instance, if the light at objective 1 is blocked, the system can be operated in a confocal reflection mode with reflection from the bottom of the sample and analyzer 1 perpendicular to the direction of polarization of the incident light and analyzer 2 aligned at 90° to analyzer 1. Finally, by leaving objective lens 1 unblocked, with the analyzers in the same position, the sum of the reflected signals $B_{1R} + B_{2R}$, from the top and bottom of the sample can be obtained.

The transmission system has been designed with the rooftop prism and lenses 3 and 4 so that a point on the center of the back pupil of objective lens 1 is imaged onto the point A. This makes it possible to use a beam scanner and thus obtain a relatively fast scan. A transmission image taken with this microscope along with a similar image taken without the pinhole present (a standard transmission microscope image) is shown in Fig. 2.19. The sample is an 80-μm-thick, Golgi stained specimen containing neurons imaged using a matched pair of 100×/0.7 *N.A.* objective lenses. Figure 2.19(a) shows a transmission image of the specimen obtained with the confocal detector. Figure 2.19(b) is a nonconfocal scanned image obtained by replacing the second objective lens in the transmission arm with a large-area silicon detector. The scan area is 200 × 200 μm, at

Figure 2.19 Two images of a 80-μm-thick, Golgi stained specimen. (a) A CSTM image of the specimen obtained with the confocal detector. (b) A non-confocal scanned image obtained by replacing the second objective lens in the transmission arm with a large-area silicon detector. The scan area is 200 × 200 μm, at a depth of 40 μm into the sample. [Courtesy: A. E. Dixon and S. Damaskinos, University of Waterloo, Waterloo, Ontario, Canada.]

a depth of 40 μm into the sample. The removal of glare and improved sharpness are readily apparent in the CSTM.

Improved differential phase imaging with better contrast than in the standard microscope using a Nomarski cell has been obtained with a similar stage scanned system designed by Cogswell and Sheppard.[65] It is apparent that by making the system somewhat more complex than for a reflection mode CSOM, CSTM imaging can provide quality transmission images.

2.6 Alternative Imaging Configurations

Throughout this text we have emphasized that the requirements for a CSOM are a point source, point detector, and a confocal lens system. A more accurate statement is that spatially coherent illumination, spatially coherent detection, and a confocal lens system are needed. Coherent illumination and detection can be carried out either optically or computationally. In this section we will discuss two alternative methods of generating coherent illumination and coherent detection. The first is a hardware solution that uses the coherence properties of a laser to build a CSOM. The second uses deconvolution software to reconstruct an image with a shallow depth response.

Laser Detection The coherence properties of a laser can be used to build a CSOM without any pinholes in the optical path as described by Juskaitis et al.[66,67] In their microscope, a single-mode laser is used as both the source and detector. Coherent illumination of the sample is achieved by passing the output of the laser through a beam expander so that it uniformly fills the pupil of the objective lens as described in Section 2.2.1. The light is then focused to a single point on the sample. To use the laser as a coherent detector, the light reflected from the sample is directed back into the laser cavity and the signal is detected with a separate diode as a modulation of the laser power. Coherent detection occurs because only that portion of the reflected field from the point x_0, y_0, z_0 on the sample, $\mathbf{E}_R(x,y,z,x_0,y_0,z_0)$ within the laser cavity that matches the geometrical field variation of the lasing mode $\mathbf{E}_L(x,y,z)$ modulates the laser power. Thus the modulation in laser power ΔP is of the form

$$\Delta P(x_0,y_0,z_0) = K \int \mathbf{E}_R(x,y,z,x_0,y_0,z_0)\mathbf{E}_L^*(x,y,z)\, dxdydz \quad (2.3)$$

where K is a constant. This modulation depends on the amplitude and phase of the reflected signal, and is proportional to both the reflectivity of the sample and the defocus distance in a way similar to the response of standard CSOM.

2.6 Alternative Imaging Configurations

The image generated by this type of confocal microscope depends on the type of laser being used. For a single-mode helium-neon gas laser with a long correlation length, the modulation of the laser power is proportional to the real part of the optical field that is coupled back into a lasing mode after reflection from the sample and depends on both the amplitude and the phase of the reflected light. Since the image, in this case, is proportional to the field amplitude, it is similar to that produced by the interference microscopes described in Section 2.6.

Semiconductor diode lasers, on the other hand, have a short correlation length, so that phase correlation is random. The net result is to eliminate any phase interference that may be detected between the two signals. The output is thus proportional to the envelope of the function $\Delta P(x_0, y_0, z_0)$ given in Eq. (2.3).

In both these instruments the output signal is proportional to the amplitude of the reflected field. Traditional CSOMs using conventional detectors yield image signals proportional to the intensity of the reflected light.

The advantage of this microscope is the simplicity of the optical system. Either the laser power can be detected, or the change in current through a laser diode used as the output signal. This technique is particularly well suited for use with fiber optics, where a single diode can be used as both source and detector. Extrapolating, such a laser feedback technique would be particularly suitable if it proved possible to work with a scanning system switching from one element to another of a multiple array of semiconductor diode lasers.[68]

The nonlinear nature of the laser feedback system can also provide amplification and be made very sensitive to small signal changes. The disadvantage of nonlinear operation is that similar reflectivity changes on the sample may not appear with similar magnitude in the image. Although the pinholes have been eliminated from this microscope, the instrument, like any other CSLM, is still critically sensitive to the alignment of the components. Small changes in the alignment of the return beam can affect the modulation intensity and thus appear in the image as a defocus or reflectivity change on the sample.

Deconvolution of a Standard Microscope Image Deconvolution software can also be used to emulate the properties of a CSOM using an image taken on a standard optical microscope. A number of different algorithms are available for the deconvolution.[69] Most rely on an accurate knowledge of the *point spread function* (PSF) of the microscope. Some of the advanced algorithms are able to calculate the PSF from several images stacked together. In these systems the original image must be carefully acquired using both averaging and background subtraction to obtain the best results. Since light from out-of-focus planes is not eliminated from the image, this method cannot be used to image a weak reflecting layer adjacent (in z) to a strongly reflecting layer. The weak image will be swamped by light reflected from the out-of-focus planes. In addition, the lack of an accurate PSF will ultimately limit the instrument's measurement accuracy.

Despite these limitations, digital deconvolution had been successfully used to image a variety of biological samples. One further application has been the combination of a CSOM and digital deconvolution image analysis. This combination has been reported to improve the spatial resolution and increase the signal-to-noise ratio of the CSOM image.[69]

2.7 Interference Microscopes

Interferometric microscopes use a different approach than the CSOM to generate images with a shallow depth of focus. These microscopes detect the interference signal between a reference and a probe beam. Since this imaging method utilizes a form of coherent detection, these instruments share many of the properties of the CSOM.

A wide variety of interference optical microscopes have been developed for different applications in optical microscopy. These include laser-based CSOMs adapted for interference imaging, custom-designed instruments containing highly stable interferometers, and conventional microscopes with interferometric lenses in place of a standard microscope objective. In this section, we will discuss a number of these instruments, beginning with interference CSOMs. From there we will move to microscopes with interferometric objective lenses. Three different interferometric lens configurations will be examined: Michelson in Section 2.7.2, Linnik in Section 2.7.3, and Mirau interferometers in Section 2.7.4.[70,71,72] In the discussion of these instruments, emphases will be placed on objective lenses incorporating a Mirau interferometer. The use of this type of interferometer with high numerical aperture lenses is relatively new. Section 2.7 will close with a brief discussion of the Tolanski multiple beam interference microscope in Section 2.7.5.

2.7.1 Interference CSOMs

Brakenhoff's Interference CSOM One of the earliest examples of an interference CSOM reported in the literature was developed by Brakenhoff et al. at the University of Amsterdam.[61] In their microscope, a reference path was added to a transmission CSOM forming the Mach-Zehnder inter-

2.7 Interference Microscopes

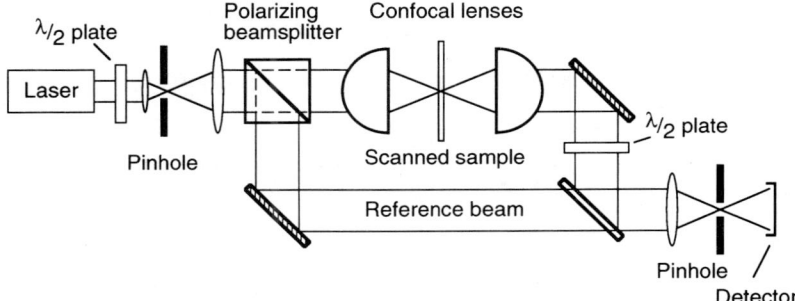

Figure 2.20 Schematic of the transmission interference CSOM which uses a Mach-Zehnder interferometer to provide the reference path for the microscope.

ferometer shown in Fig. 2.20. In addition to a reference path, the microscope contains two λ/2 plates designed to give control over the relative phase and amplitude of the light in the interferometer. These waveplates enable either a phase or amplitude image of the sample to be made. The purpose of the first waveplate is to adjust the relative power in the two arms of the interferometer. Since the light from the laser is linearly polarized, rotating this waveplate changes the polarization direction of the light incident on the polarizing beamsplitter, which, in turn, will change the amplitude and phase of the light in each arm of the interferometer.

After passing through the polarizing beamsplitter, the polarization directions of the light in each arm of the interferometer are mutually perpendicular. Orthogonally polarized light beams will not normally interfere at the detector. To generate an interference image, the polarization direction of either the sample or reference beam must be further rotated by 90° before entering the second conventional-type beamsplitter. The rotation is accomplished by the second λ/2 waveplate.

Hamilton and Sheppard's Interferometric CSOM Shortly after Brakenhoff's publication, Hamilton and Sheppard reported a reflection CSOM based on the Michelson interferometer shown schematically in Fig. 2.21.[73] Their microscope used two detectors so that either an amplitude image or an interferogram of the sample could be observed. To generate separate amplitude and phase images, Hamilton and Sheppard exploited the fact that the phase difference is $\pi/2$ between the beam passing directly through and the beam deflected by a beamsplitter. Consequently, when two beams are combined in a beamsplitter, the outputs from the two different faces of beamsplitter correspond to the sum and difference

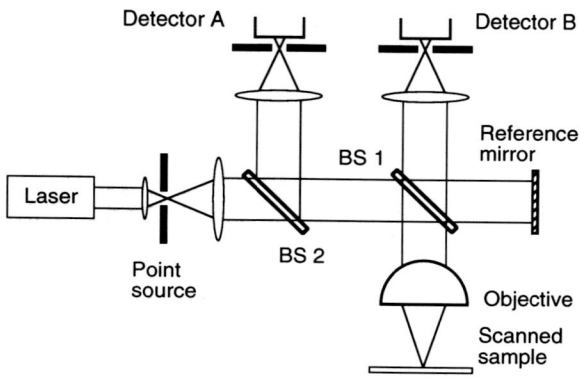

Figure 2.21 Schematic of a reflection interference CSOM which uses an interferometer and two detectors to form an image.

of the original beams. For the optical arrangement shown in Fig. 2.21, the signals at the two detectors A and B are given by

$$I_{A,B} = I_1 \pm I_2 \cos(\phi_S - \phi_R). \quad (2.4)$$

In Eq. (2.4) I_1 is the reflected intensity from the sample, I_2 is the intensity of the modulations in the interference term, and ϕ_S and ϕ_R are the phases of the light reflected from the sample and reference surfaces, respectively.

The sum and difference of the two detector signals are used to extract the phase and amplitude terms from the detected responses:

$$I_S = I_A + I_B = 2I_1 \quad (2.5)$$

and

$$I_D = I_A - I_B = 2I_2 \cos(\phi_S - \phi_R). \quad (2.6)$$

The sum signal I_S is a normal CSOM intensity image, while the difference signal, I_D, is an interferogram of the sample modulated by its reflectivity and the correlation function of the two beams. The interferogram can be analyzed to extract the phase information by using one of the techniques developed for the Mirau or Linnik interference microscope, described in Chapter 4.

The need to control the relative intensity of the light in both arms of the interferometer is a common problem for interference CSOMs. In Hamilton and Sheppard's microscope, the ratio of the intensities reflected from the sample and reference must be the same at each detector in order for the sum and difference signal to be accurate. This requirement can be

2.7 Interference Microscopes

met if the beamsplitter BS1 has equal transmittance and reflectance. Fine adjustment of the intensity in each arm can be made by using a linearly polarized laser source and rotating the laser until the two ratios become equal.

The 4Pi Microscope An interesting alternative to the phase imaging CSOMs described above has been proposed by Hell and Stelzer.[74] They used two opposing microscope objective lenses to illuminate a fluorescent sample from both sides and to collect the fluorescent emission from both sides. Interference of either the illumination or the collected wavefronts in a common detector pinhole generates the interferogram.

The confocal interference microscope has several advantages compared with a standard CSOM. It is able to image phase information while maintaining the shallow depth response and improved transverse resolution of the CSOM. Since light from out-of-focus planes does not reach the detector, the confocal interference microscope can be used to produce an interferogram of a weakly reflecting object on a highly reflecting substrate.

A major disadvantage of many of the interferometric microscopes described above is that they employ long-path interferometers with dc detection and hence are sensitive to vibrations and other environmental factors. A second disadvantage of these microscopes is that they do not directly measure the phase of the sample beam. Instead, an interferogram is produced from which the phase information can be extracted. These disadvantages can be overcome, in principle, by using an acousto-optic or electro-optic cell to phase modulate the beam so that ac detection techniques can be used. Further improvements can be made by ensuring that the reference and sample beams follow as close to a common path as possible. Some examples of systems that fulfill this criterion are the Zernike phase-contrast microscope, described in Chapter 4, and the Mirau correlation microscope (MCM) described in Section 2.7.4.

2.7.2 The Michelson Interference Microscope

An interference microscope that uses a Michelson interferometer built into the objective lens is shown in Fig. 2.22. In this microscope a focused beam from the objective is split into two components by a beamsplitter. The beamsplitter directs part of the light to a reference mirror and part to the sample. After reflection from the sample and reference mirror, the beams recombine at the beamsplitter and are relayed to the detector. There, light from the reference and sample interferes, forming a fringe pattern or interferogram, which depends on both the sample topography

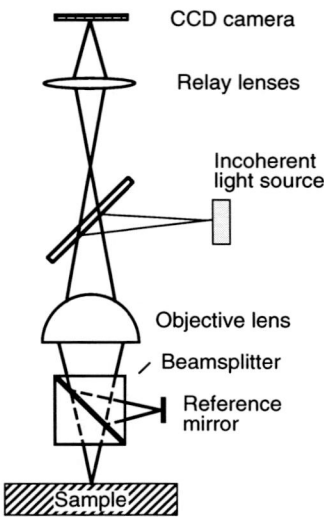

Figure 2.22 Schematic of an interference microscope based on a Michelson interferometer.

and the distance between the reference mirror and the sample. The interferogram can be processed to extract depth information or the phase difference between the two beams. The phase is used to accurately determine height changes on the sample. The signal processing techniques used to calculate the phase and amplitude of the reflected light from the interferogram are discussed in Chapter 4.

The envelope of the interference signal depends on both the illumination bandwidth and the numerical aperture of the objective lens. If a laser is used for illumination, the reference and sample beams from a low numerical aperture lens will interfere even if the path length difference is long. If a broadband arc lamp is used as the source, the short coherence length of the light will limit the width of the interference fringes to a few microns as discussed in Chapter 3. However, if the numerical aperture is large, the envelope of the interference pattern of even a laser-illuminated microscope will have much the same shape as the amplitude variation of the depth response in a confocal microscope with the same aperture.

In a Michelson interferometric microscope, the beamsplitter must be fitted underneath the objective lens. This placement severely limits the numerical aperture of the objective lens that can be used, since most high numerical aperture objectives have short working distances. In addition, the objective must be corrected for aberrations introduced by the beam-

2.7 Interference Microscopes

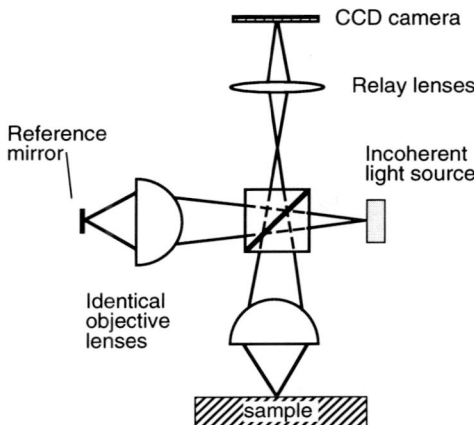

Figure 2.23 Schematic of an interference microscope based on a Linnik interferometer.

splitter so that conventional microscope objectives cannot be used to produce high-quality images. These limitations make Michelson interference microscopes best suited for low-resolution applications.

2.7.3 The Linnik Interference Microscope

Some of the disadvantages of the Michelson interference microscope can be overcome by moving to a Linnik configuration, shown in Fig. 2.23. The Linnik microscope uses two matched objective lenses and a beamsplitter located between the objectives. With this design, high numerical aperture lenses can be used to produce high-resolution, diffraction-limited images without correcting for aberrations introduced by the beamsplitter. In the Linnik microscope, however, the beam paths are longer than in the Michelson interferometer, so that both mechanical vibrations and thermal expansion must be tightly controlled to keep the interference path constant to within a fraction of a wavelength. In addition, finding two matched lenses with minimum aberrations can vastly increase the complexity of the lens sorting and testing problem. Despite some of these challenges, a successful commercial instrument has been developed by KLA Instruments in San Jose, California.

The KLA 5000 Coherence Probe Metrology System was designed specifically for semiconductor metrology.[70,75] It uses a Linnik microscope with two $100\times/0.9\,N.A.$ Leitz objective lenses. The high magnification and numerical aperture are necessary for accurate measurements of submicron

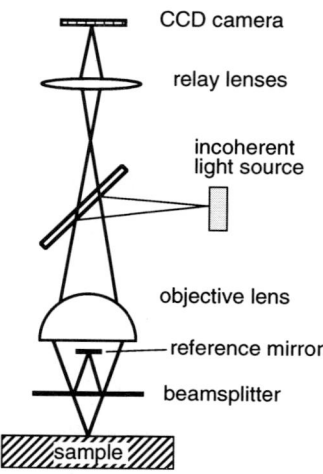

Figure 2.24 Schematic of an interference microscope based on a Mirau interferometer.

structures. In addition to high-quality objectives, the beamsplitter is another critical element in the Linnik microscope. The transmitted wavefront distortion of the beamsplitter should be as small as possible, but certainly better than $\lambda/10$. A large transmitted wavefront error can significantly change the fringe pattern of the interferogram. If the change occurs near the edge of a line that is being measured, there will be a large error in the reported linewidth.

Since the KLA 5000 uses a white light source, the path lengths in the two arms of the interferometer must be identical to within less than a wavelength. This requirement means that fine focus control must be used to adjust the position of the objectives, sample, and reference mirror. In the KLA system, automated calibration adjustments are made periodically to the positions of both the reference mirror and the beamsplitter. In addition, the system is built massively to avoid vibration problems.

2.7.4 The Mirau Interference Microscope

The Mirau interference microscope, like the Michelson interference microscope, has a beamsplitter below the objective lens. The Mirau interferometer consists of a beamsplitter and a reference mirror placed underneath a long working distance objective, as shown in Fig. 2.24. At the beamsplitter, approximately half of the incident light from the objective lens is reflected toward the reference mirror; the remainder is transmitted

2.7 Interference Microscopes

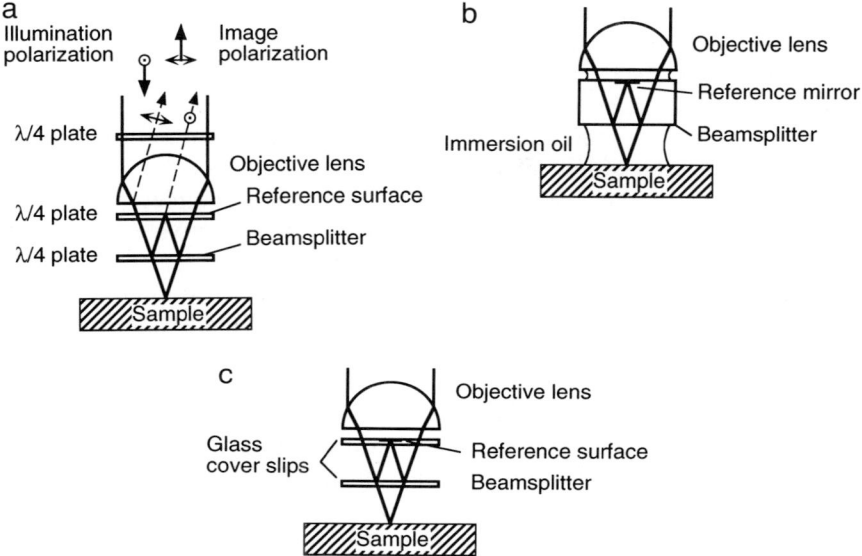

Figure 2.25 Alternative configuration for an objective lens based on a Mirau interferometer: (a) a low-magnification lens where the reference spot is replaced with an unobscuring reference surface, (b) a high numerical aperture oil immersion lens where the beamsplitter and reference surface are located on the opposite sides of a large glass block, and (c) with the beamsplitter coated on a glass coverslip.

to the sample. After reflection from the sample and the reference mirror, the two light beams recombine at the beamsplitter and pass back through the objective lens to the detector. Since only one objective is used for both the signal and reference beams, the range aberrations of the system are minimized. The Mirau interference microscope also does not have the long interference beam paths typical of the Linnik microscope and is, thus, less sensitive to vibration and temperature drift. However, the presence of a thin-film beamsplitter beneath the objective can limit the numerical aperture of the objective lens.

Unobscured Reference Surface There are several alternatives to the simplified microscope described above.[76] For low-magnification imaging the reference spot is so large that it obscures a significant portion of the objective. In these applications, the reference spot can be replaced with an unobscuring reference surface, illustrated in Fig. 2.25(a). The absence of the central reference spot improves both the light efficiency and spatial frequency response of the objective lens, thereby improving the contrast.

If the illumination bandwidth is narrow, polarization effects can be used to reduce the amount of background light reflected back to the eyepieces from the beamsplitter and reference surface. Using polarized illumination and an analyzer in front of the eyepieces, an optical isolator can be constructed which will eliminate the light reflected from the beamsplitter, shown by the dashed line in Fig. 2.25(a). To complete the optical isolator, a set of $\lambda/4$ plates above and below the objective ensures that the illumination beam is circularly polarized while the reference beam is linearly polarized at the beamsplitter. The dielectric coating of the beamsplitter must then be designed to pass the illumination beam but reflect the reference beam.

Oil Immersion System It is also possible to build a Mirau interferometer for a high numerical aperture, high-magnification microscope objective. The highest numerical aperture objectives are usually oil immersion lenses. With these lenses both the beamsplitter and reference surface can be located on the opposite sides of a large glass block that is index matched to the immersion oil, Fig. 2.25(b). In this case the presence of the glass introduces minimal aberrations into the system because the light does not refract at its surface.

In the intermediate region of numerical aperture, a Mirau interferometer can be constructed by coating a beamsplitter on a glass coverslip and suspending it below an objective that is corrected for spherical and chromatic aberration introduced by the coverslip, Fig. 2.25(c). The reference spot can be placed either directly on the last element of the objective lens or on another coverslip held just in front of the objective lens. The disadvantage of this configuration is that any thickness nonuniformity or other wavefront aberrations due to the coverslip will be observed in the interferogram and the system has to be very carefully aligned.

To minimize the aberrations introduced by the beamsplitter's plate, it should be made as thin as possible. Plastic membranes approximately 2 μm-thick have been used to produce Mirau interferometers commercially. These beamsplitters, however, are too thick to be used with high numerical aperture lenses, as shown in Chapter 3. In order to avoid image degradation, a microscope objective with a numerical aperture of 0.8 should have a beamsplitter membrane no thicker than 100 nm. Recently, therefore, research has been directed at producing high-quality, thin-film beamsplitters.[77]

2.7 Interference Microscopes

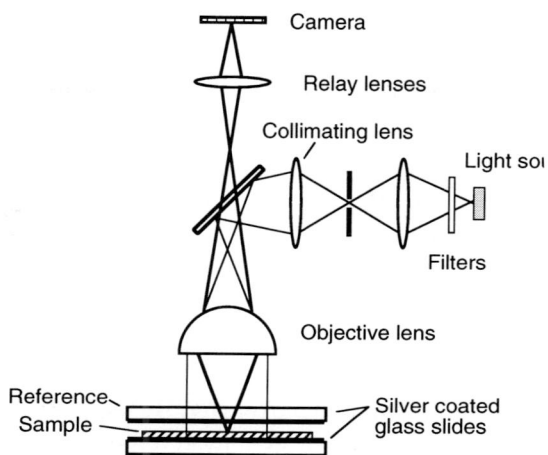

Figure 2.26 Schematic of the interferometer used in a Tolanski interference microscope.

2.7.5 The Tolanski Interference Microscope

The Tolanski multiple-beam interference microscope improves the sensitivity of the microscope to height changes by creating a multiple-beam interference pattern.[78,79,80] The physical arrangement is similar to a Mirau interferometer except that the beamsplitter and the sample are illuminated with a collimated beam of light and are positioned so that they form a wedged air gap, Fig. 2.26.[81] An interferogram of the sample is formed by light that has undergone multiple reflections between the reference and sample surface. The optical system used for multiple-beam interferometry differs slightly from a standard optical or confocal scanning optical microscope because of the requirement that the illumination be normally incident on the beamsplitter. In a multiple-beam interference microscope the light source is focused at the back focal plane of the objective, rather than on the pupil plane as is common with Köhler illumination. In addition, the illumination path should be equipped with a variety of narrowband interference filters in order to aid the formation of sharp fringes with a large number of multiply reflected beams.

In multiple-beam interferometry the spacing of the fringes is the same as in a Linnik or Mirau interferometer, $\sim(\lambda/2)$; however, the width of an individual fringe can be very narrow. Under optimum conditions the width of a single fringe can be 50 times narrower than the fringes in a Mirau or Linnik interferometer. The fringe width is determined by the number of paths that contribute to the interference pattern; the greater the number of multiple reflections, the finer and sharper the fringes. However, because the higher order beams are reduced in intensity by the multiple reflections, the width and brightness of the fringes also depend on the reflectivity of the substrate and beamsplitter. In order to achieve optimum results it is often necessary to coat the sample with a thin layer (50–100 nm) of high-reflectivity metal. For this reason the Tolanski interference microscope has not been widely adopted as a measurement tool.

2.8 Near-Field Microscopy

The minimum spot size of a standard or confocal microscope is limited by diffraction to approximately $0.5\lambda/(N.A.)$. By using an ultraviolet light source, one might hope to obtain definition almost a factor of 2 better than in the optical range. However, this approach is very difficult because most of the lens materials that are normally used in the visible range are opaque to far UV wavelengths (<250 nm). We are therefore left with a limited choice of materials like fused silica and magnesium fluoride. Such limited choices make it difficult to design high numerical aperture compound lenses with compensation for chromatic and other aberrations. An alternative is to use reflecting Schwarzchild lenses rather than refracting lenses. These lenses, however, are difficult to accurately align, and are not readily available with numerical apertures larger than 0.6. Furthermore, their characteristics are usually not adequate for precision microscopy at UV wavelengths.

In this section, we will describe an alternative approach for obtaining improved definition, near-field imaging. In Section 2.8.1, the *near-field scanning optical microscope* (NSOM) is introduced, and in Section 2.8.2 some applications of the NSOM are discussed. The microscope makes use of a small pinhole placed near the sample to enhance the definition. In Section 2.8.3, a second near-field technique, *the solid immersion microscope* (SIM), which makes use of a high-refractive-index material to obtain effectively shorter wavelengths, is described.

2.8.1 The Near-Field Scanning Optical Microscope

The NSOM is a device in which the incident or received light passes through a small pinhole and the definition is determined by the size of the pinhole rather than by diffraction. Although this technique is not strictly confocal or interference microscopy, we shall review it here because it makes possible observation of structures with resolutions

2.8 Near-Field Microscopy

of a small fraction of a wavelength, and some techniques employed in this connection follow fairly naturally from the earlier work on confocal microscopy.

The basic idea, which has been reinvented several times, was suggested by Synge in 1928,[82] rediscovered by O'Keefe in 1956,[83] and rediscovered again and demonstrated by Ash and Nicholls at microwave frequencies in 1972.[84] In Ash and Nicholls' experiment, a microwave cavity with a 0.5-mm hole drilled in it, resonant at 10 GHz corresponding to a free space wavelength of 3 cm, was used. They observed the perturbation in resonant frequency of the cavity when the fields leaking through the hole interacted with a 0.5-mm period grating located outside the cavity. By these means, they were able to observe a structure with a periodicity of only 1/60 of a wavelength.

In 1984 Pohl at IBM Zurich and Lewis et al. at Cornell University, in independent work, demonstrated the first optical microscopes using this technique.[85,86] The basic idea of these systems was to use a glass rod tapered down to a small diameter, cover it with a metal film by deposition in a vacuum, and introduce a small pinhole at its end. To obtain a definition comparable to the diameter of the pinhole, the pinhole must be placed within a distance from the sample comparable to its radius. Otherwise the fringing fields of the pinhole spread out over a relatively large area, as discussed in more detail in Chapter 3.

Several configurations are possible, as illustrated in Fig. 2.27. Figure 2.27(a), the collection mode, shows a transmission system with light incident from the bottom of the sample. Light transmitted through the sample passes through a pinhole into a metal-covered glass rod or fiber, and then on to a detector. To form an image, the fiber is scanned in an x–y raster pattern by means discussed later, while keeping the spacing between the pinhole and the sample constant. Figure 2.27(b) illustrates the transmission mode NSOM in which the sample is illuminated through the fiber and pinhole. The light transmitted by the sample is collected using a large-aperture lens which focuses it onto a detector. Typically, in the configurations described above, the incident or received light beams are focused into the neighborhood of the pinhole by a microscope objective.

It is also possible to work in a reflection mode, by focusing the incident light onto the sample from one side of the pinhole, as shown in Fig. 2.27(c). Finally, it is possible to work in a frustrated internal reflection mode by coupling the light into the sample through a prism, so that it is totally internally reflected at the top surface of the sample. The light incident on the top surface of the sample can be detected by a fiber probe and pinhole, as shown in Fig. 2.27(d).

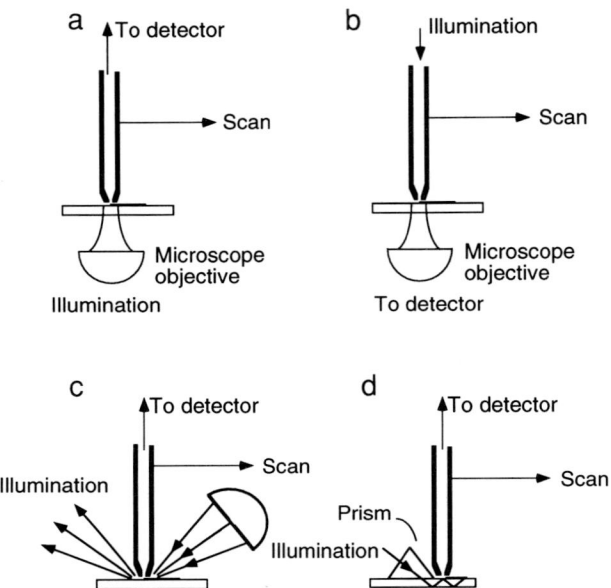

Figure 2.27 Different configurations for near-field imaging: (a) collection mode; (b) transmission mode; (c) reflection mode; (d) frustrated internal reflection mode.

In principle, the pinhole size can be decreased indefinitely to improve the definition. However, the ultimate limit of the definition is, in fact, determined by penetration of the fields into the metal in the region around the pinhole. With aluminum, the penetration depth due to skin effect at a wavelength of 500 nm is approximately 6.5 nm, so that the ultimate resolution is of the order of 13 nm. As will be discussed below, one approach that has been tried to circumvent this limit is to use a metal probe to reflect light rather than a hole to transmit light.

Pipette System   Pohl made his first devices by etching a small quartz pipette in hydrofluoric acid to a conical shape and covering it with metal.[87] The end was scraped against a hard surface to remove metal and form a pinhole. Betzig et al. drew down a glass pipette with a flame, using techniques commonly employed for medical applications, and covered the pipette with metal. Since the drawn-down pipette was hollow, the metal film would have a hole with a well-controlled size of less than 0.1-μm diameter in its end.[88,89]

2.8 Near-Field Microscopy

Fiber System The technique for manufacturing the probes using optical fibers most commonly employed was developed by Betzig. In his method a CO_2 laser is used to heat the fiber while it is drawn down to a controllable size. The fiber is broken off at its thinnest point, and metal is evaporated on the fiber from the side as it is rotated in the deposition chamber, which forms the pinhole automatically.[90] A major advantage of this technique is that, since an adiabatically-tapered optical fiber is employed, light can be transmitted or received directly through the fiber. The optical waveguide diameter is below the cutoff wavelength only for the last micron or so of its length, so that some light still reaches the end of the fiber. It is desirable to use a transparent material with a high refractive index, so that the waveguide cutoff at a radius $a = 0.29 \, \lambda/n$ for the lowest order mode, the TE_{11} mode, occurs at the smallest possible radius, where λ is the optical wavelength in free space and n the refractive index of the material. Normally the standard material employed is fused silica ($n = 1.5$), although in some experiments unmetallized cleaved diamond samples ($n = 2.4$) or silicon nitride ($n = 2$) has been used. These probes lend themselves well to polarization imaging and fluorescence imaging, although the sensitivity is poor for fluorescence imaging so that collection of data is slow.

Optical Losses One major problem with the use of these near-field optical techniques is the high transmission loss of the light. The most efficient device is the adiabatically tapered optical fiber because the coupling into the fiber is relatively efficient. Even for this device, however, the attenuation in the cutoff region near the end of the fiber will be of the order of 40–50 dBs, as is discussed in Chapter 3. In addition, since the definition of the system is of the order of 50 nm and the spot size of the beam formed by the objective lens used to focus the transmitted or received light will be of the order of 1 μm, there will be at least an additional 20 dBs of coupling loss from the pinhole to the objective. In Betzig's experiments, with 15 mW of input power and an 80-nm pinhole, he obtained nanowatts at the detector, corresponding to about 70 dBs loss.[96] The pipette systems are even lossier, since a multimode fiber is used to introduce the signal into a relatively large core; there are large-mode conversion losses at this stage—typically, an additional 20 dBs. Because of the high one-way attenuation, it is not possible to operate either of these systems in a direct reflection mode with the light introduced into and received from the fiber.

Positioning The scanning systems commonly employed for near-field optics are based on those developed for scanning probe microscopes such as the *scanning tunneling microscope* (STM) and *atomic force micro-*

scope (AFM). The antivibration mounts are also similar to those developed for the STM/AFM. It is also advisable to keep the NSOM small in order to minimize vibration problems.

To accommodate a coarse x,y,z movement, stepping motors or an inchworm movement which is capable of a slow scan over several inches of travel with steps of 2.5 nm is used.[91] The fine movement in the x, y, and z directions is carried out with a piezoelectric tube scanner, a device now generally available for use with scanning probe microscopes and capable of 10-μm movement in the x and y directions with 1-μm movement in the z direction.

In the NSOM it is necessary to maintain the fiber at a fixed distance of less than half the pinhole diameter of the order of 20 nm, away from the sample. Atomic force microscope techniques are well developed and have proved to be the most useful in this application.[92,93,94,95,96] In practice, either the normal component or transverse component of force between the probe itself and the sample is used to produce a separate force microscope image that controls the distance between the probe and the sample.

Normal forces are difficult to employ because the fiber or pipette is rigid in the direction perpendicular to the sample. Shalom et al. overcame this difficulty by using a pipette bent near its tip and detected the change in vibration frequency of the pipette as the tip approached the sample.[92] Tortonese et al. manufactured small silicon nitride apertures which were supported on a thin piezoresistive silicon membrane to measure the force between the probe and the sample.[93] However, in this first attempt, the resolution was limited to 0.25 μm.

In the system preferred by Betzig et al., the probe is moved parallel to the surface and its motion detected with a shear force sensor.[96] A simplified schematic of this system is shown in Fig. 2.28. Light is transmitted through a tapered fiber and the sample and observed through an objective lens. Part of the light leaving the objective lens passes through a beamsplitter and a pinhole into photomultiplier detector (PMT). Small transverse vibrations of the fiber will cause the signal received by the PMT to vary periodically. This periodic signal is detected with a lock-in amplifier and divided by the direct signal to remove intensity variations. The fiber itself is vibrated at its shear resonant frequency by a "dither piezo." As the fiber approaches the sample, the shear or frictional force on it changes, and the damping of the vibration resonance is affected. This change in vibration amplitude is used in a control circuit that moves the "scan piezo" to keep the vibration signal at a preset value determined by the required spacing between the fiber and the sample. By measuring the voltage applied to the control circuit, the same technique can be used to make a shear force image as in an AFM.

2.8 Near-Field Microscopy   125

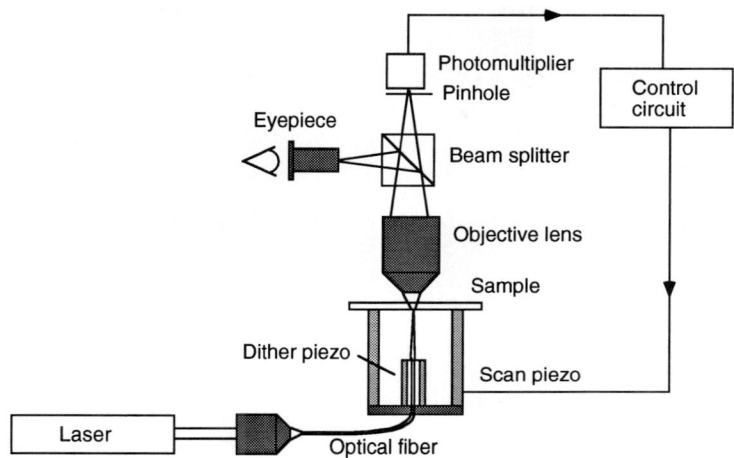

Figure 2.28 A simplified schematic of an NSOM using transverse force to regulate the probe-sample spacing.

Similar results were obtained independently at about the same time by R. Toledo-Crow et al.[94] who used two separate lasers, one for near-field imaging and the other with a Nomarski system, as described in Chapter 4, to measure the vibration of the cantilever formed by a pipette. Differential optical images were obtained at the oscillation frequency, along with *shear force microscope* (SFM) images and direct NSOM images at dc. An example of an image of latex spheres, using this technique, is shown in Fig. 2.29.

Probes for Near-Field Optical Imaging The NSOM systems using the fiber optical and pipette probes have several disadvantages: (1) the definition is limited to about 13 nm by penetration of fields into the metal; (2) the probes must be made individually and the parallel processing advantages of conventional optical imaging are lost; (3) the optical losses are very high. Various stratagems have been tried to decrease some of these difficulties.

To improve the efficiency and definition of the probe, small fluorescent particles have been placed on the end of a pipette and used as a light source,[97] the fiber has been doped with neodymium so that it lases,[98] and field enhancements using the plasmon resonances of small silver particles placed at the end of the fiber have been tried. None of these techniques has, as yet, been fully developed.[99,100]

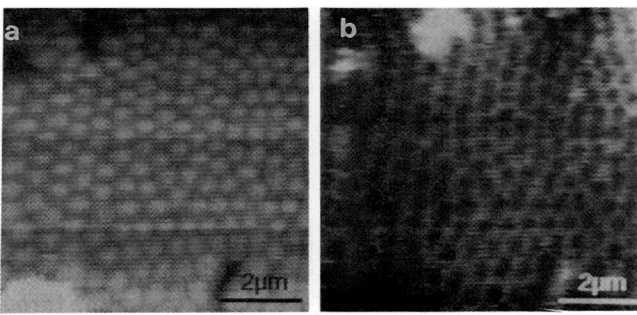

Figure 2.29 Latex spheres measured with (a) an NSOM; (b) a scanning force microscope. [From R. Toledo-Crow, P. C. Yang, Y. Chen, and M. Vaez-Iravani, "Near-field differential scanning optical microscope with atomic force regulation," Appl. Phys. Lett. **60**, 2957–2959 (1992), with permission.]

The standard probes discussed above can be destroyed if they touch the sample, and they do not work well on samples with rough surfaces. A different approach to making the probe is to use the technology already available for the AFM to build the probe on the end of a cantilever made with semiconductor microfabrication techniques. The basic idea, demonstrated by van Hulst et al., is to use a commercially available microfabricated silicon nitride probe with an integrated cantilever of silicon nitride (Park Scientific Instruments).[101] The hollow SiN pyramid-shaped probe, illustrated in Fig. 2.30, has a high refractive index ($n = 2$) with a 20–50-nm radius apex (compared to 250-nm radius for the fiber and pipette probes). The cantilever probe is flexible and can be operated in close contact with the sample to produce both AFM and NSOM images.

The experimental arrangement shown in Fig. 2.27(d) is used to generate an evanescent wave in the sample substrate with a HeNe laser beam focused to a 200-μm-diameter spot. The cantilever is held at an angle of 15° to the sample normal with the probe in contact with the surface. A 10-mm-long working distance objective (40×/0.5 *N.A.*) is used to collect the light from the tip of the probe and pass it through a pinhole to a photomultiplier detector, so as to eliminate other scattered light. In addition, the light passes through a beamsplitter to a CCD camera so that the probe tip can be observed.

In operation, as the probe approaches the substrate, the optical signal increases exponentially in amplitude, corresponding to the expected coupling to the exponentially decaying evanescent wave, until the tip jumps into contact with the sample. Due to adhesion in the presence of a water

2.8 Near-Field Microscopy

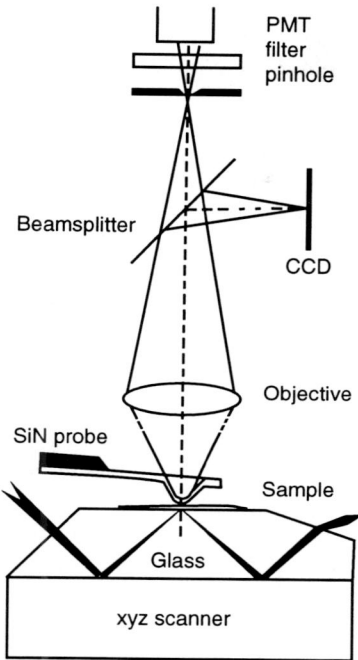

Figure 2.30 Near-field optical microscope with a silicon nitride cantilever probe. [From N. F. van Hulst, M. H. P. Moers, O. F. J. Noordman, R. G. Tack, F. B. Segensk, and B. Bölger, "Near-field optical microscope using a silicon-nitride probe," Appl. Phys. Lett. **62**, 461–463 (1993), with permission.]

film, the tip remains in contact with the sample as the probe is withdrawn and a constant optical signal of about 0.1 nW is observed. A plot of intensity versus movement of the probe toward the sample and then away again is shown in Fig. 2.31. It will be observed that when the tip is withdrawn, the cantilever bends by as much as 300 nm.

Optical images of gratings and other samples yield an estimated transverse definition of 50 nm, as compared to a 10–40-nm definition when used in a constant force mode AFM with a similar SiN cantilever probe. The advantage of this form of NSOM is that the scanning can be carried out with the probe in contact with the sample, since the force between the probe and the sample is only 20 nN. In addition, the technology lends itself well to the building of multiple probes made on one silicon substrate and thus to parallel processing of images. Eventually, it is anticipated by those working in the field that the probes can be excited from optical waveguides laid down on the same substrate.

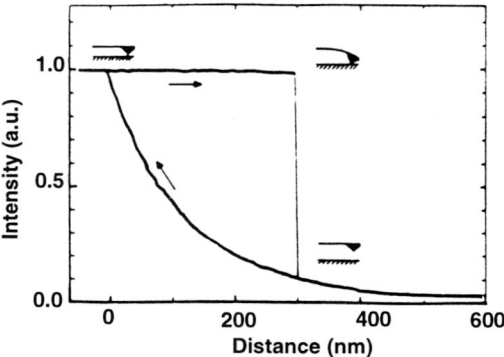

Figure 2.31 Optical signal as a function of tip-substrate distance on approaching and retracting the SiN probe. [From N. F. van Hulst, M. H. P. Moers, O. F. J. Noordman, R. G. Tack, F. B. Segensk, and B. Bölger, "Near-field optical microscope using a silicon-nitride probe," Appl. Phys. Lett. **62**, 461–463 (1993), with permission.]

Apertureless NSOM Zenhausern, O'Boyle, and Wickramasinghe have demonstrated an NSOM that operates by reflecting light from the tip of a scanning probe rather than passing light through a small pinhole.[102] The definition is now comparable to the radius of the probe tip, which can be of atomic dimensions, far smaller than a pinhole. However, extreme sensitivity is required to observe the weakly scattered light from the probe in the presence of other strong reflectors. The system shown in Fig. 2.32 uses a Nomarski objective (see Chapter 4), illuminated with a HeNe laser, to form two diffraction-limited spots about 100 μm apart on the bottom side of a transparent sample, in this case a microscope coverslip. The sharp silicon tip of an AFM cantilever is moved toward one of the spots. The tip is used as part of an AFM. Resonance frequency vibration of the cantilever at 239 kHz is used to control the tip-sample spacing.

The reflected electric field consists of two components, E_r from the back surface of the coverslip and a relatively weak component E_S from the tip. These two components are slightly phase shifted from each other due to the Gouy shift through a focused Gaussian beam.[103] Therefore, the total phase shift of the total field, $E = E_r + E_s$, relative to the phase of the second spot, is $\Delta\phi = KE_s/E_r$, where K is a constant.

The field E_s will contain not only components from the tip but also scattered light from the shank of the probe and the cantilever. To emphasize the tip component, several stratagems have been adopted by this group. First, they use a confocal microscope for detection, which limits the detected signal to within a few hundred nanometers of the tip. Second,

2.8 Near-Field Microscopy

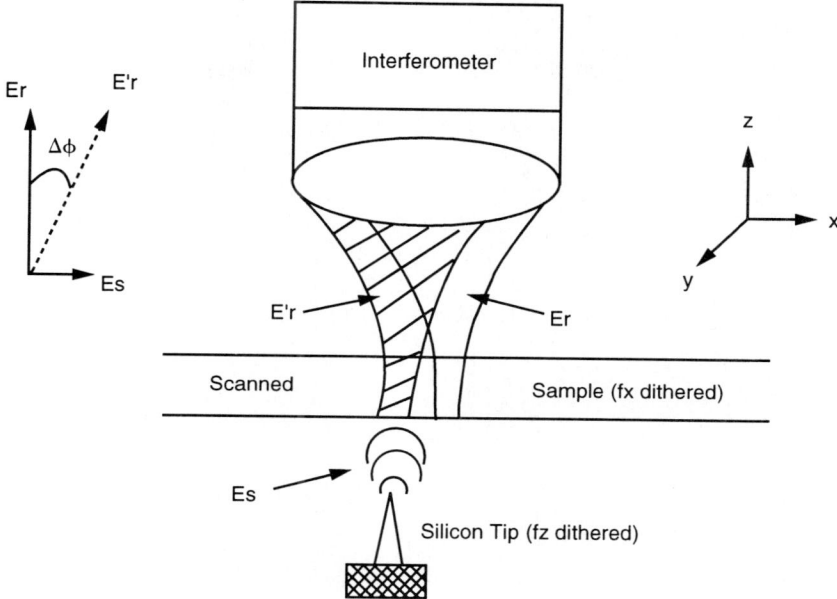

Figure 2.32 Concept for detecting the near-field from a sharp tip. [From F. Zenhausern, M. P. O'Boyle, and H. K. Wickramasinghe, "Apertureless near-field optical microscope," Appl. Phys. Lett. **65**, 1623–1625 (1994), with permission.]

they vibrate the tip in the z direction at a frequency $f_z = 1$ kHz, so that only the tip signal will be emphasized at this frequency. Third, they vibrate the sample laterally by approximately the tip radius at a frequency $f_x = 3$ kHz. The signal from the two spots is then passed into an interferometer, and detection is carried out at the sum frequency $f_x + f_z$ or 4 kHz. With the interferometer system used, the smallest detectable phase difference $\Delta\phi$ was of the order of 10^{-8} radians/$\sqrt{Hz}$.

The first experiments were performed using the back of the coverslip itself with a 2-nm-radius silicon tip. In later experiments, with an improved system, oil droplets on a Mica substrate were imaged. Figure 2.33 compares a interferometric near-field and AFM image of dispersed oil droplets. Figure 2.33(a) is an attractive mode AFM topographic image displayed in 3-D with correct scaling between the x, y, and z magnifications. Most of the droplets are imaged as bumps by the AFM; however, some are imaged as dips due to electrostatic charging effects. Figure 2.33(b) is a simultaneously recorded optical image in which the oil droplets appear as bright scattering regions. The smallest resolvable features in this image are

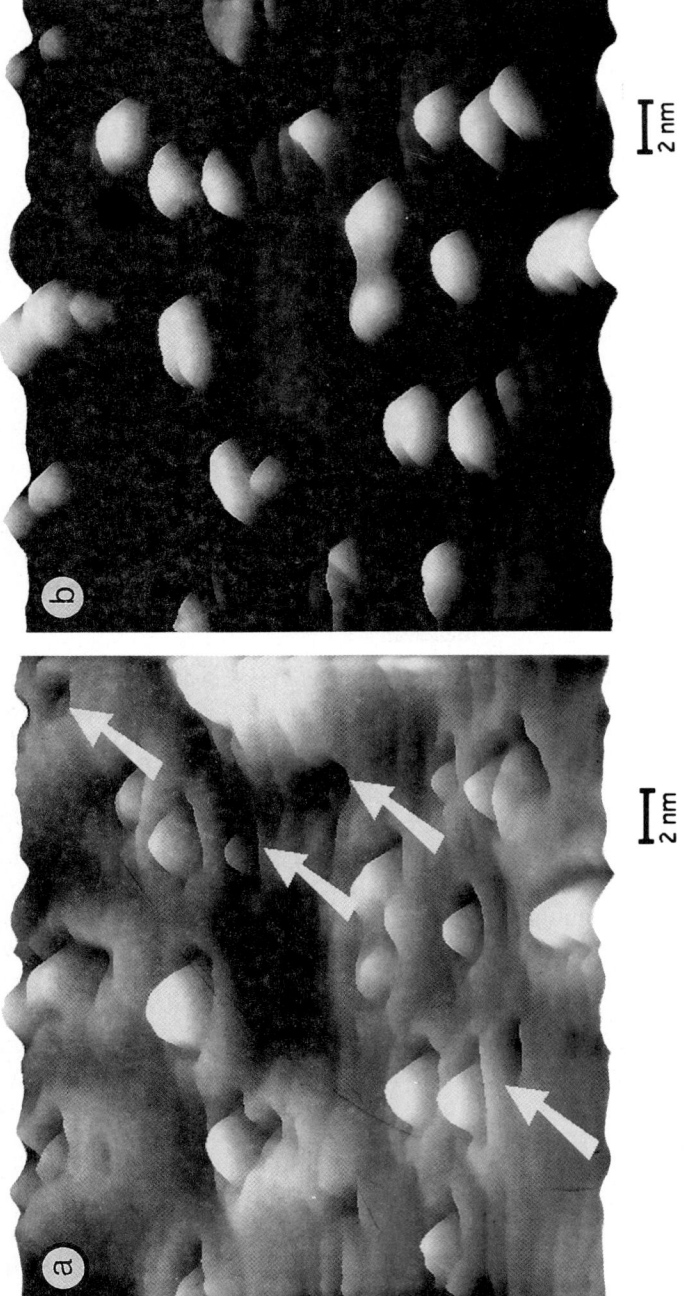

Figure 2.33 Comparison of interferometric near-field and AFM images of dispersed oil droplets on mica. (a) Attractive mode AFM topographic image. (b) Simultaneously recorded optical NSOM image showing the oil droplets as bright scattering regions; smallest resolvable feature ≈1 nm. [Courtesy: K. Wickramasinghe, IBM T. J. Watson Research Center, Yorktown Heights, New York, USA.]

2.8 Near-Field Microscopy

≈1 nm. These results are comparable in definition to those obtained with an AFM.

2.8.2 Applications of the NSOM

In addition to simple imaging of semiconductors, fibers, crystals, and biological materials, there are many possible applications of the NSOM. These applications include optical spectroscopy, optical lithography, studies of GaAs/AlGaAs quantum wells, measurement of surface plasmons, and magneto-optical storage. Brief descriptions of some but not all of these techniques are given below.

Spectroscopy of Individual Molecules   One recent exciting development in near-field imaging is optical spectroscopy of individual molecules carried out both at room temperature and at liquid He temperatures.[104,105] The technique is so sensitive that it is possible to determine from optical polarization measurements the direction of the dipoles giving rise to the fluorescence of the individual molecules.

Optical Lithography   It is difficult to carry out lithography with dimensions below 0.1 μm. One approach that has been tried recently is to use an NSOM with an ArF excimer laser operating at a wavelength of 193 nm. The laser light is guided through a hollow glass pipette to an NSOM probe, which directs it onto the sample. There it is used to ablate photoresist without heating effects. Linewidths as small as 70 nm have been fabricated with this technique, although, at the present time, the rate of exposure is too slow to be of practical importance.[106]

Magneto-optical Storage   When polarized light passes through a magnetic medium, its polarization may be rotated by the magnetic material. The effect, which can be exploited for magneto-optical storage, has been used by Betzig et al. with the near-field optical microscope for reading in transmission and writing on magneto-optical material.[107] The disk is subject to a magnetic field perpendicular to it surface. If a small region of the disk is heated by a light spot, the magnetic domains will rotate to line up with the field. If polarized light is then transmitted through this region, the polarization will be rotated by the Kerr effect. The polarization rotation can be detected with good sensitivity through a tapered fiber which maintains the polarization. Using a scheme similar to that shown in Fig. 2.28, with the addition of the polarization detection components, Betzig et al. were able to obtain a reading resolution of 30–50 nm consistently and in the writing mode a definition of 60 nm, corresponding to a density of

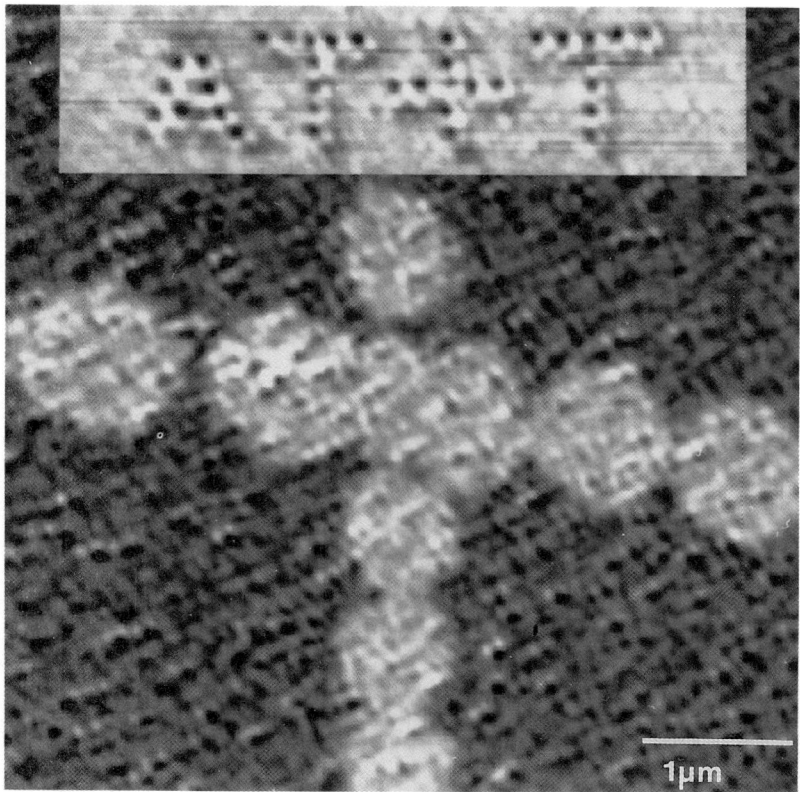

Figure 2.34 Near-field magneto-optic image of conventionally written domains. Inset: domains of ~100 nm diameter both written and imaged with near-field techniques. [From E. Betzig, J. K. Trautman, R. Wolfe, P. L. Finn, M. H. Kryder, and C. H. Chang, "Near-field magneto-optics and high density data storage," Appl. Phys. Lett. **61**, 142–144 (1992), with permission.]

45 Gbits/inch^2. Since the power passing through the fiber is so low, writing was most likely due to heating of the fiber itself with corresponding proximity heating of the disk. A near-field image of spots read into the magneto-optical disk with normal methods is shown in Fig. 2.34. The inset shows an image of spots, approximately 100 nm in diameter, written into and read from a magneto-optic material using near-field techniques.

We have described here just a few examples of the possible applications of the NSOM. Before leaving the subject, it is also worthwhile to

2.8 Near-Field Microscopy

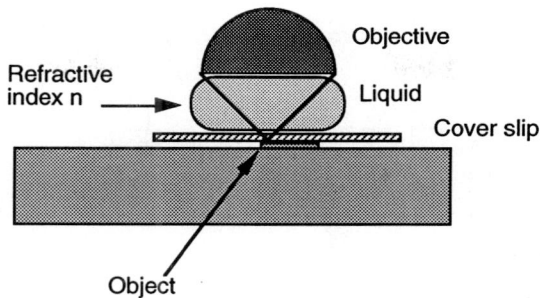

Figure 2.35 An illustration of Guerra's photon tunneling microscope. [From G. S. Kino, and S. M. Mansfield, "Solid immersion lens photon tunneling microscope," *Scanning Microscopy Instrumentation,* G. S. Kino, editor, SPIE **1556**, 2–10 (1991), with permission.]

consider an alternative near-field system which has higher efficiency but poorer resolution, the *solid immersion microscope* (SIM).

2.8.3 The Solid Immersion Microscope

Alternative systems for near-field microscopy based on the use of evanescent fields from internally reflected light have been demonstrated by Guerra[108] and by Mansfield and Kino.[109]

Guerra used a liquid immersion objective lens with the liquid in contact with a microscope coverslip placed close to the object being observed, as shown in Fig. 2.35. The wavelength in the liquid and the coverslip is reduced by a factor $1/n$, where n is the refractive index of the liquid. Hence, the resolution obtained in the coverslip material is reduced by the same factor as from an air immersion lens. However, rays that impinge upon the coverslip/air interface from inside the coverslip at an angle, θ, greater than the cutoff angle ($\sin\theta > 1/n$) will be internally reflected and have only exponentially-decaying fields on the air side of the interface. Therefore, the improved resolution can only be obtained near the surface of the object being observed when it is placed close to the coverslip. The advantage of this system is that a directly observable image is obtained without scanning. The disadvantage of the device is that the improvement in resolution is limited by the refractive index of the liquids commonly used with liquid immersion lenses ($n = 1.33$–1.45).

Mansfield and Kino constructed a related type of device, which they called the solid immersion microscope, that makes use of the principles of the liquid immersion microscope. However the SIM uses an additional *solid immersion lens* (SIL) element instead of a liquid as illustrated in

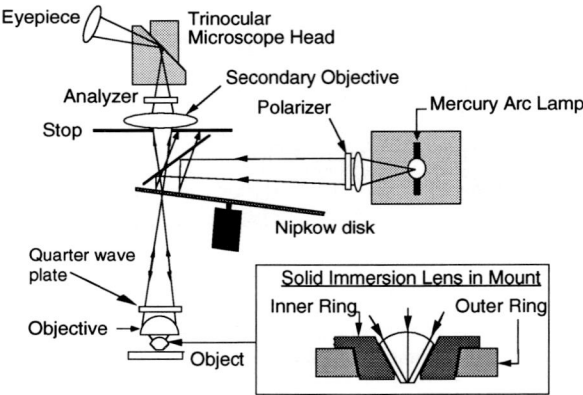

Figure 2.36 Solid immersion microscope. [From G. S. Kino and S. M. Mansfield, "Solid immersion lens photon tunneling microscope," *Scanning Microscopy Instrumentation*, G. S. Kino, editor, SPIE **1556**, 2–10 (1991), with permission.]

Fig. 2.36. Like Guerra's microscope, the SIM has a definition that is reduced by a factor of $1/n$ from that of a standard or confocal optical microscope, where n is the index of refraction of the SIL. Typically SILs are made out of glasses with indices up to about 2.0. Thus, with an objective of $N.A. = 0.8$, the effective $N.A.$ of the SIM would be 1.6, and the theoretical half-power width of a confocal microscope would be $0.37\lambda/(N.A.)$ or about 93 nm at a wavelength $\lambda = 405$ nm.

These resolution limits are not as small as those predicted for the NSOM but the SIM has several advantages over those systems. It provides a real-time image directly observable by eye or CCD camera, requires no mechanical scanning, has a better light budget, operates in reflection, and can be easily added on to an existing optical system like the RSOM.

The SIM uses a solid immersion lens placed between a long working distance objective and the sample. In the original configuration, the SIL was a hemisphere that had been ground into a cone, leaving about a 100-μm-diameter flat spot on the bottom surface. This flat surface was placed precisely on the focal plane of the objective lens and then centered, so that rays that focus on the optic axis would pass unrefracted through the spherical surface of the SIL and continue to focus directly onto the flat surface. Rays focusing off axis will make only small angles with the surface normal and will not be strongly aberrated. The system magnification is also increased by a factor of n and the field of view is decreased to a diameter w/n, where w is the field of view without the SIL. Simple theory and ray tracing show that for a $100\times/0.8$ $N.A.$ objective with 2-mm working

2.8 Near-Field Microscopy

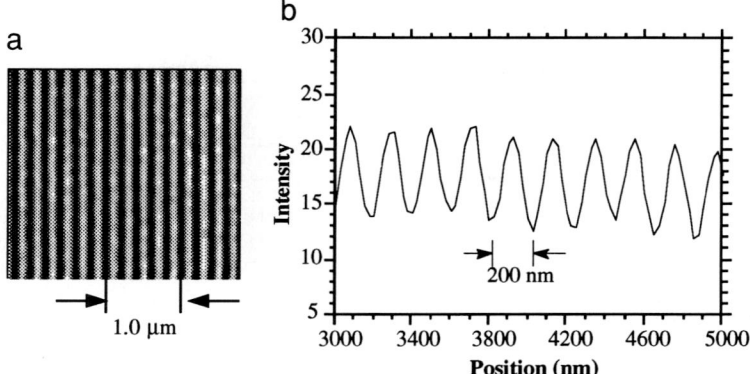

Figure 2.37 Measurement of a 100-nm linewidth grating with the SIM. (a) Gray scale image. (b) Line scan of intensity. [From G. S. Kino and S. M. Mansfield, "Solid immersion lens photon tunneling microscope," *Scanning Microscopy Instrumentation*, G. S. Kino, editor, SPIE **1556**, 2–10 (1991), with permission.]

distance, a full field of view of 30–40 μm in diameter can be achieved without significant degradation of the image. As with Guerra's system, the SIM uses evanescent fields that decay away from the bottom of the SIL and the sample must, therefore, be placed very close to the surface of the SIL.

The SIL has been incorporated in an RSOM as illustrated in Fig. 2.36. The RSOM was used because its range definition diminishes glare from out-of-focus objects, allows for easier focusing and tilt adjustment of the SIL, and has amplitude components in the image at twice the spatial frequency of a standard microscope. The SIL was placed in a floating mount so that it would not be damaged if it came in contact with the sample. The mount was held in a Vernier screw system that provided precise positioning of the SIL in the x, y, and z directions along with a tilt adjustment to get the SIL flat on the focal plane of the objective. The objective was also placed in a PZT tube to provide fine focusing and a method to adjust the focus while leaving the sample and SIL fixed. The sample could be precisely positioned under the SIL using piezoelectric transducers in the stage.

With an SIL of refractive index 1.9 and with a 0.8 $N.A.$ long working distance (2 mm) objective, the SIL had an effective $N.A.$ of 1.52. This SIL was used to image various small structures at a wavelength of 436 nm. The smallest structures imaged were 100-nm lines and spaces in photoresist shown in Fig. 2.37. In comparison, the smallest structures that were

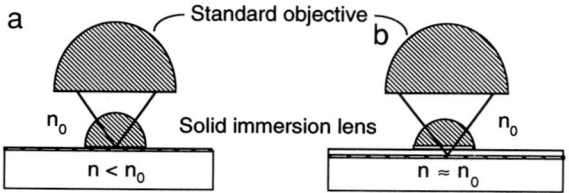

Figure 2.38 Illustration of a solid immersion lens. (a) Hemispherical lens with focusing at the bottom surface; (b) lens of same refractive index as the CD disk. [From S. Hayashi and G. S. Kino, "Solid immersion lens for optical storage," *Three-Dimensional Microscopy: Image Acquisition and Processing II*, T. Wilson and C. J. Cogswell, editors, SPIE **2412**, 80–87 (1995), with permission.]

observed with the confocal microscope, with a 0.9 *N.A.* objective under the same conditions, were 150-nm lines and spaces with a calculated minimum detectable periodicity of 242 nm (120-nm lines). The calculated minimum detectable periodicity for the SIL is 144 nm (72-nm lines). In Chapter 3, measurements of edge response for confocal microscopes and SIMs are shown which are in excellent agreement with theory.

Magneto-optical Storage Like the NSOM, this type of microscope can be used for magneto-optical storage, as well as for reading compact disks of higher density than the presently available systems. The SIL has the advantage that information on neighboring tracks can be read for tracking along with the image required, it can be read in reflection, and the power efficiency is high. The standard CD or magneto-optical disk is covered with a 1.2-mm-thick layer of plastic ($n = 1.5$) for protection. Normally, the beam is focused by a lens through the protective layer to the bottom surface of the disk, and control circuitry is employed with a magnetic actuator to move the objective up and down to follow variations in flatness and thickness of the protective layer. Since the beam is relatively large at the top surface, particles of dust do not present a severe problem, and the disk can be made removable. Aberration limit the maximum numerical aperture of the lens to about 0.55.

For operation with an SIL, there are two possible approaches: (1) as illustrated in Fig. 2.38(a), an SIL is used to focus on an active region at the surface of the disk. In this case, if the numerical aperture is greater than 1, the spacing of the lens from the disk must be kept to less than 100 nm to minimize aberrations; (2) an SIL can be used which has the same refractive index as the protective layer covering the disk, as illustrated in Fig. 2.38(b). In this case, if the numerical aperture is less than 1, there

2.8 Near-Field Microscopy

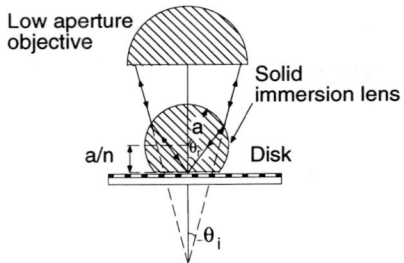

Figure 2.39 The stigmatic solid immersion lens. [From S. Hayashi and G. S. Kino, "Solid immersion lens for optical storage," *Three-Dimensional Microscopy: Image Acquisition and Processing II*, T. Wilson and C.J. Cogswell, editors, SPIE **2412**, 80–87 (1995), with permission.]

will be no totally reflected rays, and the air gap is not critical. By using an aspherical objective, it is possible to work with an air gap of as much as 100 μm with an effective *N.A.* of 0.8, although keeping the spacing constant is critical. A major advantage of this configuration is that it is not too sensitive to dust and can be operated in an unsealed environment. Its disadvantage is that a numerical aperture greater than 1 is not used so the increase in area density is limited.

By using a solid immersion lens with an effective numerical aperture of 0.83 and green light from a doubled semiconductor laser at 532 nm, the theoretical density should be ~5 Gbs/inch^2, about 10 times the density presently used in commercially available magneto-optic disks. So far, densities of the order of 3–4 Gbs/inch^2 have been observed with this configuration in our laboratory at Stanford in cooperative work with Sony.

A further improvement is to use a stigmatic focusing SIL, illustrated in Fig. 2.39. In this case, the beam emerging from the objective is focused to a point distant a/n from the center of the lens, and the effective *N.A.* of the objective is increased by a factor n^2, where a is the radius of the SIL.[110] The major advantage of this approach is that the numerical aperture of the objective can be relatively small so that the objective can have a relatively large working distance which allows room for the SIL. Just as with the hemispherical SIL, the beam can be focused through a transparent layer of the disk if it is chosen to be of the same refractive index as the lens. Alternatively, the beam can be focused onto the top surface of the disk, which makes it possible to obtain a spot size of the order of 120 nm, with a wavelength $\lambda = 450$ nm and a refractive index of 2. This could lead to a storage density greater than 30 Gbs/inch^2.

In this system, the lens must be placed very close to the disk (typically less than 100 nm), and it is obviously advantageous to work with materials with a high refractive index. Calculations have been made of the air gap which indicate that if the air gap is kept to less than 100 nm, the aberrations are not severe enough to affect the spot size. The conclusion is that it is preferable to use an SIL in the form of a floating head, much as is used in standard magnetic disk storage. Following this line of reasoning, it is possible to make suitable lenses with a refractive index of 1.96 of the order of 0.5–1 mm in diameter.

Experiments and theory carried out by Terris et al. at IBM Almaden and at Stanford have shown that loss of polarization information is not a major problem.[111,112] Terris et al. have been able to read spots in a magneto-optical disk made for this purpose without difficulty. Furthermore, with the stigmatic lens, they were able to obtain a measured Gaussian beam profile with a half-width of 317 nm, in good agreement with the calculated value of 304 nm, using a 780-nm laser source with an objective with $N.A. = 0.55$ and an $n = 1.9$ index glass for a stigmatic SIL. The potential storage density was, therefore, increased from that of the objective alone by an order of magnitude with the use of the SIL.

A floating air bearing can be constructed for use with the SIL so that it can float at a distance of less than 100 nm from the disk as it rotates. The control system for the objective can be very similar to that presently used, since the critical focus distance is magnified by n^2 at the objective. Furthermore, since this system can provide images over a finite field of view of the order of 100 spot widths in diameter, it should be possible to use standard tracking methods and thus make a viable and practical high-density optical storage system.

2.9 Conclusion

A wide variety of confocal scanning and interference optical microscopes are available today. Each microscope has its advantages and disadvantages relating to the many samples available to the researcher. For maximum flexibility in a research tool, or for fluorescent imaging, a laser-based system may be best. These microscopes can be easily modified and provide the researcher the opportunity to experiment with a variety of different configurations. For quantitative measurements on dielectric or semiconductor materials, it is necessary to use broadband light to reduce coherent interference effects. For real-time imaging, ease of use, and sample flexibility, one of the many Nipkow disk systems may be the appropriate choice. For phase imaging capability, one of the many varieties of interference microscopes can be employed. Finally, if the ultimate resolution is required, an NSOM or SIM can be used to obtain definitions below the normal diffraction limits.

References

1. To cite an example, SiScan instruments of Campbell California tried to build a CSLM using a HeCd laser operating at 440 nm for measuring the dimensions of small structures on semiconductor substrates. They found, however, that intensity fluctuations in the laser prevented them from achieving the desired repeatability of the measurements.
2. S. Hell, S. Witting, M. V. Schickfus, R. W. Wijnaendts van Resandt, S. Hunklinger, E. Smolka, and M. Neiger, "A confocal beam scanning white light microscope," Journal of Microscopy **163** Pt. 2, 179–187 (1991).
3. T. R. Corle, G. Q. Xiao, G. S. Kino, and N. S. Levine, "Characterization of a real-time confocal scanning optical microscope," *Integrated Circuit Metrology, Inspection, and Process Control III*, K. M. Monahan, editor, SPIE **1087**, 138–145 (1989).
4. H. J. Tiziani and H.-M. Uhde, "Three-dimensional image sensing by chromatic confocal microscopy," Appl. Opt. **33**, 1838–1843 (1994).
5. Scanning '92, Atlantic City New Jersey (March 30–April 3, 1992).
6. Theoretically, the tube length is the distance between the back principal plane of the lens and the image plane. In practice, however, another distance called the mechanical tube length is often used. This distance is measured between the shoulder that acts as the stop when threading the objective into the turret mount and the image plane. These two distances, the tube length and the mechanical tube length are usually not very different and so are often used interchangeably.
7. T. R. Corle, "Studies in confocal scanning optical microscopy," Ph.D. Dissertation, Applied Physics Department, Stanford University, Stanford, California (1989).
8. C. J. R. Sheppard and T. Wilson, "Multiple traversing of the object in the scanning microscope," Optica Acta **27**, 611–624 (1980).
9. Camera lenses have the advantage that they are affordable and of high quality.
10. T. Wilson and A. Carlini, "Size of the detector in confocal imaging systems," Opt. Lett. **12**, 227–229 (1987).
11. J. S. Ploem, "Laser scanning fluorescent microscope," Appl. Opt. **26**, 3226–3231 (1987).
12. S. R. Goldstein, T. Hubin, S. Rosenthal, and C. Washburn, "A confocal video rate laser-beam scanning reflected-light microscope with no moving parts," Journal of Microscopy **157** Pt 1, 29–38 (1990).
13. T. Takamatsu and S. Fujita, "Microscopic tomography by scanning microscopy and its three-dimensional reconstruction," Jorn. Microscopy **149** Pt. 3, 167–174 (1988).
14. J. J. Art and M. B. Goodman, "Rapid scanning confocal microscopy," Methods in Cell Biology **38**, 47–77 (1994).

15. P. M. Houpt and A. Draaijer, "A real-time confocal laser scanning microscope for fluorescence and reflection," Inst. Phys. Conf. Ser. **98**: Chapter 14, 639–642 (1989).
16. A. Draaijer and P. M. Houpt, "High scan-rate confocal laser scanning microscopy," *Electronic Light Microscopy*, 273–287 (Wiley-Liss, 1993).
17. A. Draaijer and P. M. Houpt, "A standard video-rate confocal laser-scanning reflection and fluorescence microscope," Scanning **10**, 139–145 (1988).
18. G. S. Kino, *Acoustic Waves, Devices, Imaging, and Analog Signal Processing* (Prentice-Hall, 1987).
19. E. H. K. Stelzer, "The intermediate optical system of laser-scanning confocal microscopes," *Handbook of Biological Confocal Microscopy*, J. B. Pawley, editor, 93–103, (Plenum Press, 1990).
20. A. Draaijer and P. M. Houpt, "A real-time confocal laser scanning microscope (CSLM)," *Scanning Imaging Technology*, Tony Wilson and Ludwig Balk, editors, SPIE **809**, 85–88 (1987).
21. G. M. Svishchev, "Microscope for the study of transparent light-scattering objects in incident light" Opt. Spectrosc. **26**, 171–172 (1969).
22. G. J. Brakenhoff and K. Visscher, "Confocal imaging with bilateral scanning and array detectors," Jorn. Microscopy **165** Pt. 1, 139–146 (1992).
23. B. R. Masters and A. A. Thaer, "Real-time scanning slit confocal microscopy of the *in vivo* cornea," Appl. Opt. **33**, 695–701 (1994).
24. B. R. Masters, "Scanning slit confocal microscopy of the *in vivo* cornea," Opt. Eng. **34**, 684–692 (1995).
25. A. E. Dixon, S. Damaskinos, A. Ribes, E. Seto, M.-C. Beland, T. Uesaka, B. Dalrympl, S. P. Duttagupta, and P. M. Fauchet, "Confocal scanning beam laser microscope/macroscope: applications requring large data sets," *Three-Dimensional Microscopy: Image Acquisition and Processing II*, SPIE **2412**, 12–20 (1995).
26. A. C. Ribes, S. Damaskinos, A. E. Dixon, G. E. Carver, C. Peng, P. M. Fauchet, T. K. Sham, and I. Coulthard, "Photoluminescence imaging of porous silicon using a confocal scanning laser macroscope/microscope." Appl. Phys. Lett. **66**, 2321–2323 (1995).
27. J. B. Pawley, A. G. Wright, and C. C. Garrard, "Optical enhancement and pulse-counting improve the quality of confocal data," Proc. 1993 Int. Conf. on Confocal Microscopy and 3D Image Processing, Sydney, Australia, C. J. R. Sheppard, editor, 69–75 (1993).
27. L. Giniunas, R. Juskaitis, and S. U. Shatalin "Scanning fiber optic microscope," Elect. Lett. **27**, 724–726 (1991).
29. S. Kimura and T. Wilson, "Confocal scanning optical microscopes using single mode fiber for signal detection," Appl. Opt. **30**, 2143–2150 (1991).
30. T. Dabbs and M. Glass, "Fiber-optic confocal microscope—FOCON," Appl. Opt. **31**, 3030–3035 (1992).
31. L. Giniunas, R. Juskaitis, and S. V. Shatalin, "Endoscope with optical sectioning capability," Appl. Opt. **32**, 2888–2890 (1993).
32. T. Wilson, "Image formation in two-mode fiber-based confocal microscopes," J. Opt. Soc. Am. A **10**, 1535–1543 (1993).

References

33. M. Gu and C. J. R. Sheppard, "Three-dimensional optical transfer function in a fiber-optical confocal fluorescent microscope using annular lenses," J. Opt. Soc. Am. A **9**, 1991–1999 (1992).
34. D. Dickensheets and G. S. Kino, "A scanned optical fiber confocal microscope," Proc. SPIE, **2184**, 39–47 (1994).
35. R. Juskaitis and T. Wilson, "Differential confocal scanning microscope with a two-mode optical fiber," Appl. Opt. **31**, 898–902 (1992).
36. R. Juskaitis and T. Wilson, "Surface profiling with scanning optical microscopes using two-mode optical fibers," Appl. Opt. **31**, 4569–4574 (1992).
37. R. Juskaitus and T. Wilson, "Direct-view fiber-optic confocal microscope," Opt. Lett. **19**, 1906–1908 (1994).
38. P. M. Delaney, M. R. Harris, and R. G. King, "Fiber optic laser scanning confocal microscope suitable for fluorescence imaging," Appl. Opt. **33**, 573–577 (1994).
39. A. F. Gmitro and D. Aziz, "Confocal microscopy through a fiber-optic imaging bundle," Opt. Lett. **18**, 565–567 (1993).
40. M. Petran and M. Hadravsky, "Tandem-scanning reflected-light microscope," JOSA **58**, 661–664 (1968).
41. M. D. Egger and M. Petran, "New reflected-light microscope for viewing unstained brain and ganglion cells," Science **157**, 305–307 (1967).
42. M. Petran, A. Boyde, and M. Hadravsky "Direct view confocal microscopy" *Confocal Microscopy*, T. Wilson, editor (Academic Press, 1990).
43. A. Frosch and H. E. Korth, "Method of increasing the depth of focus and or the resolution of light microscopes by illuminating and imaging through a diaphragm with pinhole apertures," U.S. Patent 3,926,500, December 16, 1975.
44. G. Q. Xiao and G. S. Kino, "A real-time confocal scanning optical microscope," *Scanning Imaging Technology*, Tony Wilson and Ludwig Balk, editors, SPIE **809**, 107–113 (1987).
45. G. Q. Xiao, T. R. Corle and G. S. Kino, "Real-time Confocal Scanning Optical Microscope," Appl. Phys. Lett. **53**, 716–718 (1988).
46. J. J. McCarthy, J. D. Fairing, and J. C. Bucholz, "Confocal tandem scanning reflected light microscope," U.S. Patent 4,802,748, February 7, 1989.
47. J. D. Hill, "Scanning disks for use in tandem scanning reflected light microscopes and other optical systems," U.S. Patent 5,083,220, January 21, 1992.
48. G. Q. Xiao, "Confocal optical imaging systems and their applications in microscopy and range sensing" Ph.D. Dissertation, Physics Department, Stanford University, Stanford, California (December 1989).
49. G. Q. Xiao and G. S. Kino, "A real-time confocal scanning optical microscope," *Scanning Imaging Technology*, Tony Wilson and Ludwig Balk, editors, SPIE **809**, 107–113 (1987).
50. J. H. Wayland, "Scanning microscopy," U.S. Patent 4,806,004, February 21, 1989.
51. J. H. Wayland and W. G. Freasher Jr., "Intravital microscopy on the basis and application of an intravital microscope for microvascular and neurophysi-

ological studies," Modern Tech. in Physiological Sciences, 125–153 (Academic Press, 1978).
52. W. E. Westell, "Apodization filters," U.S. Patent 4,030,817, June 21, 1977.
53. J. L. Jewell, G. R. Olbright, R. P. Bryan, and A. Scher, "Surface-emitting lasers break the resistance barrier," Photonics Spectra 126–130 (November 1992).
54. T. R. Corle "Confocal scanning optical microscope," U.S. Patent 5,067,805, November 26, 1991.
55. M. D. Egger, W. Gezari, P. Davidovits, M. Hadravsky, and M. Petran, "Observation of nerve fibers in incident light," Experiential (Basel), **25**, 1225–1226 (1969).
56. C. J. R. Sheppard and X. Q. Mao, "Confocal microscopes with slit apertures," J. Mod Opt. **35**, 1169–1185 (1988).
57. B. R. Masters, "In vivo confocal microscopy of the human eye," USA Microscopy and Analysis, 15–17 (November 1994).
58. D. M. Maurice, "A scanning slit optical microscope," Investigative Ophthalmology **13**, 1033–1037 (1973).
59. C. J. Koester, "A comparison of various optical sectioning methods: the scanning slit confocal microscope," *Handbook of Biological Confocal Microscopy*, J. B. Pawley, editor, 207–214 (Plenum Publishing, 1990).
60. D. Awamura and T. Ode, "Optical properties of type1 + type2 microscopes," *New Trends on Scanning Optical Microscopy*, S. Minami, editor, Proceedings of the Second International Forum of the Optoelectronic Industry and Technology Development Association, Okinawa, Japan, 43–48 (1992).
61. G. J. Brakenhoff, "Imaging modes in confocal light microscopy CSLM," Journal of Microscopy **117**, 232–242 (1979).
62. C. J. R. Sheppard and T. Wilson, "Multiple traversing of the object in the scanning microscope," Opt. Acta. **27**, 611–624 (1980).
63. A. E. Dixon, S. Damaskinos, and M. R. Atkinson, "A scanning confocal microscope for reflection and transmission imaging," Nature **351**, 551–553 (1991).
64. A. E. Dixon and C. Cogswell, "Confocal microscopy with transmitted light," *Handbook of Biological and Confocal Microscopy, 3rd edition,* J. B. Pawley, editor, 479–490 (Plenum Press 1995).
65. C. J. Cogswell and C. J. R. Sheppard, "Confocal differential interference contrast (DIC) microscopy: including a theoretical analysis of conventional and confocal DIC imaging," Journal of Microscopy **165**, 81–101 (1992).
66. R. Juskaitis, T. Wilson, and F. Reinholz, "Spatial filtering by laser detection in confocal microscopy," Opt. Lett. **18**, 1135–1137 (1993).
67. R. Juskaitis, N. P. Rea, and T. Wilson, "Semiconductor laser confocal microscopy," Appl. Opt. **33**, 578–584 (1994).
68. R. H. Webb and F. J. Rogemich, "Microlaser microscope using self detection for confocality," Opt. Lett. **20**, 533–536 (1995).
69. J. Kesterson and M. Richardson, "Confocal microscope capability with desktop affordability," Advanced Imaging 23–26 (October 1991).

70. M. Davidson, K. Kaufman, I. Mazor, and F. Cohen, "An application of interference microscopy to integrated circuit inspection and metrology," *Integrated Circuit Metrology, Inspection, and Process Control* Kevin M. Monahan, editor, SPIE **775**, 233–247 (1987).
71. M. Davidson, K. Kaufman, I. Mazor, and F. Cohen, "The coherence probe microscope," Solid State Technology, 57–59 (Sept. 1987).
72. S. S. C. Chim and G. S. Kino, "Mirau correlation microscope," App. Opt. **29**, 3775–3783 (1990).
73. D. K. Hamilton, and C. J. R. Sheppard, "A confocal interference microscope," Optica Acta **29**, 1573–1577 (1982).
74. S. Hell and E. H. K. Stelzer, "Properties of a 4Pi confocal fluorescent microscope," J. Opt. Soc. Am. A **9**, 2159–2166 (1992).
75. J. W. Dockrey and D. Hendricks, "The application of coherence probe microscopy for submicron critical dimension linewidth measurements," *Integrated Circuit Metrology, Insepction, and Process Control III*, Kevin M. Monahan, editor, SPIE **1087**, 120–137 (1989).
76. J. F. Biegen, "New developments in Mirau interferometry," Presented at the Annual Meeting of the Optical Society of America (October 31, 1988).
77. S. S. C. Chim, "The Mirau correlation microscope—a new tool for optical metrology," Ph.D. Dissertation, Electrical Engineering Department, Stanford University, Stanford California (June 1991).
78. S. Tolanski, *Multiple-Beam Interferometry of Surfaces and Films* (Clarendon Press, 1948).
79. S. Tolanski, *An Introduction to Interferometry* (Longmans, 1955).
80. S. Tolanski, *Multiple-Beam Interference Microscopy of Metals* (Academic Press, 1970).
81. H. Komatsu, "Interferometry: principles and applications of two-beam and multiple-beam interferometry," Nikon Technical Bulletin, Nikon Instrument Group (1990).
82. E. H. Synge, "A suggested method for extending microscopic resolution into the ultra-microscopic region," Phil. Mag. **6**, 356–362 (1928).
83. J. A. O'Keefe, "Resolving power of visible light," J. Opt. Soc. Am. **46**, 359 (1956).
84. E. A. Ash and G. Nicholls, "Super-resolution aperture scanning microscope," Nature **237**, 510–512 (1972).
85. D. W. Pohl, W. Denk, and M. Lanz, "Optical stethoscopy image recording with resolution $\lambda/20$," Appl. Phys. Lett. **44**, 651–653 (1984).
86. A. Lewis, M. Isaacson, A. Harootunian, and A. Muray, "Development of a 500 Å spatial resolution light microscope," Ultramicroscopy **13**, 227 (1984).
87. D. W. Pohl, W. Denk, and M. Lanz, "Optical stethoscopy image recording with resolution $\lambda/20$," Appl. Phys. Lett. **44**, 651–653 (1984).
88. A. Lewis, M. Isaacson, A. Harootunian and A. Muray, "Development of a 500 Å spatial resolution light microscope," Ultramicroscopy **13**, 227 (1984).
89. E. Betzig, A. Lewis, A. Harootunian, M. Isaacson, and E. Kratschmer, "Near-field scanning optical microscopy (NSOM)," Biophys. J. **49**, 269–279 (1986).

90. E. Betzig, J. K. Trautman, T. D. Harris, J. S. Weiner, and R. L. Kostelak, "Breaking the diffraction barrier," Science **251**, 1468–1470 (1991).
91. Burleigh Instruments Inc., Burleigh Park, Fishers, New York 14453 USA.
92. S. Shalom, K. Lieberman, A. Lewis, and S. R. Cohen, "A micropipette force probe suitable for near-field scanning optical microscopy," Rev. Sci. Instrum. **63**, 4061–4065 (1992).
93. M. Tortonese, H. Yamada, and C. F. Quate, presented at the *International Conference on Scanning Tunneling Microscopy*, Interlaken, August 12–16, 1991, unpublished personal communication.
94. R. Toledo-Crow, P. C. Yang, Y. Chen, and M. Vaez-Iravani, "Near-field differential scanning optical microscope with atomic force regulation," Appl. Phys. Lett. **60**, 2957–2959 (1992).
95. A. Shchemelin, M. Rudman, K. Lieberman, and A. Lewis, "A simple lateral force sensing technique for near-field micropattern generation," Rev. Sci. Instrum. **64**, 3538–3541 (1993).
96. E. Betzig, P. L. Finn, and J. S. Weiner, "Combined shear force and near-field scanning optical microscope," Appl. Phys. Lett. **60**, 2484–2486 (1992).
97. K. Lieberman, S. Harush, A. Lewis, and R. Kopelman, "A light source smaller than the optical wavelength," Science **247**, 59–61 (1990).
98. E. Betzig, S. G. Grubb, R. J. Chichester, D. J. DiGiovanni, and J. S. Wiener, "Fiber laser probe for near-field scanning optical microscopy," Appl. Phys. Lett. **63**, 3550–3552 (1993).
99. U. C. Fischer and D. W. Pohl, Phys. Rev. Lett. **62**, 458–461 (1989).
100. T. J. Silva and S. Schultz, *Proceedings of the 2nd Conference on Near-Field Optics*, Raleigh NC, 20–22 October 1993, to be published in Ultramicroscopy.
101. N. F. van Hulst, M. H. P. Moers, O. F. J. Noordman, R. G. Tack, F. B. Segensk, and B. Bölger, "Near-field optical microscope using a silicon-nitride probe," Appl. Phys. Lett. **62**, 461–463 (1993).
102. F. Zenhausern, M. P. O'Boyle, and H. K. Wickramasinghe, "Apertureless near-field optical microscope," Appl. Phys. Lett. **65**, 1623–1625 (1994).
103. A. E. Siegman, *Lasers,* (University Science Books 1986).
104. E. Betzig and R. J. Chichester, "Single molecules observed by near-field scanning optical microscopy," Science **262**, 1422–1427 (1993).
105. W. E. Moerner, T. Plakhotnik, T. Irngartinger, U. P. Wild, D. W. Pohl, and B. Hecht, "Near-field optical spectroscopy of single molecules in solids," Phys. Rev. Lett. **73**, 2764–2767 (1994).
106. P. Dawson, F. de Fornel, and J. P. Goudonnet, *Proceedings of the 2nd Conference on Near-Field Optics,* Raleigh, NC, 20–22 October 1993, Ultramicroscopy **57**, (2–3) (February 1995).
107. E. Betzig, J. K. Trautman, R. Wolfe, P. L. Finn, M. H. Kryder, and C. H. Chang, "Near-field magneto-optics and high density data storage," Appl. Phys. Lett. **61**, 142–144 (1992).
108. J. M. Guerra, "Photon tunneling microscope," Appl. Opt. **29**, 3741–3752 (1990).
109. S. M. Mansfield and G. S. Kino, "Solid immersion microscope," Appl. Phys. Lett. **57**, 2615–2616 (1990).

110. M. Born and E. Wolf, *Principles of Optics* (Pergamon Press, 1975).
111. B. D. Terris, H. J. Mamin, D. Rugar, W. R. Studenmund, and G. S. Kino, "Near-field optical data storage using a solid immersion lens," Appl. Phys. Lett. **65**, 388–390 (1994).
112. S. Hayashi and G. S. Kino, "Solid immersion lens for optical storage," *Three-Dimensional Microscopy: Image Acquisition and Processing II,* T. Wilson and C. J. Cogswell, editors, SPIE **2412**, 80–87 (1995).

CHAPTER 3

Depth and Transverse Resolution

3.1 Introduction

The basic formulae for the depth and transverse response of the *confocal scanning optical microscope* (CSOM) and the correlation microscope were introduced in Chapter 1. The construction of these microscopes was described in Chapter 2 along with a description of various types of near-field optical microscopy; without much theory. In this chapter, we shall derive the formulae for the depth and transverse resolution of the confocal scanning, interference, and near-field microscopes, discuss their limitations, and show how the resolution and efficiency are affected by factors such as lens aberrations and the pinhole size.

The definition of resolution depends, to a large extent, on the type of sample, and which measurements are important to the observer. For integrated circuits, we are often interested in measuring the height differences of planar surfaces and the profiles of structures containing sharp edges. Thus, the formulae for depth resolution with a plane mirror reflector and the edge response of the microscope are the relevant definitions. For biological samples, however, the scattering from individual cells is important. A definition of depth resolution based on the use of point reflectors is the most applicable. The transverse resolution in biological applications is best defined in terms of when two closely spaced neighboring points can be distinguished, using the Rayleigh or Sparrow criteria discussed in Section 1.3.4.

This chapter is divided into several sections which discuss these different aspects of depth and transverse resolution. Section 3.2, Depth Response of the Confocal Microscope with Infinitesimal Pinholes and Slits, discusses the range resolution of the confocal microscope for both plane reflectors and point reflectors in brightfield and fluorescence imaging. The effect of aberration on the range resolution is also examined.

Section 3.3, Depth Response of the Confocal Microscope with Finite-Sized Pinholes, extends the theory to deal with the effect of finite pinhole and slit size on the range resolution. Approximate theories and formulae are first given, followed by a more exact theory for the effect of finite pinhole size.

Section 3.4, Transverse Response of the Confocal Microscope, deals with the theory of the transverse response for both infinitesimal size pinholes and finite size pinholes. The various types of definition for resolution are dealt with in some detail, and the half-power width, Sparrow, and Rayleigh two-point resolutions are defined for the confocal microscope.

Section 3.5, Depth and Transverse Resolution of the Interferometric Microscope, examines the response of the interferometric microscope. It is shown that the depth and transverse responses of the interferometric microscope to narrowband illumination are the same as for a confocal microscope using an objective of the same numerical aperture. In the interference microscope, aberrations have less effect than in the confocal microscope, but the beamsplitter in a *Mirau correlation microscope* (MCM) can still significantly degrade the image. In this section we shall show how the lens and beamsplitter aberrations affect the point spread function and depth response, but leave the question of accuracy of phase measurements to Chapter 4. We shall also consider the effect of the illumination bandwidth.

Section 3.6, The Near-Field Scanning Optical Microscope, examines the range resolution, transverse resolution, and efficiency of different types of *near-field scanning optical microscopes* (NSOMs). Section 3.6.1 calculates the attenuation in the cutoff mode of an optical fiber demonstrating that the efficiency of the NSOM is dominated by the light loss in the tapered probe. In Section 3.6.2 it is shown how the fields fall off away from the pinhole and how the resolution is affected by the distance from the pinhole.

Section 3.7, The Solid Immersion Microscope, carries out the analysis required for these types of instruments, and compare the theoretical performance with experimental data. Section 3.7.1 derives the transverse and longitudinal magnification of the *solid immersion lens* (SIL), and Sections 3.7.2 and 3.7.3 examine the depth and transverse responses, respectively. These sections demonstrate that the efficiency and frame rate of the *solid immersion microscope* (SIM) are much higher than that of the NSOM, but the resolution is worse.

3.2 Depth Response of the Confocal Microscope with Infinitesimal Pinholes and Slits

3.2.1 Scalar Theory for a Plane Reflector

The distinguishing feature of scanning optical microscopes is their shallow depth response, meaning that out-of-focus portions of the sample are not imaged. The first theoretical treatment of the depth response of the CSOM had its origin in acoustic microscopy. The acoustic theory was first derived in 1978[1] and since then refined by many authors.[2,3] Cox, Hamilton, and Sheppard adapted it to develop a scalar form for the scanning optical microscope in 1982.[4] Corle and Kino[5] adapted the full electromagnetic wave vector theory of Richards and Wolf[6] to the CSOM imaging a plane reflector and obtained the same result for an infinitesimal pinhole as that derived from scalar theory.[7] A corrected form of their treatment is given in Appendix A.

In this section, we shall review the derivation of the depth response characteristic of a CSOM using nonparaxial scalar theory with an infinitesimally small pinhole and a plane reflector. The effects of finite-sized pinholes and of point reflectors will be considered later in the chapter.

For this derivation it is convenient to consider a CSOM with a collimated beam illuminating the objective as illustrated in Fig. 3.1. The beam reflected from the sample passes back through the objective and is reflected by the beamsplitter to a pinhole relay lens, which focuses it onto an infinitesimally small detector pinhole. We will assume that the objective

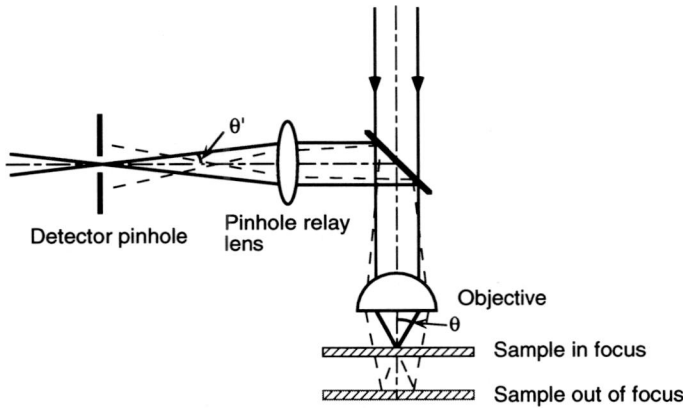

Figure 3.1 The CSOM configuration used in the derivation of the depth response.

lens is characterized by a pupil function $P(\theta)$ whose amplitude and phase may vary with θ, the angle between a ray from the pupil plane to the focal point and the lens axis. The aberrations of the entire lens system can be combined mathematically into a term represented by the pupil function of the objective. Finally, it is assumed that the linear dimensions of the lens pupil are large compared to the wavelength of light and that the sine condition or stigmatic focusing condition for a perfect lens is obeyed,[8]

$$n \sin \theta = M \sin \theta'. \qquad (3.1)$$

In Eq. (3.1), M is the magnification of the optical system, n is the index of refraction of the material between the sample and objective, θ is the angle between a ray to the objective pupil and the optical axis of the system, and θ' is the angle formed by the corresponding ray at the pinhole. The sine condition is applicable to a large-angle system but requires that a perfect image is obtained at the image plane for points in the object plane that are a small distance apart.

To calculate the depth response of the microscope it is assumed that the amplitude of the input beam at the pupil plane of the objective lens is uniform with a value of $\Phi^I(r,\theta,\phi) = \Phi_0$. When a perfect plane mirror is placed at the focal plane of the objective lens, the amplitude of the reflected signal associated with the ray passing through the pupil plane at r,θ,ϕ is $\Phi^R(r,\theta,\phi)$, where, by symmetry,

$$\Phi^R(r,\theta,\phi) = P^2(\theta)\Phi^I(r,\theta,-\phi) = P^2(\theta)\Phi_0. \qquad (3.2)$$

The pupil function is squared in Eq. (3.2) because the beam passes through the lens twice. This equation does not account for any nonidealized behavior of the reflecting mirror. Those contributions are covered in Section 3.2.5.

After reflecting off the beamsplitter and pinhole lens, the rays converge to the pinhole forming an angle θ' with the optical axis. The signal $V(0)$ received at the on-axis pinhole in front of the detector is proportional to the integral of $\Phi^R(r,\theta,\phi)$ over the angle θ', or

$$V(0)_{\text{plane}} = \Phi_0 \int_0^{\theta_0'} P^2(\theta) \sin \theta' \, d\theta'. \qquad (3.3)$$

In Eq. (3.3) θ_0' is the maximum angle subtended by the focused beam at the pinhole.

We now consider the situation when the mirror is moved a distance z from the focal plane and repeat the calculation. When the mirror moves a distance z, the image of the focused spot in the mirror will move by a distance $2z$ from the focal plane, as discussed in Section 1.3.3. Conse-

3.2 Depth Response of the Confocal Microscope with Infinitesimal Pinholes and Slits

quently, the fields along a reflected ray in the pupil plane of the lens passing through (r,θ,ϕ) pick up an additional phase shift of $2knz \cos \theta$, where $k = 2\pi/\lambda$. The amplitude of the reflected field on the ray at the pupil of the objective is, proportional to the incident field times the phase factor $\exp(-2jknz \cos \theta)$,

$$\Phi^R(r,\theta,\phi) = P^2(\theta)e^{-2jknz\cos\theta}\Phi^I(r,\theta,-\phi) = \Phi_0 P^2(\theta)e^{-2jknz\cos\theta}. \quad (3.4)$$

The normalized electric field amplitude, $V(z)$, as a function of the defocus distance is calculated by integrating the reflected field over the angles in the focused beam. We assumed, as before, that $\Phi^I(r,\theta,\phi) = \Phi_0$, so that $V(z)$ normalized to its value for the mirror located at the focal plane is

$$V(z)_{\text{plane}} = \frac{\int_0^{\theta_0} P^2(\theta)e^{-2jknz\cos\theta} \sin \theta' \, d\theta'}{\int_0^{\theta_0} P^2(\theta) \sin \theta' \, d\theta'}, \quad (3.5)$$

where θ_0 is the half-angle subtended by the focused beam at the objective lens. As discussed in Chapter 1, the notation $V(z)$ is a term from acoustic microscopy, which is also commonly used in the literature of confocal microscopy.

Equation (3.5) can be simplified if it is written entirely in terms of the objective coordinate θ rather than pinhole coordinate θ'. Differentiating Eq. (3.1), the sine condition for a perfect lens gives the transformation

$$n \cos \theta \, d\theta = M \cos \theta' \, d\theta'. \quad (3.6)$$

Substituting Eqs. (3.1) and (3.6) into Eq. (3.5), it follows that

$$V(z) = \frac{\int_0^{\theta_0} P^2(\theta)e^{-2jknz\cos\theta} \sin \theta \cos \theta \, d\theta}{\int_0^{\theta_0} P^2(\theta) \sin \theta \cos \theta \, d\theta}. \quad (3.7)$$

This expression is identical to the results obtained by Sheppard and Wilson for the acoustic microscope[9] and the expression derived by Kino and Xiao for the *real-time scanning optical microscope* (RSOM), in the limit of infinitesimal pinhole size.[10] It should be emphasized that since the square of the pupil function occurs in Eq. (3.7), the depth response is very sensitive to the system's optical aberrations.

Paraxial Equation A simplified paraxial solution for the depth response can be obtained by direct integration of Eq. (3.7) making the assumptions that $P^2(\theta) = 1$, and $\cos \theta \approx 1$ in the amplitude term of the integrand. In this case, after some algebra, the analytic form of the depth response becomes

$$V(z)_{\text{plane}} = e^{-jknz(1+\cos\theta_0)} \frac{\sin knz(1 - \cos \theta_0)}{knz(1 - \cos \theta_0)}. \tag{3.8}$$

It will be noted that in addition to the variation in amplitude derived as Eq. (1.14) of Chapter 1, there is also a phase change $knz(1 + \cos \theta_0)$, a term of great importance in the closely related interferometric microscopes, which are able to directly measure the phase. The presence of the phase term also implies that there can be interference between two objects at different depths, for instance, the top and bottom of a transparent layer of material. This interference phenomenon has been used to measure the thickness of thin transparent films laid down on semiconductors and by biologists to help elucidate the structure of cells. However, it can also give rise to speckle which degrades the optical image.

A photodiode detects the intensity of the depth response signal, $I(z)_{\text{plane}}$, which is proportional to $|V(z)|^2$,

$$I(z)_{\text{plane}} = |V(z)_{\text{plane}}|^2 = \left(\frac{\sin knz(1 - \cos \theta_0)}{knz(1 - \cos \theta_0)}\right)^2. \tag{3.9}$$

Depth of Focus The depth of focus of the microscope is commonly defined as the *full width at half maximum* (FWHM) distance or 3-dB resolution of the intensity response,

$$d_z(3\text{dB})_{\text{plane}} = \frac{0.45\lambda}{n(1 - \cos \theta_0)}. \tag{3.10}$$

If the angle θ_0 is small, Eq. (3.10) can be written in the convenient form

$$d_z(3\text{dB})_{\text{plane}} \approx \frac{0.90 n\lambda}{(N.A.)^2}. \tag{3.11}$$

Equations (3.10) and (3.11) are also formulae for the distance between the 0.707 points of the central lobe of the amplitude depth response.

Equation (3.11) shows that the depth response varies inversely as the square of the numerical aperture. However, it is accurate only for values of $N.A.$ less than ~ 0.5. The form of Eq. (3.10) can be derived by setting

3.2 Depth Response of the Confocal Microscope with Infinitesimal Pinholes and Slits

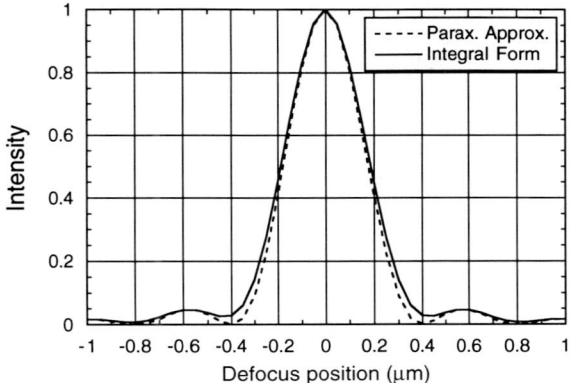

Figure 3.2 The form of the intensity depth response for a plane reflector calculated using both the paraxial theory and the nonparaxial scalar theory for $N.A. = 0.95$.

Eq. (3.9) $= 1/2$ and solving for the constants using the expansion $\sin x = x - x^3/3! + \cdots$. While this method will yield a constant that is slightly in error due to the $\sin x$ expansion, it is nonetheless useful for understanding the origin of Eq. (3.10). The exact constant is derived by using a more exact expression for the sinusoid.

The intensity depth response calculated using both paraxial, (Eq. 3.9), and nonparaxial scalar theory, (Eq. 3.7), are shown in Fig. 3.2 for a lens with numerical aperture $N.A. = 0.95$ and a wavelength $\lambda = 546$ nm. The depth resolution given by Eq. (3.10) is $d_z = 353$ nm. The more exact calculation using Eq. (3.7) yields $d_z = 375$ nm. There is a 6% error in the use of Eq. (3.10), which reduces to a 2% error for $N.A. = 0.9$. Thus the analytic forms of Eqs. (3.8) through (3.10) are very convenient and sufficiently accurate for most purposes.[11]

The analysis given above presents useful guidelines for determining the FWHM of the intensity depth response. Unless the experiment is made very carefully, however, lens aberrations will broaden the depth response and change the sidelobe levels from that predicted by the theory. The most common aberration is spherical aberration that can be introduced when the focused beam passes through the objective lens or a plane of transparent material such as a microscope coverslip. Other effects that degrade the depth response include nonuniform illumination of the lens pupil and chromatic aberration. These aberrations will be discussed in more detail in Section 3.2.5.

3.2.2 Scalar Theory for Depth Response of a Point Reflector

If the sample can be regarded as being composed of point reflectors, such as widely scattered cells suspended in a fluid, then a more useful definition of depth response is based on that for a point reflector. To modify the depth response equations, the phase and amplitude of the light at the point reflector are first calculated assuming that the objective is illuminated with a collimated beam, as before. This light is reflected by the point reflector back to the pupil of the objective. From there, a procedure similar to that given above for the plane reflector can be followed.

If the point reflector is located on the optical axis of the objective lens a distance z away from the focal point of the lens, the normalized complex field amplitude of the incident beam at this point, $\Phi^I(0,z)$, is given, in the paraxial approximation, by

$$\Phi^I(0,z,\theta) = \frac{\int_0^{\theta_0} P(\theta) e^{-jknz\cos\theta} \sin\theta \, d\theta}{\int_0^{\theta_0} P(\theta) \sin\theta \, d\theta}. \tag{3.12}$$

where the superscript I denotes the incident field. Taking $P(\theta) = 1$, Eq. (3.12) can be directly integrated to obtain the on-axis field at the sample,

$$\Phi^I(0,z,\theta) = e^{-j\frac{knz}{2}(1+\cos\theta_0)} \frac{\sin\frac{knz}{2}(1-\cos\theta_0)}{\frac{knz}{2}(1-\cos\theta_0)}. \tag{3.13}$$

A reflected ray at angle θ will undergo an additional phase shift $e^{-jknz\cos\theta}$ while traveling back to the objective so that the reflected field at the pupil plane of the objective is

$$\Phi^R(0,z,\theta) = e^{-j\frac{knz}{2}(1+\cos\theta_0)} \frac{\sin\frac{knz}{2}(1-\cos\theta_0)}{\frac{knz}{2}(1-\cos\theta_0)} P(\theta) e^{-jknz\cos\theta}. \tag{3.14}$$

Following the procedure outlined in Section 3.2.1, this field can be integrated over all angles of the focused beam to yield the amplitude depth response for a point reflector.

An alternative approach is to use the reciprocity theorem and take the receiver response to be the same as the transmitter response. In the

3.2 Depth Response of the Confocal Microscope with Infinitesimal Pinholes and Slits

paraxial approximation, if the objective is illuminated by a pinhole, the response is given by Eq. (3.13). It follows from reciprocity that the receiver response is also given by Eq. (3.13). The signal received at the pinhole is the product of the illumination and the receiver responses, or

$$V(z)_{\text{point}} = e^{-jknz(1+\cos\theta_0)} \left[\frac{\sin\frac{knz}{2}(1-\cos\theta_0)}{\frac{knz}{2}(1-\cos\theta_0)} \right]^2. \quad (3.15)$$

Squaring the amplitude depth response gives the point intensity response,

$$I(z)_{\text{point}} = |V(z)_{\text{point}}|^2 = \left(\frac{\sin\frac{knz}{2}(1-\cos\theta_0)}{\frac{knz}{2}(1-\cos\theta_0)} \right)^4. \quad (3.16)$$

The range resolution for a point reflector or the FWHM distance in the intensity depth response for this sample is

$$d_z(3\text{dB})_{\text{point}} = \frac{0.62\lambda}{n(1-\cos\theta_0)}. \quad (3.17)$$

The single-point depth resolution for a point reflector is, therefore, 1.38 times greater than for a plane mirror. For a wavelength of $\lambda = 546$ nm and a lens with numerical aperture of $N.A. = 0.95$, the range resolution calculated from Eq (3.17) is 482 nm. With a liquid immersion lens and a larger aperture, the resolution would be correspondingly smaller. A detailed nonparaxial analysis for a pinhole of finite size is given in Section 3.3.3.

Response Far from Focus In a CSOM observing a sample made up of a collection of point reflectors, the intensity of the signal due to individual small scatterers far from the focus falls off as $1/z^4$ rather than the $1/z^2$ rate for a plane reflector. This result follows from the mathematical formulae but may also be understood physically from ray optics as illustrated in Fig. 3.1. In the ray optic model the focused beam from the objective forms a point image. If the beam is defocused after reflecting from a mirror, the reflected light forms a conical set of rays converging to a point in front of, or behind, the pinhole. As the sample is defocused, the area of the cone at the pinhole plane increases as z^2. The intensity of light received by a detector behind the pinhole thus varies as $1/z^2$ for a plane mirror sample. Following the same reasoning for point scatterers leads to the conclusion that the illumination intensity at a point scatter falls off as $1/z^2$ and that the receiver response also decreases as $1/z^2$. The signal received at the pinhole, therefore, decreases as $1/z^4$. Consequently, a large number of small scatterers far from the focus produces very little glare in the image.

Sparrow Depth Resolution When imaging samples containing point reflectors, it is appropriate to define the Sparrow depth criterion to determine when two points a distance z apart can be distinguished. If the sample is illuminated by a laser beam or other coherent source, the reflected light from the two points is added coherently. If the two points are fluorescent emitters, then the reflected light adds incoherently as discussed in the next section.

The Sparrow criterion states that two points at $z = \pm d/2$ can just be distinguished if the amplitude of the signal received from $z = 0$ is equal to the signals received from $z = \pm d/2$. Since in the coherent case the amplitudes of the signals add, as two points are separated from each other, the relative phase of their reflected signals will vary rapidly. The signal at $z = 0$ will then be given by

$$V(0) = e^{j\frac{knz}{2}(1+\cos\theta_0)} \left[\frac{\sin\frac{knz}{4}(1 - \cos\theta_0)}{\frac{knz}{4}(1 - \cos\theta_0)} \right]^2$$

$$+ e^{-j\frac{knz}{2}(1+\cos\theta_0)} \left[\frac{\sin\frac{knz}{4}(1 - \cos\theta_0)}{\frac{knz}{4}(1 - \cos\theta_0)} \right]^2 \quad (3.18)$$

$$= 2\cos\left(\frac{knz}{2}(1 + \cos\theta_0)\right) \left[\frac{\sin\frac{knz}{4}(1 - \cos\theta_0)}{\frac{knz}{4}(1 - \cos\theta_0)} \right]^2.$$

Equation (3.18) has the form of a rapidly oscillating carrier wave with an envelope that falls off more slowly as $\sin X/X$. The amplitude of the signal halfway between the points is dominated by the coherent interference of the reflected light signals from the two points, the cosine term in Eq. (3.18). For a lens with $N.A. = 0.95$ and a wavelength of $\lambda = 546$ nm, the separation in the z direction at which two points can be distinguished according to the Sparrow criterion is only $d_z = 0.11$ μm. At this spacing the round-trip path difference is approximately 0.8π so that the light reflected from the two points adds almost out of phase. The phenomenon

3.2 Depth Response of the Confocal Microscope with Infinitesimal Pinholes and Slits

is responsible for interference fringes appearing in the images of biological samples.

In many imaging applications it is useful to consider only the envelope of the total signal from the points at $z = \pm d/2$. The distance d_z(Sparrow) at which $V(0) < 1$ for the envelope signal is

$$d_z(\text{Sparrow}) = \frac{0.89\lambda}{n(1 - \cos\theta_0)}. \quad (3.19)$$

For a lens with $N.A. = 0.95$ and a wavelength of $\lambda = 546$ nm, the separation in the z direction at which the envelope of the two points can just be distinguished is $d_z(\text{Sparrow}) = 0.71$ μm.

3.2.3 Scalar Theory for Fluorescent Reflectors

In biological studies it is common to image samples to which a fluorescent dye has been added. Fluorescence imaging modifies the depth response and Sparrow depth resolution formulae in two ways. First, because the illumination and emission wavelengths are different, the formulae for depth resolution will change. Second, since the light emitted from each point is incoherent, the two-point range resolution will be smaller than in the coherent envelope case.

The general equation for fluorescent imaging in the CSOM has been derived by Kimura and Munakata,[12] Wilson,[13] and Conchello and Lichtman.[14] Their procedures used Fourier optics to calculate the intensity of light on the sample and then the intensity of the light received at the pinhole. The intensity *point spread function* (PSF) is used in these calculations because the strength of the fluorescent signal is proportional to the intensity of the light on the sample, rather than the amplitude. The calculations can be further complicated by the fact that the fluorescent image signal may not be linearly proportional to the intensity on the sample. Following the method adopted by these authors, we denote the complex amplitude of the PSF of the objective lens at the illumination wavelength λ_1 as $h_1(x,y,z)$ and the PSF at the fluorescent wavelength λ_2 as $h_2(x/\beta, y/\beta, z/\beta)$, where $\beta = \lambda_2/\lambda_1$. We further assume that the induced fluorescence signal is proportional to the incident illumination intensity and to the efficiency of the fluorescence excitation, $f(x,y,z)$, at a point x,y,z on the sample. If the incident beam is focused on the point x_0, y_0, z_0, the fluorescence intensity at a point x,y,z on the sample can then be expressed as

$$i_f(x,y,z) = |h_1(x - x_0, y - y_0, z - z_0)|^2 f(x,y,z). \quad (3.20)$$

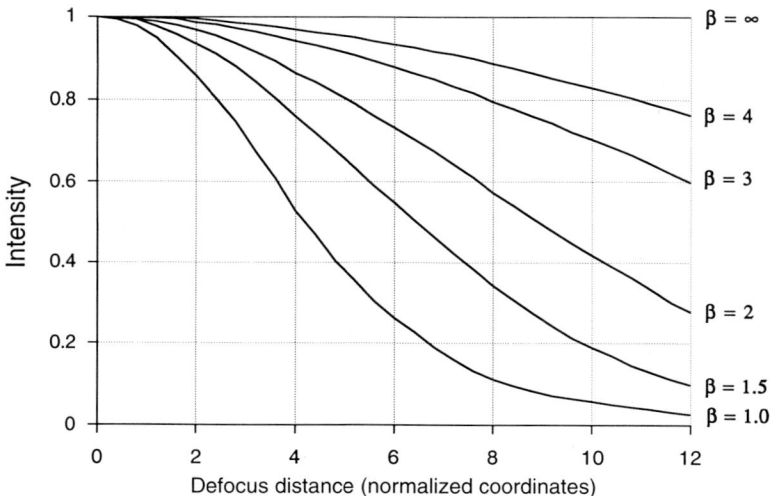

Figure 3.3 Theoretical depth response as a function of normalized defocus distance u for a variety of fluorescent wavelengths as a fluorescent plate is scanned axially through the focal plane of the lens. [From T. Wilson, "Optical sectioning in confocal fluorescent microscopes," J. Microsc. **154**, 143–156 (1989), with permission.]

With the illuminating beam and a point detector focused on the point $x_0 = 0, y_0 = 0, z_0 = 0$, the detected signal is proportional to the total fluorescent energy at the pinhole, which we can write,

$$I = \int_{-\infty}^{\infty} \int_{-\infty}^{\infty} \int_{-\infty}^{\infty} |h_1(x,y,z)h_2(x/\beta,y/\beta,z/\beta)|^2 f(x,y,z)\,dx\,dy\,dz. \quad (3.21)$$

If a uniformly fluorescing plate reflector is scanned axially through the focal plane, we can take $f(x,y,z) = 1$. An equation for the depth response of the microscope is given by performing the integral in Eq. (3.21) only over x and y,[15]

$$I^F(z)_{\text{plane}} = \int_{-\infty}^{\infty} \int_{-\infty}^{\infty} |h_1(x,y,z)h_2(x/\beta,y/\beta,z/\beta)|^2\,dx\,dy. \quad (3.22)$$

Due to the complexity of this equation, no simple analytical formula has been obtained for the depth response of a fluorescent plate reflector. In Fig. 3.3 we show a plot by Wilson of the theoretical depth response as a function of defocus distance in normalized optical coordinates for a

3.2 Depth Response of the Confocal Microscope with Infinitesimal Pinholes and Slits

variety of fluorescent wavelengths.[15] In the figure the axial distance, $u = 4kz \sin^2(\theta_0/2)$, has been normalized to the illumination wavelength λ_1. It will be observed from the figure that, as expected, the sectioning becomes coarser as the fluorescent wavelength increases. The FWHM of the depth response for an infinitesimal pinhole and equal illumination and emission wavelengths can be measured from these results. For equal illuminating and fluorescence wavelengths, the fluorescence depth resolution is

$$d_z^F(3\text{dB})_{\text{plane}}^{\lambda_1=\lambda_2} = \frac{0.67\lambda}{n(1 - \cos\theta_0)}. \tag{3.23}$$

This width is 1.5 times larger than the FWHM for a mirror reflector in brightfield confocal imaging. If the fluorescence wavelength is double that of the illumination wavelength ($\lambda_2 = 2\lambda_1$), the factor 0.67 in the numerator increases to 1.40 and the depth of focus is increased by a factor of 2.09 over the brightfield CSOM response for a plane mirror.

For a fluorescent point reflector the situation is somewhat simplified. The range resolution can then be derived from the results of Section 3.2.2. For an excitation wavelength λ^1 and fluorescent wavelength λ^2, by analogy to Eq. (3.16), the normalized intensity as a function of defocus distance z is

$$I^F(z)_{\text{point}} = \left\{ \frac{\sin\left[\frac{k_1 n_1 z}{2}(1 - \cos\theta_0)\right]}{\frac{k_1 n_1 z}{2}(1 - \cos\theta_0)} \right\}^2 \left\{ \frac{\sin\left[\frac{k_2 n_2 z}{2}(1 - \cos\theta_0)\right]}{\frac{k_2 n_2 z}{2}(1 - \cos\theta_0)} \right\}^2. \tag{3.24}$$

For the simplest case when $k_1 = k_2$, the single-point resolution is given by Eq. (3.16).

Sparrow Depth Resolution The formula for the Sparrow depth resolution is also simple to obtain for two fluorescent points. Since the light generated by the two points is incoherent, the intensities of the two signals, rather than their amplitudes, will add at the detector. In this case the signal halfway between them will equal 1 when the intensity value of each point is 0.5. The required separation for two points is thus equal to the FWHM distance of a single point,

$$d_z^F(\text{Sparrow})_{\text{point}}^{\lambda_1=\lambda_2} = \frac{0.62\lambda}{n(1 - \cos\theta_0)}. \tag{3.25}$$

The incoherent Sparrow resolution is smaller than the coherent envelope value because it is the sum of the intensities rather than the amplitudes of the two points that dictates the strength of the signal.

3.2.4 Scalar Theory for Confocal Slit Microscopes

The efficiency of a scanning optical microscope can be increased by using a slit rather than a pinhole in front of the detector or the source.[16] Slits allow more light to pass through the aperture plane, thus producing higher intensity images. They also make it possible to scan in only one direction, thus decreasing the frame time radically. However, the disadvantages of using a slit are that the depth resolution is a little worse than for a pinhole and the response far from focus in the z direction and glare from far-out structures are considerably worse. In addition, the transverse response along the direction of the slit is worse than in the direction perpendicular to the slit.

The response of a slit microscope to an object with an amplitude reflectance $r(x_0, y_0, z_0)$ has been calculated by Wilson and Hewlett,[17]

$$I(x_s, y_s, z_s)$$
$$= \int_{-\infty}^{\infty}\int_{-\infty}^{\infty}\int_{-\infty}^{\infty} \left| \int_{-\infty}^{\infty}\int_{-\infty}^{\infty}\int_{-\infty}^{\infty} [h(x_0, y_0, z_0) h(x_0 + x_2/M, y_0 + y_2/M, z_0 + z_2/M)] \right.$$
$$\left. \times r(x_0 - x_s, y_0 - y_s, z_0 - z_s) \, dx_0 \, dy_0 \, dz_0 \right|^2 D(x_2, y_2, z_2) \, dx_2 \, dy_2 \, dz_2. \tag{3.26}$$

In Eq. (3.26) M is the magnification of the system, the coordinates (x_0, y_0, z_0) are in the sample plane, (x_2, y_2, z_2) are in the detector plane, (x_s, y_s, z_s) indicates the point on which the microscope is focused, and $D(x_2, y_2, z_2)$ is the response at the detector. For a slip-shaped detector parallel to the x-axis $D(x_2, y_2, z_2) = \delta(x_2)\delta(z_2)$. As with the fluorescent plate reflector there is no simple analytical solution for the depth response of a slit detector.

The depth response for a plate may be calculated from Eq. (3.26), however, Wilson has found it easier to use a transfer function approach.[18] Figure 3.4 compares the depth response of a CSOM to a perfect plane reflector for both fluorescent and nonfluorescent microscopes using slit and point detectors. The fluorescent curves are for the case $\beta = 1$. For an infinitesimally thin slit, the FWHM of the depth response can be calculated from Wilson and Hewlett's results to be

$$d_z(3\text{dB})_{\text{plane}}^{\text{slit}} = \frac{0.54\lambda}{(1 - \cos\theta_0)}, \tag{3.27}$$

for a plane mirror reflector, and

$$d_z^F(3\text{dB})_{\text{plane}}^{\text{slit}\,\lambda_1 = \lambda_2} = \frac{0.95\lambda}{(1 - \cos\theta_0)}, \tag{3.28}$$

for a fluorescent plate.

3.2 Depth Response of the Confocal Microscope with Infinitesimal Pinholes and Slits

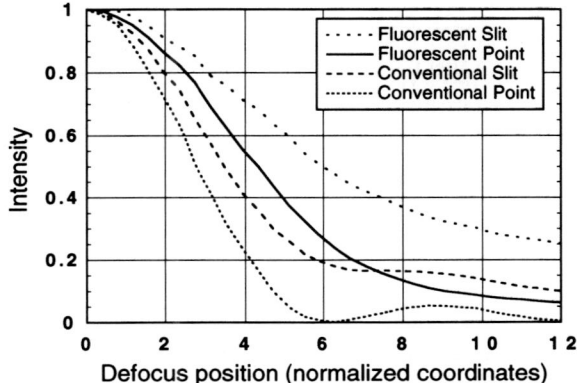

Figure 3.4 The depth response of a CSOM to perfect plane reflectors for both fluorescent and nonfluorescent samples using slit and point detectors. [From T. Wilson and S. J. Hewlett, "Imaging in scanning microscopes with slit-shaped detectors," J. Microsc. **160**, 115–139 (1990), with permission.]

As might be expected, the FWHM range resolution is slightly worse in a slit microscope than in a pinhole system. More important, for a plane reflector located far from the focal point, the image intensity falls off as $1/z$ rather than as $1/z^2$ as it does with a pinhole system. The reason is simple. Applying our previous ray optic model to the slit system, the objective forms a conical set of rays, which after reflection from the sample forms a second set of conical rays around the slit. As the sample is defocused, the area of the cone at the slit increases as z^2. However, the area of the slit through which light passes also increases with defocus distance in proportion to z. Thus the intensity of the light received at a detector behind the pinhole varies as $1/z$.

The range resolutions and rate of fall off in the out-of-focus regions for the various microscope and sample configurations discussed above are tabulated in Table 3.1.

3.2.5 The Effect of Sample and Lens Aberrations on the Depth Response

Experimental Results The first measurements of the depth response with a scanning optical microscope were made by Hamilton et al.[4,19,20] The technique involved scanning a front surface mirror axially through the focal plane of the objective lens while simultaneously measuring the position of the sample and the light intensity at the detector. We have repeated these experiments with the results shown in Fig. 3.5. The dotted

Table 3.1 Range Resolution and Rate of Intensity Decrease Far from the Focus

Type of detector	Criterion/Type of reflector	Value of K for depth resolution where $d_z = K\lambda/(1 - \cos\theta_0)$	Rate of intensity decrease far from the focus
Pinhole detector	FWHM resolution Plane reflector	$K = 0.45$	$1/z^2$
	FWHM resolution Fluorescent plane reflector	$K = 0.67$	$1/z^2$
	FWHM resolution Point reflector	$K = 0.62$	$1/z^4$
	Coherent Sparrow depth resolution	$K = 0.89^*$	$1/z^4$
	Fluorescent Sparrow depth resolution $\lambda_1 = \lambda_2$	$K = 0.62$	$1/z^4$
Slit detector	FWHM resolution Plane reflector	$K = 0.54$	$1/z$
	FWHM resolution Fluorescent plane reflector	$K = 0.95$	$1/z$

* Formula altered by coherent interference effects; $K = 0.89$ is for envelope.

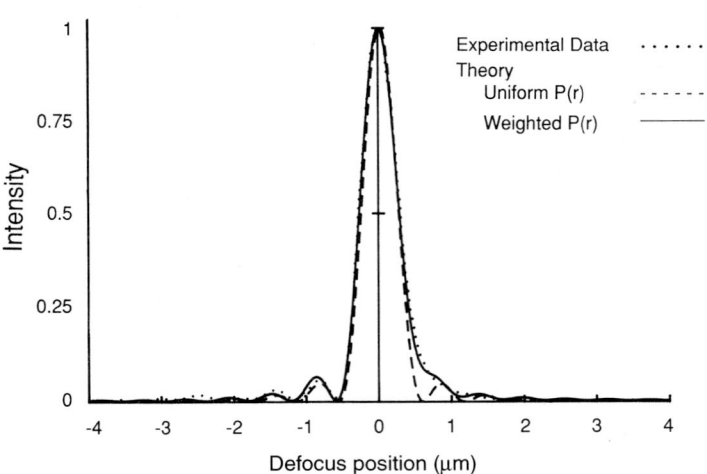

Figure 3.5 Experimental and theoretical depth responses for an Olympus 80×/0.9 N.A. lens at $\lambda = 0.633$ μm. Dotted line: experimental depth response. Dashed and solid lines: theoretical depth response with fitting coefficients added to the pupil function. [From T. R. Corle, C.-H. Chou, and G. S. Kino, "Depth response of confocal optical microscopes," Opt. Lett. **11**, 770–772 (1986), with permission.]

3.2 Depth Response of the Confocal Microscope with Infinitesimal Pinholes and Slits

line is data for an Olympus MD plan 80×/0.90 *N.A.* objective.[5] The z-axis position is taken to be zero at best focus, with positive values of z representing increased objective-sample separation.

Seidel Aberrations It will be observed that the experimental depth response is not as symmetrical as the theory in Eq. (3.8). Spherical aberration in the objective and amplitude apodization of the illumination can account for this non-ideal behavior. To account for the lens aberrations in the theory, the pupil function can be written in the form

$$P(r_1,\phi_1) = \exp 2j\pi(\alpha r_1^4 + \beta r_1^3 \cos \phi_1 + \gamma r_1^2 \cos^2 \phi_1), \quad (3.29)$$

where r_1 is the radial distance from the axis at the pupil, and α, β, and γ are the Seidel coefficients of spherical aberration, primary coma, and primary astigmatism, respectively.[8,21,22] The spherical aberration term primarily affects the symmetry of the depth response curve, while the other two parameters can affect its width.

To model spherical aberration and account for the change in amplitude through a compound lens system, Corle et al. found it convenient to use a trial apodization function of the form[5]

$$\begin{aligned} P(\theta) &= e^{A\sin^2\theta} e^{-jB\sin^4\theta} \quad (\theta < \theta_0), \\ P(\theta) &= 0 \quad (\theta > \theta_0), \end{aligned} \quad (3.30)$$

where A is an amplitude coefficient and $B \approx \alpha/z_0^2$ is related to the Seidel coefficient for spherical aberration with z_0 being the distance from the reflecting plane to the focus. The parameter $\sin \theta$ rather than θ was used in these approximations because the aperture is large. Different values of A and B were tried to obtain the best fit to the experimental results. The calculated result for $A = 0.2$ and $B = 6$ is shown by the solid line in Fig. 3.5 and can be compared to the theory for $P(\theta) = 1$, shown as a dashed line, and the experimental data, shown as a dotted line.

The phase error at the edge of the lens for this value of B corresponds to 1.25π, well within the expected values for primary spherical aberration. The value of A indicates that the amplitude is increased by 17% at the edge of the lens. These values are typical of a lens system designed for broadband illumination, when it is used at a single wavelength. These curves also illustrate that a relatively small amount of aberration can give rise to asymmetry in the depth response. With no aberrations the depth response curves are symmetric. However, with spherical aberration, the depth response curves become asymmetric around $z = 0$ and the asymmetry reverses its shape with the sign of the aberration.[23]

Changing the Shape of the Pupil Function The sensitivity of the depth response to the pupil function suggests the possibility of modifying the depth response by changing the pupil function of the lens. As we have seen, a phase change like spherical aberration affects the symmetry of the sidelobes. Amplitude apodization, on the other hand, increases the width of the main lobe. If the amplitude apodization causes the illumination intensity to decrease near the edge of the lens it will also decrease the effective numerical aperture of the lens, thus increasing the width of the central lobe. In the converse situation, if the amplitude apodization increases the transmission near the edges of the lens such that the illumination takes the form of a thin ring, all rays reaching a point on the axis will be in phase, and the depth of focus is, in principle, infinite. In practice, the penalty paid is a high sidelobe level in the transverse direction and poor illumination efficiency.

Chromatic Aberration A number of other factors can influence the shape of the depth response curve.[24] These include chromatic aberration, nonideal reflectors, and sample roughness. When broadband illumination is used in an RSOM, chromatic aberration within the objective lens will cause light of different wavelengths to focus to different depths. This effect, known as axial chromatic aberration, broadens the depth response. In a visual inspection system, chromatic aberration can be an asset because it allows height differences to be resolved by color differences.[25,26] However, in a metrology system, chromatic aberration can reduce the ability of the instrument to make measurements for it increases the effective range resolution. The use of an apochromatic objective that is color corrected at three wavelengths significantly reduces this effect; typically for narrow bandwidths, the range resolution is not much affected by chromatic aberrations, providing that the wavelength of the light is close to the design wavelength of the objective lens.

Effect on $V(z)$ of the Reflectivity Changing with Angle To account for the influence of the nonidealized behavior of the sample reflection coefficient on the depth response a term of the form $R(\theta,\phi)$ can be added into the integral of Eq. (3.7),

$$V(z)_{\text{plane}} = \frac{\int_0^{2\pi}\int_0^{\theta_0} R(\theta,\phi)P^2(\theta)e^{-2jknz\cos\theta}\sin\theta\cos\theta\,d\theta\,d\phi}{2\pi\int_0^{\theta_0} P^2(\theta)\sin\theta\cos\theta\,d\theta}. \quad (3.31)$$

As might be expected, $R(\theta,\phi)$ for a metal mirror or semi-infinite dielectric substrate varies slowly with angle, so the depth response is only slightly affected by its inclusion.[4] Furthermore, the contributions of a significant change in reflectivity at one angle, such as the Brewster angle, are small compared with the total integrated power in the rest of the beam. In these cases, the shape of the depth response is typically changed much less by the reflectivity of the sample than by the contribution of lens aberrations.[27] However, if there is a thin transparent film several wavelengths thick deposited on a substrate, as discussed in Chapter 5, $R(\theta,\phi)$ can vary rapidly with the angle θ. In this case, as would be expected, the $V(z)$ characteristic can change quite radically. Thus by knowing the reflectivity of a plane wave as a function of angle, it is possible to calculate $V(z)$ accurately. For maximum accuracy, Mansfield and Kino have carried out this type of calculation using the full vector field theory.[28]

Effect of Surface Roughness An interesting and important feature of the depth response curve for range-sensing applications is its insensitivity to surface roughness and sample tilt. This insensitivity occurs because a point illuminated on the reflector is still a point, even if the reflector has a rough surface. In this case, the reflected light is scattered over a wide angular range, so only a small portion of the reflected light reaches the objective, but an image of the focal point, albeit reduced in intensity, is obtained at the pinhole.[29] The insensitivity to roughness has been demonstrated by Corle et al., who replaced a mirror reflector with a quartz flat that had been roughened using 5-μm grit and coated with 150 nm of gold.[5] The sample was tilted at an angle of 9° from a plane normal to the z-axis of the system to eliminate specular reflections.

A depth response curve taken using a 5×/0.07 $N.A.$ lens, is shown as the solid line in Fig. 3.6. This lens was chosen because it had a spot size considerably larger than the average surface roughness to ensure operation in the diffuse reflection regime. In addition, the aperture angle of this lens is only 4°, so that it was easy to tilt the sample and exclude specular reflection. The central lobe of the curve is unchanged from the central lobe obtained using an aluminum mirror, as shown by the dashed curve in Fig. 3.6. The peak value is, however, 66 dB below the value obtained with a mirror, indicating that the return signal is generated primarily by diffuse reflections. Similar results have also been obtained with reflection from a ground aluminum surface.[30]

3.3 Depth Response of the Confocal Microscope with Finite-Sized Pinholes

The choice of the optimum pinhole size is an important criterion in the design of a confocal microscope. If the pinhole is too small, there is

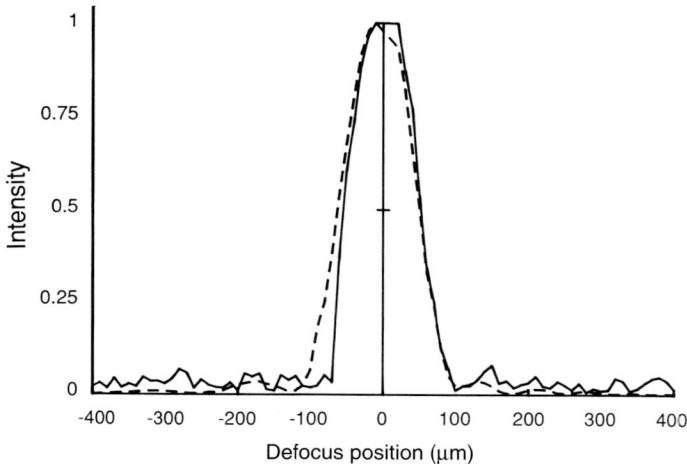

Figure 3.6 Experimental depth response curves for a 5×/0.07 *N.A.* objective. Dashed line: depth response from a mirror sample. Solid line: depth response from a rough tilted sample. [From T. R. Corle, C.-H. Chou, and G. S. Kino, "Depth response of confocal optical microscopes," Opt. Lett. **11**, 770–772 (1986), with permission.]

a significant loss of light intensity; if it is too large, the range and transverse definitions suffer.[31] The first treatment of the effect of finite pinhole size on range resolution was published by Wilson and Carlini.[32,33,34] They considered the problem of a collimated beam entering the objective with a pinhole of finite size in front of the detector. Shuman has developed a simple theory based on a Gaussian beam approximation.[35] Kino and Xiao, who were interested in the problem of the Nipkow disk in the RSOM, dealt with the situation when the transmitted and reflected beam both pass through the same pinhole.[7] This problem has also been analyzed by Wilson and Hewlett.[36] As might be expected, in the RSOM, since both the illuminating and reflected beams are affected, the performance falls off more rapidly with increase in pinhole size than it does with a CSLM illuminated with a rectilinear beam or from a pinhole of infinitesimal diameter.

3.3.1 Approximate Theory for Optimum Pinhole Size

To derive a simple formula for the optimum pinhole size we will consider the confocal microscope illustrated in Fig. 3.1. In this instrument, the light reflected by the sample is focused by the pinhole relay lens onto

3.3 Depth Response of the Confocal Microscope with Finite-Sized Pinholes

a finite-diameter pinhole detector. Assuming that the objective lens is uniformly illuminated and an ideal sample is located at the focal plane of the objective, it can be shown from Fourier optics that the normalized light intensity at the pinhole is

$$I^R = \left[\frac{J_1(kr \sin \theta_0')}{kr \sin \theta_0'}\right]^2, \quad (3.32)$$

where θ_0' is the angle between the outer ray and the axis at the pinhole relay lens.

Using this formula, we may estimate the optimum pinhole radius by choosing the radius of the pinhole, $a(\text{opt})$, to be at the half-power point of the beam,

$$a(\text{opt}) = 0.25\lambda/\sin \theta_0'. \quad (3.33)$$

From Eq. (3.1), the sine condition for a perfect lens, we can take the magnification M of the system into account and obtain an approximate formula for the optimum pinhole radius,

$$a(\text{opt}) = \frac{0.25\lambda M}{N.A.}, \quad (3.34)$$

where $N.A.$ is the numerical aperture of the objective. From this formula, we find that for a $100\times/0.95$ $N.A.$ objective lens and a wavelength of $\lambda = 546$ nm, $a(\text{opt}) = 14.4$ μm. As it turns out, this estimate of the radius is somewhat smaller than is needed for reasonable focusing and efficiency.

Ideally, if several different lenses are to be used in a turret mount, the pinhole size has to be changed for each lens, unless the ratio of magnification to the numerical aperture is kept constant. However, if the optimum pinhole size is chosen for the highest magnification lens, which is usually the lens with the highest numerical aperture, then the pinhole diameter will be larger than optimum for the lower $N.A.$ lenses, where the loss in resolution may not be so critical.

3.3.2 Approximate Theory for the Range Resolution vs. Pinhole Size

The Confocal Scanning Laser Microscope   When the pinhole diameter is much larger than the 3-dB width of the focused beam, geometric optics can be used to calculate the range resolution as a function of pinhole size.

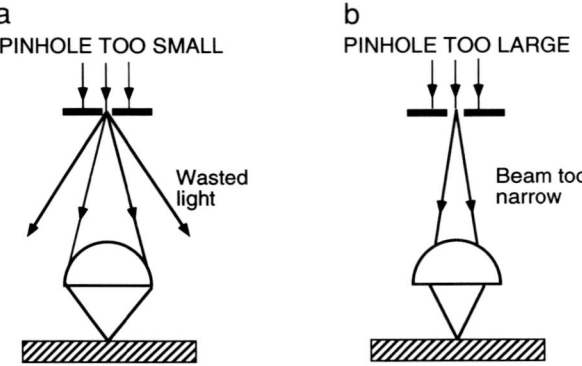

Figure 3.7 Illustration of the effect of pinhole size on the beam passing through it.

As illustrated in Fig. 3.1, when the sample is moved out of focus by a distance z, the beam passing through the pinhole is partially blocked. For a *confocal scanning laser microscope* (CSLM), we can estimate the range resolution between the half-power points in the large-pinhole limit by determining at what value of z the pinhole occupies half the area of the beam. Since the radius of the beam on the sample side, at a defocus distance z, is $r = z \sin \theta_0$, this simple assumption, with the use of the sine condition of Eq. (3.1), yields a range resolution for the CSLM of

$$d_z(3\text{dB})_{\text{plane}} = \frac{a\sqrt{2}}{M \sin \theta_0}. \tag{3.35}$$

It will be observed that the range resolution in the large-pinhole limit is linearly proportional to the radius of the hole.

The Real-Time Scanning Optical Microscope The effect of finite pinhole size on the range resolution can also be estimated for finite-size pinholes in the RSOM. We consider, as shown in Fig. 3.7, a CSOM using a single pinhole for both the illuminating and reflected beam, such as in the RSOM. In this case, we assume that the beam leaving the pinhole diverges at an angle θ'(effective). If the pinhole is too small, Fig. 3.7(a), the angle θ'(effective) will be large and thus much of the beam power is wasted. If the pinhole is too large, the angle θ'(effective) is too small for the beam to fill the pupil of the objective, and the effective numerical aperture is decreased. By using this effective numerical aperture, we can calculate the range resolution.

3.3 Depth Response of the Confocal Microscope with Finite-Sized Pinholes

It follows from Fraunhofer diffraction theory that for a pinhole of radius a, the transmitted beam will diverge by an angle θ'(effective) from the axis, where[37]

$$\sin \theta'(\text{effective}) = \frac{K\lambda}{a}. \tag{3.36}$$

If the divergence angle is assumed to be the angle to the first zero of the Airy function, then $K = 0.61$. By comparison with the more exact theory given in the next section, however, we have found that the angle at which the light intensity drops to 25% of its peak value is the best choice of divergence angle. In this case $K = 0.35$. The effective pupil size, and corresponding objective numerical aperture through which most of the beam power passes, is then given by combining the sine condition of Eq. (3.1) with Eq. (3.36) and writing

$$N.A. = n \sin \theta_0 = M \sin \theta' = \frac{KM\lambda}{a}. \tag{3.37}$$

The approximate range resolution of the microscope can be calculated by substituting Eq. (3.37) into Eq. (3.11) to give the result

$$d_z(3\text{dB})_{\text{plane}} = \frac{7.4na^2}{\lambda M^2}. \tag{3.38}$$

In Eq. (3.38), we have explicitly set $K = 0.35$.

In the large-pinhole limit, the range resolution is proportional to the square of the pinhole radius. The range resolution is inversely proportional to the wavelength in Eq. (3.38) because as the wavelength decreases, the divergence angle decreases as λ/a and the effective numerical aperture decreases.

3.3.3 Exact Theory for the Range Resolution vs. Pinhole Size

A scalar theory for the variation of the depth response as a function of pinhole size has been derived by Wilson and Carlini for the CSLM[32] and by Kino and Xiao for the RSOM.[7] We will repeat these derivations here and will not attempt to use a full vector theory because of its inherent complexity and because in most cases the polarization is not maintained as the light passes through the various lenses in the system. In the derivation, it will be assumed that the field incident on the pinhole is uniform with a value Φ_0^i and that paraxial conditions are satisfied on the pinhole side, but not on the sample side of the objective.

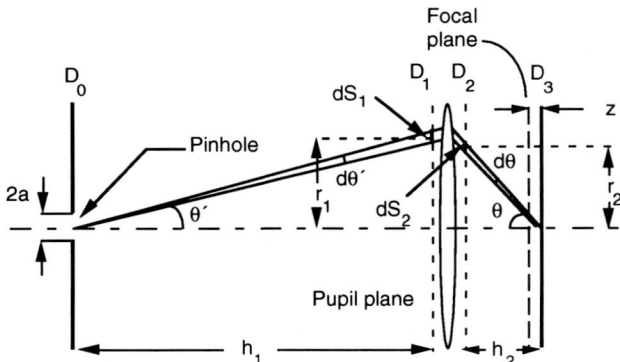

Figure 3.8 Schematic used for the calculation of the range and transverse resolution as a function of pinhole size.

We will use the Kirchhoff approximation that the field at the pinhole follows closely the form of the incident wave field, except very close to the edge of the pinhole. The Fraunhofer approximation to the Rayleigh-Sommerfeld diffraction theory will then be employed to calculate the incident field at the pupil plane of the lens, D_1, located a distance h_1 from the pinhole in the configuration illustrated in Fig. 3.8. If the field of the transmitted light passing through a pinhole of radius a is given by Φ_0^I, the form of the field, $\Phi_1^I(\theta')$, at the pupil plane of the objective will be

$$\Phi_1^I(\theta') = \frac{\Phi_0^I a^2}{j\lambda h_1} \frac{J_1(ka \sin \theta')}{ka \sin \theta'} e^{-jk(h_1 + r_1^2/2h_1)}. \tag{3.39}$$

In Eq. (3.39) $k = 2\pi/\lambda$, θ' is the angle from the axis, and r_1 the radius in cylindrical coordinates at the plane D_1. The parameter r_1 can be stated in terms of θ' and h_1 as $r_1 = h_1 \tan \theta'$.

As in the earlier treatment for a pinhole of infinitesimal size, given in Section 3.1, the reflected field Φ_1^R at the pupil plane is calculated by multiplying the incident field Φ_1^I by a phase factor which accounts for the defocus of the lens,

$$\Phi_1^R = \Phi_1^I e^{-2jnkz\cos\theta} e^{-j\psi}. \tag{3.40}$$

In Eq. (3.40) z is the distance of the mirror from the focal plane of the objective and ψ is a constant phase term associated with the total phase delay through the lens and beam path to the focal plane.

By using Rayleigh-Sommerfeld theory, the reflected wave field is then propagated back to the pinhole to determine the scalar potential at the

3.3 Depth Response of the Confocal Microscope with Finite-Sized Pinholes

pinhole, $\Phi_0^R(r_0,\varphi_0)$.[41] The power passing back through the pinhole as a function of the focus position z can then be calculated. As a first step, we write

$$\Phi_0^R(r_0,\varphi_0) = \frac{1}{j\lambda} \int_{\phi_0=0}^{2\pi} \int_{r_1=0}^{b} \Phi_1^R(r_1,\varphi_1) \frac{\exp(-jkR')}{R'} r_1 \, dr_1 \, d\phi_1, \quad (3.41)$$

where $R' = \sqrt{r_1^2 + h_1^2 - 2r_0 r_1^2 \cos(\varphi_0 - \varphi_1)}$, r_1, φ_1 are cylindrical coordinates at the pupil plane, and r_0,φ_0 are cylindrical coordinates at the pinhole plane.

Substituting Eqs. (3.39) and (3.40) into Eq. (3.41) and applying the paraxial approximation on the pinhole side of the objective: $\cos\theta' \approx 1$ with $r_1^2 \ll h_1^2$, and $r_0^2 \ll h_1^2$. After some algebra, the use of Bessel function identities, and Eqs. (3.1) and (3.6) relating θ and θ', it follows that,

$$\Phi_0^R(r_0)_{\text{RSOM}}$$

$$= \Phi_0^i \frac{kna}{M} e^{-j\psi} \int_0^{\theta_0} J_1\left(\frac{kna\sin\theta}{M}\right) J_0\left(\frac{knr_0\sin\theta}{M}\right) e^{-2jknz\cos\theta} \cos\theta \, d\theta. \quad (3.42)$$

Equation (3.42) is a corrected form of the equation published by Kino and Xiao.[7]

The depth response of the microscope as a function of pinhole radius is calculated by integrating the reflected field intensity over the area of the pinhole and normalizing to the intensity at the best focus position,

$$I(z) = \frac{\int_0^a |\Phi_0^R(r_0,z)|^2 r_0 \, dr_0}{\int_0^a |\Phi_0^R(r_0,0)|^2 r_0 \, dr_0}. \quad (3.43)$$

Equation (3.43) gives the depth response of an RSOM or a CSOM, with equal diameter transmitting and receiving pinholes. As expected, for pinholes of infinitesimal size, Eq. (3.43) reduces to Eq. (3.5).

The FWHM of the depth response as a function of the normalized pinhole radius $a/M\lambda$ for different numerical aperture lenses is plotted in Fig. 3.9 for the RSOM. The dotted lines in Fig. 3.9 are the FWHM widths calculated from Eq. (3.43) for different numerical aperture lenses. The solid line is the approximate FWHM calculated using Eq. (3.38). There is surprisingly good agreement between the two results for normalized pinhole radii $R_p = a \, N.A./M\lambda > 0.6$, indicating that the approximate calculations are sufficient for most purposes. The depth response is fairly flat out to $R_p \approx 0.6$, after which it begins to increase from the theoretical

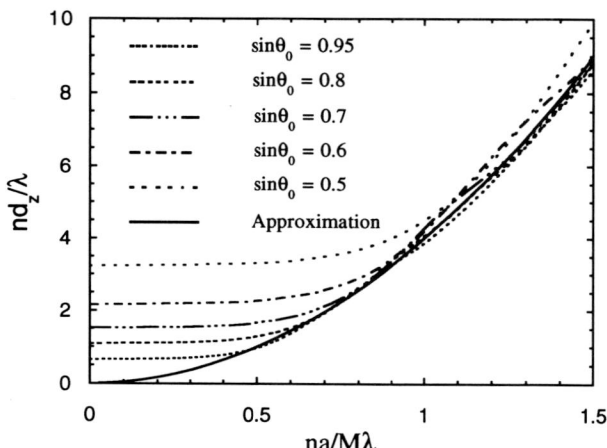

Figure 3.9 The FWHM of the depth response of an RSOM as a function of the normalized pinhole radius for different numerical aperture lenses. A comparison is made with the approximate formula Eq. (3.38).

value for an infinitesimal pinhole. For a $100\times/0.95$ $N.A.$ lens at $\lambda = 546$ nm in air, this normalized radius corresponds to an actual pinhole radius of 30 μm.

For the CSLM the calculations are modified by assuming the objective is illuminated with collimated light from a point source at infinity and that the input beam radius is b. Equation (3.42) then becomes

$$\Phi_0^R(r_0)_{CSLM} = \Phi_1^I \frac{k^2 n b^2}{2M} \int_0^{\theta_0} J_0\left(\frac{knr_0 \sin\theta}{M}\right) e^{-2jknz\cos\theta} \sin\theta \cos\theta \, d\theta. \quad (3.44)$$

The depth response for the CSLM can be calculated by substituting Eq. (3.44) into Eq. (3.43).

The FWHM of the depth response of the CSLM as a function of the normalized pinhole radius $a/M\lambda$ for an $N.A. = 0.5$ lens is plotted in Fig. 3.10. The FWHM of the approximation, Eq. (3.35), is also plotted. Just as in Fig. 3.9, the approximate and the exact result tend to converge for a normalized pinhole radius of $R_p = a\,N.A./M\lambda > 0.6$.

Efficiency The reflected fields at the pinhole can also be used to calculate the single-pinhole light efficiency of the microscope. The efficiency, η_{RSOM}, is defined as the ratio of the reflected power from a perfectly in-focus plane mirror passing through the pinhole to the incident power

3.3 Depth Response of the Confocal Microscope with Finite-Sized Pinholes

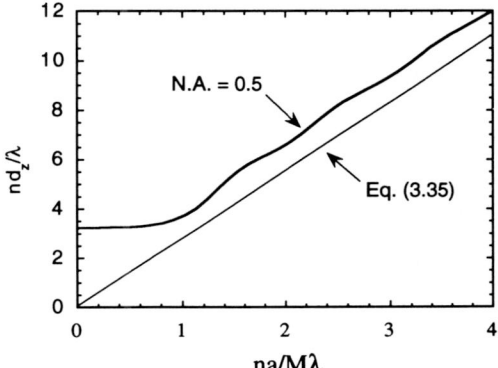

Figure 3.10 The FWHM of the depth response of a CSLM as a function of the normalized pinhole radius compared to the approximate formula Eq. (3.35).

on the pinhole. It is independent of any light loss due to the lenses or the reflection coefficient of the sample. It is also independent of the illuminator efficiency and the pinhole spacing in a Nipkow disk type of microscope. These contributions must be multiplied together with the single-pinhole efficiency to determine the overall light efficiency of the microscope. For an input signal of power P_0^I and a reflected wave power P_{refl}, the single-pinhole efficiency of the RSOM is

$$\eta_{RSOM} = \frac{P_{refl}}{P_0^I} = \frac{2 \int_0^a |\Phi_0^R|^2 r_0 \, dr_0}{\pi a^2 (\Phi_0^I)^2}. \tag{3.45}$$

For the CSLM the normalized single-pinhole power efficiency η_{CSLM} is defined as the ratio of the reflected power passing through the pinhole to the power incident on the objective lens. The efficiency is given by

$$\eta_{CSLM} = \frac{P_{refl}}{P_0^I} = \frac{2 \int_0^a |\Phi_0^R|^2 r_0 \, dr_0}{\pi b^2 (\Phi_0^I)^2}. \tag{3.46}$$

The normalized single-pinhole power efficiencies at best focus for an RSOM and CSLM as a function of normalized pinhole radius $R_p = a \, N.A./M\lambda$ are shown in Fig. 3.11. As might be expected, the efficiency is worse for small pinholes in the RSOM than in the CSLM, since the light must pass through the pinhole twice in the RSOM. In addition, the range resolution of the RSOM deteriorates more rapidly with increasing pinhole size than with the CSLM.

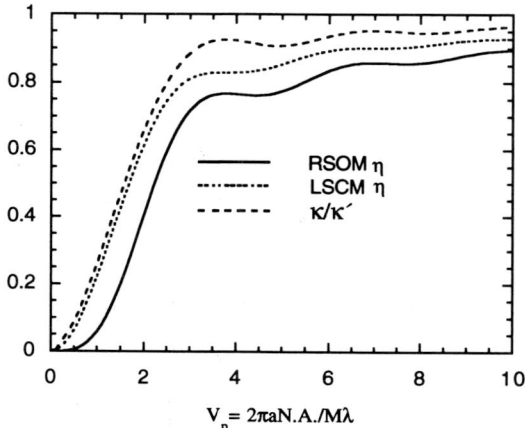

Figure 3.11 The normalized single-pinhole power efficiency, η, for an RSOM and CSLM as a function of the normalized pinhole radius. The value of κ/κ', the increase in far-out sidelobe level of the RSOM due to loss of light at the pinhole while in focus, is also plotted. [From G. S. Kino and G. Q. Xiao, "Real-time scanning optical microscopes," *Confocal Microscopy*, T. Wilson, editor (Academic Press, 1990), with permission.]

For an RSOM with a pinhole size close to the calculated optimum value from Eq. (3.34) of $a = 14.4$ μm, using a 100×/0.95 N.A. lens at $\lambda = 546$ nm, 20% of the reflected power from a plane mirror at the focus of the objective passes through the pinhole. For a CSLM with the same lens and pinhole size, the efficiency is 54%. Although the single-pinhole efficiency of the CSLM is somewhat greater than for the RSOM, this additional power may be lost in the beam expander, so that general conclusions about the performance of the two instruments cannot be drawn without considering the entire optical train. For an RSOM, it should be realized that the light beam transmitted through a single pinhole spreads to a larger diameter than the pupil of the objective, and so some light is also lost in this process.

When the system is badly defocused, as is discussed in Chapter 2, the defocused light reflected from the sample returns through the objective lens and forms a large enough spot on the disk so that some light passes through alternative pinholes to the eyepiece. As the defocusing becomes more extreme, the proportion of reflected light passing back through the pinholes approaches $\kappa = \pi a^2/A_d$, where κ is defined as the fraction of pinhole area to total illuminated area, and A_d is the surrounding disk area per pinhole. Thus, the far-out sidelobes reach a floor level determined by κ. But as we have seen, even the in-focus system is not perfectly efficient.

In order to obtain the relative sidelobe level of the badly defocused system, we must calculate the ratio of the badly defocused signal reaching the detector through all the pinholes to the power reaching the detector through a single pinhole when the system is perfectly focused. The ratio of the badly defocused signal received at the detector to the in-focus signal is called κ', which must be larger than κ. By using calculations similar to those given above, we have plotted κ/κ' as a function of $V_p = 2\pi a\, N.A./M\lambda$ in Fig. 3.11. The implication is that the relative level or sidelobe level of the badly defocused signal is higher than the fractional area of the pinholes κ.

3.4 Transverse Response of the Confocal Microscope

3.4.1 Transverse Response for Infinitesimal Pinholes

We shall now compare the transverse imaging characteristics of a standard optical microscope with a CSOM. The main difference lies in the method of illumination and detection. The standard microscope uses a condenser lens to illuminate the entire sample uniformly from an incoherent light source. A large-area detector, such as the human eye or a TV camera, detects the image as shown in Fig. 3.12(a). In a confocal microscope, illustrated in Fig. 3.12(b), only one point on the sample is illuminated at a time. In ideal operation of these two instruments a point on the sample is imaged as a diffraction-limited spot on the detector by the objective lens. The amplitude of the spot on the detector is given by the amplitude PSF of the objective lens $h(r)$, which in the paraxial approximation takes the form of the Airy pattern,[41]

$$h(r) = \frac{2J_1(kr \sin\theta)}{kr \sin\theta}. \qquad (3.47)$$

Imaging Characteristics of the Standard Microscope In the standard optical microscope, since the illumination is essentially incoherent, only the intensities of two neighboring points will add at the detector. It is therefore necessary to use the intensity PSF to calculate the imaging properties of the instrument. The intensity PSF of a standard microscope is just the square of the amplitude PSF of the objective lens, i.e., $I_S(r) = |h(r)|^2$, where the subscript S denotes a standard microscope.

For a general object with amplitude reflectance r, the intensity at any point (x, y) in the image is given by the convolution

$$I_S(x, y) = |h|^2 * |r|^2, \qquad (3.48)$$

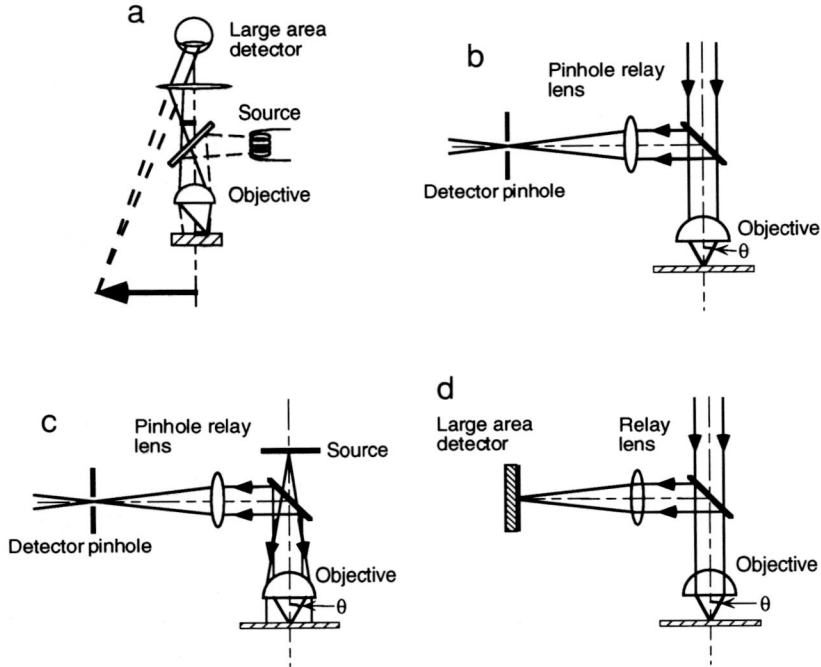

Figure 3.12 Four types of microscopes: (a) a standard optical microscope; (b) a CSOM with a parallel beam source and a point detector; (c) a microscope with uniform illumination and a point detector; (d) a microscope with a parallel beam source and a large-area detector.

where ∗ denotes the convolution operation. The standard microscope, in simplified form, is thus linear in intensity.

Partially Coherent Illumination In a standard microscope which uses a condenser to image an incoherent source onto the sample, diffraction effects in the condenser will cause the image of neighboring points on the source to overlap on the sample, producing partially coherent illumination. Optimum performance is obtained when the numerical aperture of the condenser is equal to that of the objective. This optimum condition is realized only if the apertures in the condenser of a standard microscope are adjusted carefully. In general, Fourier optics does not strictly apply to a poorly adjusted standard microscope, since the images are not linear in amplitude or intensity. Imaging in partially coherent standard microscopes has been analyzed by Hopkins,[38,39] Kintner,[40] and is discussed in the text by Born and Wolf.[8]

3.4 Transverse Response of the Confocal Microscope

Imaging Characteristics of the Confocal Microscope In a CSOM the sample is illuminated one point at a time by the objective. For analysis, we consider the configuration shown in Fig. 3.12(b). Following the arguments of Wilson and Shepard,[41] we calculate the field $U(x,y)$ reflected by a point x,y on the sample when the input beam is focused on the point $(x = 0, y = 0)$. Taking the amplitude PSF of the objective and source relay lens to be $h_1(x,y)$, $U(x,y)$ is defined as the product of the amplitude PSF of the objective, $h_1(x,y)$, and the sample reflectance, $r(x,y)$, or

$$U(x,y) = h_1(x,y)r(x,y). \tag{3.49}$$

In turn, the field at a point x_0, y_0 on the detector $\Phi(x_0, y_0)$ is the convolution of the field at the sample and the PSF of the pinhole lens, $h_2(x,y)$ (a combination of the objective and the pinhole lens responses),

$$\begin{aligned}\Phi(x_0,y_0) &= \int_{-\infty}^{\infty}\int_{-\infty}^{\infty} h_2\left(\frac{x_0}{M} - x, \frac{y_0}{M} - y\right) U(x,y)\,dx\,dy \\ &= \int_{-\infty}^{\infty}\int_{-\infty}^{\infty} h_1(x,y)r(x,y)h_2\left(\frac{x_0}{M} - x, \frac{y_0}{M} - y\right) dx\,dy,\end{aligned} \tag{3.50}$$

where M is the magnification from the sample to the detector plane.

If a point detector or infinitesimal pinhole is used at $x_0 = y_0 = 0$, as illustrated in Fig. 3.12(b), the detected intensity is

$$I = \left| \int_{-\infty}^{\infty}\int_{-\infty}^{\infty} h_1(x,y)r(x,y)h_2(-x,-y)\,dx\,dy \right|^2. \tag{3.51}$$

Assuming that the amplitude PSFs of the two lenses are even in x and y and equal, $h_1 = h_2 = h$, it follows from Eq. (3.51) that

$$I_C = |h^2 * r|^2. \tag{3.52}$$

It is clear from the above calculations that in the CSOM the amplitude PSF of the microscope is given by $h^2(r)$. The imaging properties are derived by convolving the amplitude PSF of the microscope with the amplitude response of the sample. The CSOM is thus linear in intensity. This result should be compared to the result obtained for a uniformly illuminated system with a point detector, as illustrated in Fig. 3.12(c), for which $U(x,y) = r(x,y)$. In this case, it follows from Eq. (3.50) that

$$I = |h * r|^2. \tag{3.53}$$

The same result is also obtained on a system with a focused input beam and a large-area detector as shown in Fig. 3.12(d).

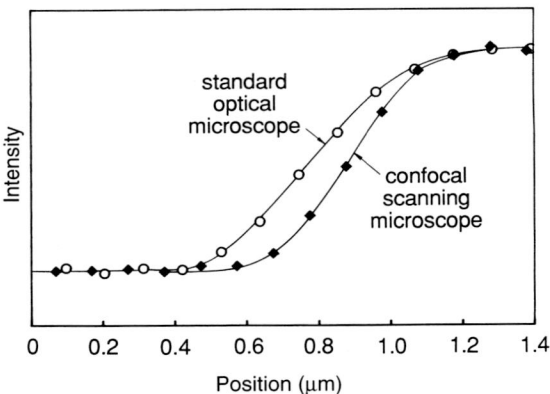

Figure 3.13 Comparison of the edge response of a standard microscope with the response of a CSOM.

The confocal microscope with an infinitesimal pinhole at the detector and focused illumination has an amplitude PSF of $h^2(r)$ and a transverse intensity response to a point reflector of $|h^2(r)|^2$. The response of the microscope to edges and points is thus sharper than the alternative configurations of Fig. 3.12. Like the standard microscope, these alternative techniques have an intensity response to a point reflector of $|h(r)|^2$. However, their response to a more general object, given by Eq. (3.53), is different from that for the standard microscope, Eq. (3.48).

For simple objects, such as edges and points, the CSOM intensity image is the square of a standard microscope's intensity image. The spatial frequency range in the amplitude response of the CSOM is equal to that for the intensity response of the standard microscope. However, confocal microscopes have more contrast and, thus, are better able to visualize the image. This principle is illustrated in Fig. 3.13, which compares the experimentally measured edge response of a CSOM with a standard optical microscope. The edge responses were measured using the same $100\times/0.9$ $N.A.$ objective lens on the two different instruments. It will be observed in the figure that the 50% intensity point in the standard microscope image is approximately equal to the 25% intensity point in the CSOM image.

Single-Point Resolution The single-point resolution for the confocal microscope is defined as the half-power width of the main lobe of the PSF,

$$d_C(3\text{dB}) = \frac{0.37\lambda}{n \sin \theta_0} = \frac{0.37\lambda}{N.A.}. \tag{3.54}$$

3.4 Transverse Response of the Confocal Microscope

The single-point definition of the CSOM is 0.73 of the single-point definition of the standard microscope. However, this fact does not mean that the CSOM has better two-point resolution than the standard optical microscope; the better confinement of the spot in a CSOM mainly allows it to produce images with better contrast.

Reduction of Speckle A major advantage of the confocal microscope, in comparison to the standard microscope is its low sidelobe level. The first sidelobe of the Airy pattern in a standard optical microscope is reduced by 18 dB from the peak of the main lobe, while for the confocal microscope, the first sidelobe is reduced by 36 dB. Since, as discussed in Chapter 1, speckle is caused by addition of the sidelobes from a large number of neighboring points, the effect of speckle tends to be almost imperceptible in the confocal microscope. Put another way, since only one point is being illuminated at a time, there is no interference from the illumination of neighboring points. Consequently, it is possible to work with temporally coherent illumination and still eliminate speckle in the image.

Speckle-like effects can also occur in transparent materials in which there is interference between two objects displaced in depth. The speckle, in this case, is not normally very noticeable. However, precision measurements of distance may be affected by distortions in the shape of the depth response curve. As with other coherent effects, the use of broadband illumination will reduce the level of the interference.

Another major advantage of the CSOM is that very little adjustment is required to achieve optimum performance, whereas, as we have seen, a standard microscope tends to require careful adjustment of the condenser apertures to operate at an optimum level.

3.4.2 Two-Point Resolution

Although the CSOM produces images with better contrast than the standard optical microscope, it is not always better for differentiating between two neighboring points in an image. Since the PSFs of the confocal microscope and standard microscopes have their zeros at the same points, it is better to use the results of Section 1.2.4 and define the Rayleigh criterion as the distance between the two points when the intensity midway between them has dropped to 26.5%.[42] On this basis, the distance between two points which can just be resolved is, for the standard microscope,[43]

$$d_S(\text{Rayleigh}) = \frac{9.61\lambda}{N.A.}, \tag{3.55}$$

and for the confocal microscope,

$$d_C(\text{Rayleigh}) = \frac{0.56\lambda}{N.A.}. \qquad (3.56)$$

Thus, the distance at which two points can be distinguished in the CSOM is only 8% closer than for the standard microscope.

The Sparrow two-point resolution for both the standard optical microscope using incoherent illumination and CSOM is

$$d_S(\text{Sparrow}) = d_C(\text{Sparrow}) = \frac{0.51\lambda}{N.A.}. \qquad (3.57)$$

The results for the two microscopes are identical because the Sparrow criterion states that two equally bright points can be distinguished when the intensity "dip" between them is equal to the intensity at each point. In the normalized response of the standard microscope, squaring the peak to find intensity of the CSOM returns the same value. So there is no improvement in resolution as measured by this criterion.

3.4.3 Edge and Line Response

Another useful definition of resolution is the edge response of the microscope.[44] This definition is often more useful than the point response because most samples, especially in the field of microelectronics, are composed of groups of lines rather than collections of points. We characterize the edge response by the distance between the 20% and 80% intensity points rather than the more conventional 10%–90% response. Spherical aberration often gives rise to some ripple in the edge response above the 80% point and below the 20% point. These ripples can, in turn, cause serious errors in the 10%–90% response measurement due to relatively minor changes in the characteristics of the sample or microscope. The 20%–80% edge response of a CSOM is given by the formula

$$d_C(\text{Edge}) = \frac{0.33\lambda}{N.A.}. \qquad (3.58)$$

Another useful measure of resolution related to the edge response is the two-line response. When measuring critical dimensions on semiconductors the minimum resolvable separation of two lines is often a number of interest. This dimension can be calculated from the *line spread function* (LSF) of the imaging system. The LSF is the response of the imaging

3.4 Transverse Response of the Confocal Microscope

system to a δ-line function. It is the integral of the point spread function in one direction. For a PSF of $h(x,y)$, the coherent LSF is

$$l(x) = \int_{-\infty}^{\infty} h(x,y)\, dy. \tag{3.59}$$

From paraxial theory, it may be shown that,[43]

$$h(x,y) = \int_{-\infty}^{\infty}\int_{-\infty}^{\infty} P(x',y') e^{jk(xx'+yy')/f}\, dx'\, dy'. \tag{3.60}$$

It follows by substitution of Eq. (3.60) into Eq. (3.59) and integrating over dy and dy' that for a radially symmetric system the coherent LSF is the one-dimensional Fourier transform of the pupil function $P(x,0)$,

$$l(x) = \int_{-\infty}^{\infty} P(x',0) e^{jkxx'/f}\, dx'. \tag{3.61}$$

In Eqs. (3.60) and (3.61) f is the focal length of the lens and x is the distance across the sample. By way of comparison, the PSF is the Fourier–Bessel, or two-dimensional Fourier transform of the pupil function.

For coherent imaging, with $P(x',0) = 1$ for $-a < x' < a$, with a lens of radius a in the paraxial approximation, N.A. $= a/f$, the line spread function is,

$$l_c(x) = \frac{2a \sin kx(N.A.)}{kx(N.A.)}, \tag{3.62}$$

where the subscript c stands for coherent imaging.

For the standard incoherent microscope, it is generally not possible to express the integral simply in terms of the pupil function of the lens. Instead the PSF can be directly integrated for a circularly symmetric lens to yield

$$\begin{aligned} l_S(x) &= \int_{-\infty}^{\infty} |h(x,y)|^2\, dy \\ &= \frac{3\pi}{8} \frac{H_1[2kx(N.A.)]}{[2kx(N.A.)]^2}, \end{aligned} \tag{3.63}$$

where H_1 is a first-order Struve function.[39] As with the PSF, the amplitude LSF of a confocal scanning microscope is equal to the intensity LSF of a standard microscope using incoherent illumination, so $l_C(x) = |l_S(x)|^2$.

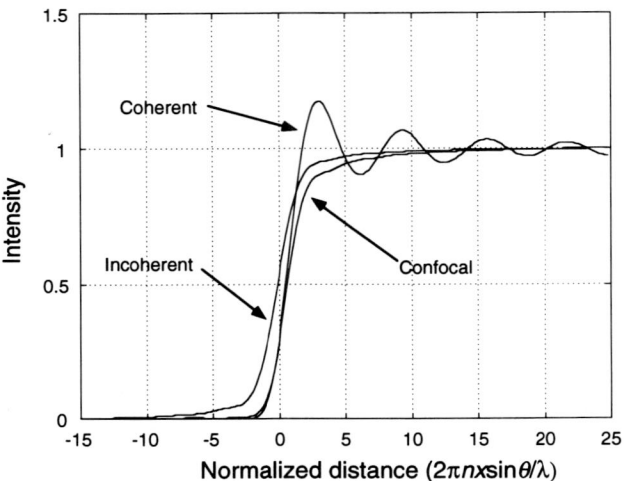

Figure 3.14 The theoretical edge responses for coherent illumination, a standard microscope with incoherent illumination, and a CSOM.

The theoretical edge responses have been calculated by integrating the LSFs for the three cases shown above and are plotted are in Fig. 3.14. The plots are normalized and plotted in terms of $2\pi x N.A./\lambda$. It will be seen that the confocal microscope gives sharper edge definition than the standard incoherent microscope, but no sharper than with coherent illumination. However, coherent illumination gives rise to fringes near the edge.

A Rayleigh two-line criterion can be developed from the LSF response by calculating the separation required for a 26.5% dip in intensity to be observed in the image between the two lines.[45] These distances are

$$d_S(\text{Line}) = \frac{0.66\lambda}{N.A.}, \qquad (3.64)$$

and

$$d_C(\text{Line}) = \frac{0.63\lambda}{N.A.}. \qquad (3.65)$$

The difference in resolution between the standard incoherent microscope and the CSOM when measuring two closely spaced lines is 5%.

The results presented in this section indicate that the improvement in two-point transverse resolution between a standard optical microscope and a confocal microscope is negligible. The real strength of the CSOM

3.4.4 The Effect of Finite Pinhole Size on the Transverse Resolution

The effect of the pinhole size on the transverse resolution of the CSLM has been calculated by Wilson and Carlini[33,46] and for the RSOM by Kino and Xiao.[7] Here, we shall use an extended form of the depth response calculation in Section 3.3.3 to derive a nonparaxial scalar theory for the transverse resolution. The procedure we will follow is summarized below. We assume that the beam is focused on the sample plane D_3, as shown in Fig. 3.8. For the RSOM shown in Fig. 3.8, the lens pupil is illuminated by light passing through the pinhole of radius a. To obtain the PSF of the system, the field amplitude at a point reflector located at a distance x from the axis on the object plane D_3 is then calculated. We determine the field amplitude at the pinhole, $\Phi^R(r_0)$, when the light is reflected from this point. The power passing back through the pinhole of radius a is then determined, and the point spread function is calculated from this result.

To calculate the field amplitude at the sample, we first assume that the pinhole is located with its center on the axis of the objective lens. The field amplitude $\Phi^I(r_1, \theta')$ at the input pupil D_1 of the objective is given by Eq. (3.39). Conservation of power along the ray path and the relation between ray angles θ' and θ on each side of the lens determine the potential at the point r_2, θ on the exit plane D_2 of the objective.

To express power in terms of the scalar field Φ, we identify it with the electric field $\mathbf{E}$ perpendicular to a ray. It then follows that the power density of the wave propagating in the θ direction is $nE^2/2Z_0$ or $n\Phi^2/2Z_0$, where Z_0 is the impedance of free space. We take the elemental areas in a plane perpendicular to the z-axis included by the rays passing through the points r_1, θ' and r_2, θ to be dS_1 and dS_2. These waves propagate at angles θ' and θ to the axis, respectively. It follows from power conservation in the z direction that

$$\Phi_1 \Phi_1^* \cos \theta' \, dS_1 = n\Phi_2 \Phi_2^* \cos \theta \, dS_2. \tag{3.66}$$

We can also write

$$dS_1 = h_1^2 \sec^3 \theta' \sin \theta' \, d\theta' \, d\varphi', \tag{3.67}$$

and

$$dS_2 = h_2^2 \sec^3 \theta \sin \theta \, d\theta \, d\varphi. \tag{3.68}$$

It follows from the stigmatic focusing relation, Eq. (3.1) and Eqs. (3.66)–(3.68), with $d\varphi' = d\varphi$, that

$$\left|\frac{\Phi_2^I}{\Phi_1^I}\right| = \frac{n^{1/2}h_1}{Mh_2}\left(\frac{\cos\theta}{\cos\theta'}\right)^{3/2}. \tag{3.69}$$

Combining Eqs. (3.39) and (3.69), it will be seen that the field at the plane D_2 due to the incident light passing through a pinhole of radius a is

$$\Phi_2^I = \frac{n^{1/2}a^2\Phi_0^I}{jMh_2\lambda}\frac{J_1(ka\sin\theta')}{ka\sin\theta'}e^{jknR_{20}}\cos^{3/2}\theta. \tag{3.70}$$

In Eq. (3.70), we have taken $\cos\theta' \approx 1$ to be consistent with the paraxial assumption on the pinhole side of the lens, and we have neglected phase terms that are uniform over the cross section. We have defined $R_{20} = \sqrt{h_2^2 + r_2^2}$ and have chosen the phase term at the plane D_2 to be $\exp(jnkR_{20})$, corresponding to focusing on a point on the axis at the plane D_3.

We may determine the value of $\Phi_3^I(r_3, \varphi_3)$ at the plane D_3 by putting $\varphi_3 = 0$, and $r_3 = x$. In this case, the distance from this point to the point r_2, φ on the plane D_2 is

$$\begin{aligned}R_2 &= \sqrt{h_2^2 + (r_2\cos\varphi_2 - x)^2 + r_2^2\sin^2\varphi_2} \\ &= \sqrt{h_2^2 + r_2^2 + x^2 - 2r_2x\cos\varphi_2}.\end{aligned} \tag{3.71}$$

Assuming that $x^2 \ll h_2^2$, but making no approximations to the r_2 term, it follows from a first-order expansion in x of Eq. (3.71) that

$$R_2 \approx R_{20} - x\sin\theta_2\cos\varphi_2. \tag{3.72}$$

The field at the plane D_3 at the point a distance x from the axis is therefore of the form

$$\Phi_3^I(x) = \frac{n}{j\lambda}\int_0^{r_{\max}(\theta)}\int_0^{2\pi}\Phi_2^I\frac{e^{-jknR_{20}}}{R_{20}}e^{jknx\sin\theta\cos\varphi_2}\cos\theta\,d\varphi_2\,r_2\,dr_2, \tag{3.73}$$

where it should be noted that r_2 is a function of θ. This expression may be integrated to yield the relation

$$\Phi_3^I(x) = -jkn\int_0^{r_{\max}(\theta)}\frac{e^{-jknR_{20}}}{R_{20}}\Phi_2 J_0(knx\sin\theta)\cos\theta\,r_2\,dr_2. \tag{3.74}$$

3.4 Transverse Response of the Confocal Microscope

Substitution of Eq. (3.70) into Eq. (3.74) yields the result

$$\Phi_3^I(x) = A\Phi_0^I \int_0^{\theta_0} J_0(knx \sin \theta) J_1(kna \sin \theta/M) \cos^{1/2} \theta \, d\theta, \quad (3.75)$$

where A is a constant.

The CSLM For the CSLM, the field amplitude at a point reflector on the sample located at the plane D_3 is calculated by assuming that the pinhole is an infinite distance from the input pupil plane, i.e., $M \to \infty$. In this case the illuminating beam is collimated at the pupil plane of the lens, and it follows from Eq. (3.75) that

$$\Phi_3^I(x) = B \int_0^{\theta_0} \Phi_1^I J_0(knx \sin \theta) \sin \theta \cos^{1/2} \theta \, d\theta. \quad (3.76)$$

where B is a constant and Φ_1^I is the field at the input pupil, D_1, of the lens. Note that if the $\cos^{1/2} \theta$ term is omitted, Eq. (3.76) can be directly integrated into the familiar paraxial form for the PSF of a simple lens.

We assume that the potential of the signal reflected from the point reflector is proportional to the amplitude of the incident signal. It can then be shown by an analysis like the one given above that the potential at a point r_0, φ_0 on the pinhole plane is of the form

$$\Phi_0^R(r_0, \varphi_0, x) = C\Phi_3^I(x)U(\rho), \quad (3.77)$$

where C is a constant and ρ is the distance between the image of the point reflector and the point r_0, φ_0 in the pinhole plane, defined by the relations

$$\rho = \sqrt{x^2 + (r_0/M)^2 - 2(xr_0/M) \cos \varphi_0}, \quad (3.78)$$

and

$$U(\rho) = \int_0^{\theta_0} J_0(kn\rho \sin \theta) \sin \theta \cos^{1/2} \theta \, d\theta. \quad (3.79)$$

The transverse resolution may now be found by calculating the normalized power, $g_{CSLM}(x) = P_{tot}(x)/P_{tot}(0)$, passing through the pinhole. It is convenient to integrate over an arc of constant radius ρ. Figure 3.15 illustrates the region of integration for two cases, (a) $x \geq a/M$ and (b) $x \leq a/M$. The intensity point spread function of the CSLM, $g_{CSLM}(x)$, is

$$g_{CSLM}(x) = \frac{|\Phi_3^I(x)|^2 \int_0^{r_0+a/M} |U(\rho)|^2 \beta(\rho,x)\rho \, d\rho}{\pi |\Phi_3^I(0)|^2 \int_0^{r_0+a/M} |U(r_0)|^2 r_0 \, dr_0}, \quad (3.80)$$

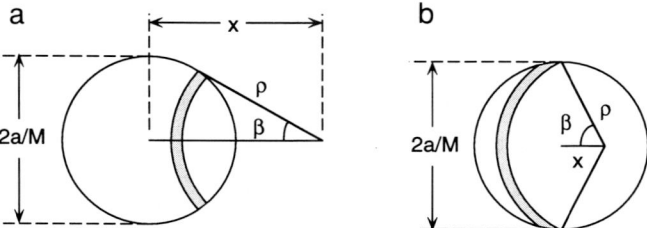

Figure 3.15 The region of integration for a finite pinhole radius a.

where

$$\beta(\rho,x) = \begin{bmatrix} \arccos\left[\dfrac{x^2 + \rho^2 - \left(\dfrac{a}{M}\right)^2}{2\rho x}\right] & x + \dfrac{a}{M} \geq \rho \geq \left|x - \dfrac{a}{M}\right| \\ \pi & \pi - \dfrac{a}{M} - x > \rho \geq 0 \\ 0 & \text{otherwise.} \end{bmatrix} \quad (3.81)$$

The RSOM The formulation developed for the CSLM has also been adapted to the RSOM. The analysis is made somewhat more complex by the fact that the pinholes in the Nipkow disk are rotating. Thus, with the RSOM it is necessary to determine the image seen by a stationary observer through a rotating set of pinholes. The analysis required is different from either the case for a beam entering and leaving through a stationary pinhole or for a collimated transmitting beam and a stationary receiving pinhole of finite size. For the RSOM, we define a *time-averaged point spread function* to allow for this effect. To derive the time-averaged PSF it is assumed that the observer on the left-hand side of the pinhole plane looks at only one point in space and measures the signal arriving at this point as the point reflector is moved a distance x from the axis. The point reflector is illuminated by a set of rotating pinholes, and the image is observed through the same set of moving pinholes. In the calculation, we first determine how the illumination of the point object varies with the position of the illuminating pinhole and then calculate the average signal received at an observation point on the axis in the plane of the rotating Nipkow disk. This signal is the time-averaged point spread function.

3.4 Transverse Response of the Confocal Microscope

When the pinhole is located with its center on the optical axis of the lens, the illumination field at the focus of the objective is given by Eq. (3.75). At some time later, the center of the pinhole has moved to a point r_0, φ_0. The radial distance to the center of the demagnified image of the pinhole from a point x in the object plane is given by ρ as defined in Eq. (3.78). It follows that the general form for the field illuminating the sample becomes

$$\Phi_3^I(r_0, \varphi_0, x) = \int_0^{\theta_0} J_0(kn\rho \sin \theta) J_1(kna \sin \theta/M) \cos^{1/2} \theta \, d\theta. \quad (3.82)$$

If a point reflector is located at a position x in the sample plane, the reflected signal will be proportional to the incident field at that location. The reflected field will be imaged by the objective lens back to the same pinhole, resulting in an on-axis field at the right-hand side of plane of the Nipkow disk given by

$$\Phi_0^R(0,0,x) = D\Phi_3^I(r_0, \varphi_0, x) U(0, x), \quad (3.83)$$

where

$$U(0,x) = \int_{\theta=0}^{\theta_0} J_0(knx \sin \theta) \sin \theta \cos^{1/2} \theta \, d\theta, \quad (3.84)$$

and D is a constant. This integral cannot be directly evaluated in terms of tabulated function, unlike the equivalent paraxial form that can be expressed in terms of the familiar Airy function.

It follows from Eqs. (3.83) and (3.84) that the reflected signal intensity at a point on axis at the plane of the disk, on its right-hand side, is

$$I^R(0,x) = D^2 U^2(0,x) |\Phi_3^I(\rho, \varphi_0, x)|^2. \quad (3.85)$$

As a pinhole moves, the reflected beam associated with it moves across the axis, ρ and φ_0 vary, and the intensity of the beam on axis changes. Consequently, we need to determine the average intensity seen by an observer on the left-hand side of the Nipkow disk. To do this, it is convenient to refer to Fig. 3.15, and integrate over an arc of radius ρ whose included angle 2β varies with ρ, while keeping x constant. The average value of $|\Phi_3^I(\rho, \varphi_0, x)|^2$, as determined by the observer on axis, is then proportional to a quantity $Q(x)$ given by the relation

$$Q(x) = \left\langle \int_0^{x+a/M} [\Phi_3^I(\rho)]^2 \beta(\rho, x) \rho \, d\rho \right\rangle, \quad (3.86)$$

where $\beta(\rho,x)$ is defined in Eq. (3.81) and illustrated graphically in Fig. 3.15.

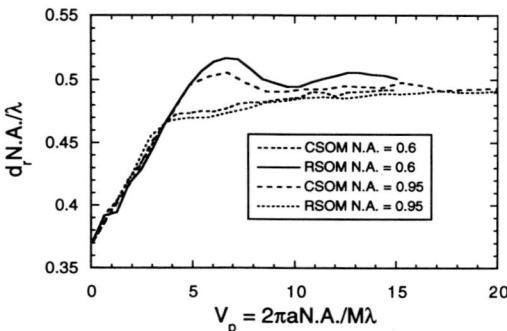

Figure 3.16 A plot of the FWHM of the point spread function of the CSOM and RSOM as a function of the normalized pinhole diameter $V_p = 2na/\lambda M$ for different numerical aperture lenses.

Finally, we determine the normalized average intensity of the signal at the observer on axis, $\langle g_{RSOM}(x) \rangle$, by dividing by the average intensity for $x = 0$, with the result,

$$\langle g_{RSOM}(x) \rangle = \frac{|U(0,x)|^2 Q(x)}{|U(0,0)|^2 Q(0)} = \frac{|U(0,x)|^2 \int_0^{x+a/M} |\Phi_3^I(\rho)|^2 \beta(\rho,x) \rho \, d\rho}{\pi |U(0,0)|^2 \int_0^{a/M} |\Phi_3^I(\rho)|^2 \rho \, d\rho}. \quad (3.87)$$

A plot of the normalized FWHM, $d_r N.A./\lambda$, of the intensity PSFs of the CSLM and RSOM for lenses with different numerical apertures as a function of the normalized pinhole radius $V_p = 2\pi a N.A./M\lambda$ is shown in Fig. 3.16. Although the curves for $N.A. = 0.6$ and 0.95 do not lie on top of each other, it will be observed that, unlike the depth response, the transverse resolution begins to increase almost immediately with increasing pinhole size. The FWHM increases by a factor of approximately 1.4 as the pinhole size increases from infinitesimal to very large. When the pinhole size is infinitesimal, the point spread response is equal to that of a perfect CSOM. When the pinhole size is large, the resolution of the system becomes equal to that of a standard optical microscope. The transverse resolution worsens by 10% from the ideal CSOM with a normalized pinhole radius of $V_p = 2\pi a N.A./M\lambda \approx 1.5$. This size corresponds to a pinhole radius of $a = 14$ μm for a $100\times/0.95$ $N.A.$ lens at $\lambda = 546$ nm, much less than the value needed for a 10% increase in the depth resolution. This value corresponds well to our approximate estimate of the optimum pinhole size from Eq. (3.34).

3.5 Depth and Transverse Resolution of the Interferometric Microscope

The confocal microscope obtains its improved resolution by passing the light through a small pinhole. Interferometric microscopes, like the *coherence probe microscope* (CPM) microscope and MCM, make use of a different principle. They rely on the fact that the correlation length of a focused beam from a spatially incoherent source is extremely small. We have shown in Chapter 1 that if the signals from the sample and the reference are of amplitudes S and R, respectively, the output current $I_{SR}(x,y,z)$ from one element of the CCD detector is of the form

$$I_{SR}(x,y,z) = |S|^2 + |R|^2 + 2|SR|g(x,y,z-z_0)\cos\phi(x,y,z-z_0), \quad (3.88)$$

where z is the distance of the reflecting point from the focal point of the lens, and z_0 is the distance of the reference mirror from its focus. The term $g(x,y,z-z_0)$ is the envelope of the correlation function. For a broadband spatially incoherent light source and a plane reflector, the envelope is equal to the depth response $V(z-z_0)$ of a confocal microscope. The $\cos\phi(x,y,z-z_0)$ term corresponds to the interference fringes between the sample and the reference.

In this section we will derive the response of the interferometric microscope for narrowband illumination and show that its point spread function is identical to that of the confocal microscope. We shall also show how the response is modified by broadband illumination.

3.5.1 Scalar Theory for the Depth Response with a Plane Reflector

We first consider the response of the microscope to a plane reflector located a distance z from the focus. Let the incident waveform at the focal plane be $u(x,y)$ and its spatial Fourier transform be

$$U(k_x,k_y) = \int_{-\infty}^{\infty} u(x,y)e^{-j(k_x x+k_y y)}\,dx\,dy. \quad (3.89)$$

If the sample is located a small distance z away from the focal plane of the lens the reflected wave component $U_S(k_x,k_y)$ from the sample can be written as

$$U_S(k_x,k_y) = S(k_x,k_y)U(-k_x,-k_y)e^{-j(2k_z z+\phi)}, \quad (3.90)$$

where ϕ is the phase change upon reflection from the object and k_z is the z-axis component of the wave vector. Similarly, the reflected wave component from the reference mirror located a distance z_0 from the reference focal plane is

$$U_R(k_x,k_y) = R(k_x,k_y)U(-k_x,-k_y)e^{-2jk_z z_0}. \tag{3.91}$$

In Eqs. (3.90) and (3.91), $S(k_x,k_y)$ and $R(k_x,k_y)$ are parameters proportional to the reflection coefficient of the sample and reference mirror, respectively.

These two signals propagate through the objective lens to the detector, where they interfere. The amplitude of the beam transmitted by the objective is also modified by the pupil function of the lens, $P(k_x,k_y)$. The detected intensity is proportional to the square of the sum of the amplitudes of the reference and signal beams, integrated over all spatial frequencies,

$$I(z) = \int |U_S(k_x,k_y) + U_R(k_x,k_y)|^2 |P(k_x,k_y)|^2 \, dk_x \, dk_y. \tag{3.92}$$

In the MCM, which uses only a single objective, one pupil function multiplies both terms in Eq. (3.92). In the CPM two objectives are used but are normally chosen to be identical, so we take the pupil functions to be the same. Since the system is radially symmetric, it is easier to work in cylindrical coordinates, where the radial and axial wave propagation constants are

$$k_z = \sqrt{k^2 - k_x^2 - k_y^2} = k\cos\theta, \tag{3.93}$$

and

$$k_r = \sqrt{k_x^2 + k_y^2} = k\sin\theta. \tag{3.94}$$

In Eqs. (3.93) and (3.94), θ is the angle the incident ray makes with the optical axis. Changing coordinate systems and integrating over the angles in the focused beam, it follows from Eqs. (3.90) and (3.91) that Eq. (3.92) becomes

$$I(z) = 2\pi \int |U_S(k_r) + U_R(k_r)|^2 |P(k_r)|^2 k_r \, dk_r$$

$$= 2\pi k^2 |U|^2 \int_0^{\theta_0} \{S^2(\theta) + R^2(\theta) + \tag{3.95}$$

$$2S(\theta)R(\theta)\cos[2k(z - z_0)\cos\theta + \phi]\} |P(\theta)|^2 \sin\theta \cos\theta \, d\theta.$$

3.5 Depth and Transverse Resolution of the Interferometric Microscope

In Equation (3.95), it is assumed that the incident illumination is uniform over all angles of the focused beam so that the $|U|^2$ term is a constant and may be removed from the integral. The first two terms in the integrand represent a constant background bias. The third term is the interference term. Since the shallow depth response arises out of the interference term, we are primarily interested in its contribution. The first two terms will, therefore, be dropped from the analysis, although it should be remembered that their contributions exist and give rise to a background signal that limits the dynamic range of the system.

When broadband illumination is used, the interference term in Eq. (3.95) must also be integrated over the spectrum of the illumination. In this case, the cross-product term is

$$I_{SR}^b(z) = 4\pi |U|^2 \int_{\substack{\text{Illumination} \\ \text{bandwidth}}} \int_0^{\theta_0} |P(\theta)|^2 S(\theta) R(\theta) \qquad (3.96)$$
$$\{\cos[2k(z - z_0) \cos \theta + \phi] \sin \theta \cos \theta \, d\theta\} k^2 F(k) \, dk$$

where $F(k)$ is the function describing the intensity distribution over the frequency range of the source and the superscript b stands for broadband.

The mathematical methods used to extract the envelope and phase terms from this signal are described in Chapter 4. A brief summary of them is given below. In one method the output is measured for several different positions of the reference plane, separated by multiples of a quarter wavelength, and the information obtained is used to determine the phase. A second method treats the output of the microscope as a modulated carrier, modulated in space rather than time. The signal is Fourier transformed and the low frequency and negative frequency terms are removed. The inverse Fourier transform is taken yielding a processed signal of the form

$$I_P(z) = g(z - z_0) e^{-j[\psi(z - z_0) + \phi]}. \qquad (3.97)$$

Narrowband Depth Response The normalized depth response of the microscope can be calculated in the narrowband case by following a procedure analogous to that of the confocal microscope. For narrowband illumination $F(k) = \delta(k_0)$, where $\delta(k_0)$ is a delta function in frequency space. In addition, it is assumed that the reflection coefficients $S(\theta)$ and $R(\theta)$ are independent of angle so that they can be removed from the integral. It can be shown that since there is a $\pi/2$ phase change between the transmitted and reflected beam at a beamsplitter, the phase change between the sample and reference beams for an ideal plane reflector is $\phi = \pi$. With these substitutions Eq. (3.96) normalized to the intensity detected by an in focus pinhole becomes

$$I_{SR}(z) = \frac{\int_0^{\theta_0} |P(\theta)|^2 \cos[2k_0(z - z_0)\cos\theta + \phi] \sin\theta \cos\theta \, d\theta}{\int_0^{\theta_0} |P(\theta)|^2 \sin\theta \cos\theta \, d\theta}. \quad (3.98)$$

This formula is similar to the normalized amplitude depth response of the confocal microscope. Following the analysis for the CSOM, Eq. (3.98) can be directly integrated in the paraxial approximation $\cos\theta \approx 1$ to yield

$$I_{SR}(z) = \frac{\sin[k_0(z - z_0)(1 - \cos\theta_0)]}{k_0(z - z_0)(1 - \cos\theta_0)} \cos[k_0(z - z_0)(1 + \cos\theta_0) - \phi]. \quad (3.99)$$

Although Eq. (3.99) is the detected intensity, it describes the amplitude depth response of the microscope, since amplitudes from the signal arm add in the interference term. Like the interference microscope, the CSOM response is linear in amplitude. By analogy with Eq. (3.7) for the confocal microscope, the integral in Eq. (3.98) is an integral over the amplitude spatial frequency response of the microscope.

The equation for the amplitude depth response of the correlation microscope consists of two terms. The first term $\sin X/X$ is the envelope of the correlation function. It is equivalent to the amplitude depth response of a CSOM. The second term represents the fringes that arise from interference between the sample and reference beams. It corresponds to the phase in Eq. (3.8) for the confocal microscope. However, because it is generated by correlation, both the positive and negative phase terms are present and must be separated to obtain phase information, as is discussed in Chapter 4. On the other hand, a major advantage over the confocal microscope is that there is a reference present, so that the phase can be measured.

It is interesting that even with a narrowband source, by employing a high numerical aperture objective, a narrow depth response can still be obtained in an interference microscope. The interference, in this case, occurs between rays of light reaching the sample at different angles and therefore suffering different phase delays from the lens to the sample and back. In addition, there is the possibility of using a broadband light source and obtaining a short correlation length due to temporal interference effects.

3.5 Depth and Transverse Resolution of the Interferometric Microscope

Range Resolution A simple paraxial formula for the range resolution between the half-power points of the intensity response in an interference microscope can be derived from Eq (3.99). As mentioned above, although Eq. (3.99) is a detected intensity, it gives the amplitude response of the instrument. The intensity response of the interference microscope is obtained by squaring the envelope of Eq. (3.99). Like the CSOM, the depth response is defined as the distance between the 3-dB points of this response. It is given by the formula

$$d_z(3\text{dB})_{\text{plane}}^{\text{narrowband}} = \frac{0.45\lambda}{1 - \cos\theta_0}. \tag{3.100}$$

Broadband Operation For a broadband source with a center frequency corresponding to k_0, and with uniform illumination over a wave number bandwidth Δk, the envelope $g(z)$ of the correlation function is given in the small numerical aperture approximation by the relation[47]

$$V(z - z_0) = \frac{\sin[\Delta k(z - z_0)(1 + \cos\theta_0)/2]}{\Delta k(z - z_0)(1 + \cos\theta_0)/2}. \tag{3.101}$$

Thus, with a broadband source, even if the aperture of the lens is small, the range resolution between half-power points is limited by the bandwidth of the source to

$$d_z(3\text{dB})_{\text{plane}}^{\text{broadband}} = \frac{1.78\pi}{\Delta k(1 + \cos\theta_0)}. \tag{3.102}$$

More generally, with finite bandwidth and a finite aperture, the depth response is somewhat smaller than its value due to either effect alone.

Figure 3.17(a) shows a comparison between the experimental and theoretical amplitude depth responses of an interference optical microscope for an illumination bandwidth from 400 to 500 nm.[47] The theory used, which is given in Section 3.5.3, takes into account the presence of a silicon nitride thin-film beamsplitter 840 nm thick. The illumination source was a broadband xenon arc lamp and the microscope used a 0.8 *N.A.* objective lens. A cosine variation for the intensity of the illumination spectrum was assumed for the theoretical calculations with zeros at 400 and 500 nm. Figure 3.17(b) shows the theoretical and experimental data after processing to determine the envelope of the intensity response.

Removal of Aberrations The amplitude depth response exhibits a central dark fringe or a local minimum at the center. The location of the minimum agrees with the prediction that a π phase shift is added to one of the beams when both the sample and reference are at their respective

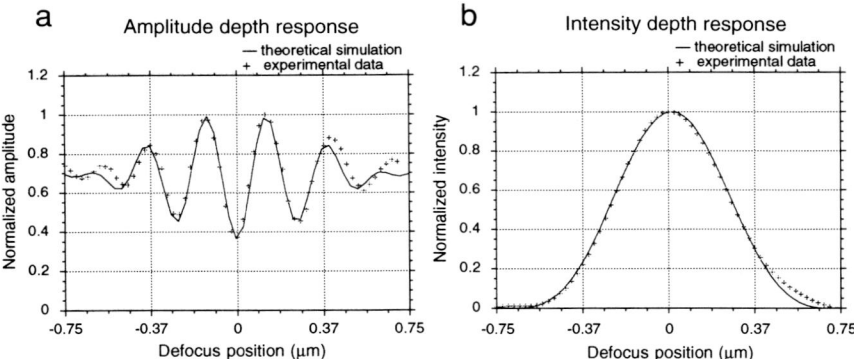

Figure 3.17 (a) A comparison of the experimental and theoretical amplitude depth responses of an interference optical microscope with an illumination bandwidth of 400–500 nm; (b) a comparison of the experimental and theoretical intensity depth responses of the processed data. [Courtesy: S. S. C. Chim, "The Mirau correlation microscope—a new tool for optical metrology," Ph.D. Dissertation, Electrical Engineering Department, Stanford University, Stanford, California, USA (June 1991).]

focal planes. As illustrated in the figure, the experimental depth responses are smooth and they agree with the simulated data. An important conclusion from this data is that aberrations are removed by the correlation process. Because both the signal and the reference beams suffer the same phase distortion in the correlation microscope, the correlation process cancels out these phase errors. This cancellation can be observed by comparing the symmetry of the sidelobe levels in the depth response curve for the CSOM, Fig. 3.5, with the interference microscope, Fig. 3.17(b). It will be observed that the depth response curve of the interference microscope is more symmetrical than that of the CSOM. Spherical aberration does not manifest itself in the depth response of the interference microscope. The phase canceling also accounts for the fact that very little aberration is introduced by the thin-film beamsplitter.

The depth resolution defined as the distance between the half-power points of the intensity depth response curves is tabulated in Table 3.2. As the bandwidth of the illumination is increased, the average optical wavelength also increases. As a result we might expect the depth resolution to increase as well. However, interference between the different frequency components of the broadband light decreases the depth resolution. It is clear from both the theoretical and experimental results, however, that the depth resolution improves with increasing bandwidth.

3.5 Depth and Transverse Resolution of the Interferometric Microscope

Table 3.2 The Effect of Bandwidth on the Range Resolution

Spectrum	d_z(FWHM) experimental	d_z(FWHM) theoretical
400–500 nm	0.55 μm	0.55 μm
400–550 nm	0.54 μm	0.53 μm
400–800 nm	0.47 μm	0.45 μm

3.5.2 Transverse Resolution

The in-focus image intensity of the interference microscope can be found by standard imaging theory as has been done by Davidson et al.[48] Let the sample, reference, and image planes be labeled by the coordinates (x_s, y_s), (x_r, y_r), and (x_i, y_i) respectively. The fields at the detector are given by the convolution of the field at the sample and reference plane with the point spread function of the objective lens,

$$u_R(x_i, y_i) = \int h_R(x_r - x_i, y_r - y_i) u_R(x_r, y_r) \, dx_r \, dy_r, \quad (3.103)$$

and

$$u_S(x_i, y_i) = \int h_S(x_s - x_i, y_s - y_i) u_S(x_s, y_s) \, dx_s \, dy. \quad (3.104)$$

In Eqs. (3.103) and (3.104) $h_R(x,y)$ and $h_S(x,y)$ are the point spread functions of the objectives in the sample and reference paths of the interferometer. In the ideal case, the two lenses are identical in the Linnik system, and in the Mirau microscope the beamsplitter is infinitesimally thin and does not introduce any aberrations. The two-point spread functions then reduce to the point spread function of an ideal objective.

The signal detected by the CCD camera is $\langle |u_R + u_S|^2 \rangle$. The cross-product or the correlation term of the detected signal is $|u_R(x_i, y_i) u_S(x_i, y_i)|$, where

$$\langle |u_R(x_i, y_i) u_S(x_i, y_i)| \rangle = \iint h_S(x_s - x_i, y_s - y_i) h_R^*(x_r - x_i, y_r - y_i)$$
$$\times \langle |u_R(x_r, y_r) u_S(x_s, y_s)| \rangle \, dx_s \, dy_s \, dx_r \, dy_r. \quad (3.105)$$

For a spatially incoherent source, $\langle |u_R(x_r, y_r) u_S(x_s, y_s)| \rangle$ is zero unless the coordinates (x_s, y_s) and (x_r, y_r) are identical. Furthermore, if a perfect reference is used, $u_R(x_r, y_r) = 1$. The correlation term for an object in the focal plane can, therefore, be simplified to

$$I_{SR}(x_i,y_i) = \iint h_S(x_s - x_i, y_s - y_i) h_R^*(x_s - x_i, y_s - y_i) u_S(x_s,y_s) \, dx_s \, dy_s$$

$$= h_S h_R^* * u_S. \qquad (3.106)$$

As with the depth response, Eq. (3.106) describes the amplitude response of the microscope. The interference microscope, like the CSOM, is linear in amplitude with an amplitude PSF $h(x,y) = h_S(x,y) h_R^*(x,y) \approx |h_R(x,y)|^2$ where $u_S(x_s,y_s) = \delta(x,y)$, and $\delta(x,y)$ is the Dirac delta function. To calculate the equivalent intensity response of the microscope, Eq. (3.106) is squared,

$$I = |I_0|^2 = |h_S h_R^* * u_S|^2. \qquad (3.107)$$

It will be observed that the transverse resolution of the interference optical microscope is identical to that of the CSOM and superior to that of the standard optical microscope. As a result, the formulae for the edge response and other imaging characteristics derived for the CSOM can also be applied to the interference microscope.

This theory can be modified to take into account the finite bandwidth, the use of a nonparaxial system, and the effect of a finite-thickness beamsplitter in the Mirau correlation microscope, as is done below. The finite bandwidth tends to lower the level of the outer sidelobes, but does not much affect the width of the main lobe if this width is calculated by using the midband frequency of the light. The elimination of the paraxial assumption tends to widen the main lobe slightly, as it does for the confocal microscope. The finite-thickness beamsplitter, if the thickness is small, has much the same effect as a finite-diameter pinhole in the confocal microscope.

3.5.3 The Effect of the Thin-Film Beamsplitter and Mirror Support of the MCM on Signal Levels, Range, and Transverse Resolution

To obtain high-quality images with a MCM the beamsplitter should satisfy four requirements. First, it should have constant transmission and reflectance over the entire illumination spectrum, as well as over a range of incident angles. Second, taking account of the multiple reflections and transmission through it, the correlation signal should be as large as possible. Third, the rays passing through the beamsplitter at different angles should not be displaced relative to one another, otherwise spherical aberration will be introduced into the focused beam. Finally, the thickness variations should be small compared to the optical wavelength, not exceed-

3.5 Depth and Transverse Resolution of the Interferometric Microscope

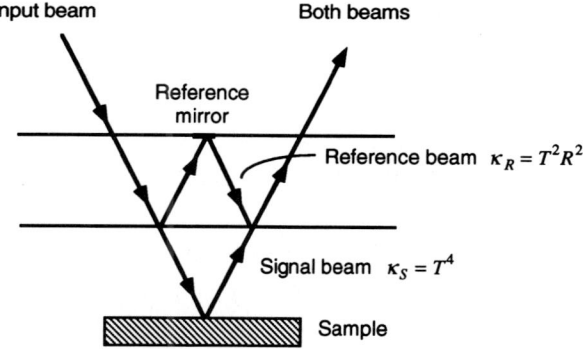

Figure 3.18 The Mirau interferometer showing the reference, signal, and beam paths.

ing $\lambda/16$. These requirements put severe limitations on the choice of membranes which can be used. When any of these requirements is not satisfied, the effect of small deviations from the ideal conditions is very similar to that of having a finite pinhole size in the CSOM.

Effect of the Beamsplitter on Signal Amplitude The finite thickness of the beamsplitter will effect the signal amplitude in a MCM. As illustrated in Fig. 3.18, when the signal reflected from the sample, u_S, passes through two films it is multiplied by an amplitude coefficient $\kappa_S = T^4$. The signal reflected from the reference, u_R, passes through only one film, is reflected from the other and hence is multiplied by a coefficient $\kappa_R = R^2T^2$. Therefore the beamsplitter introduces a multiplication factor of $\kappa = \kappa_R\kappa_S = |R^2T^6|$ into the imaging equations. Ideally, the film material should be chosen to maximize the parameter κ and thus to maximize the image intensity. Since $|T|^2 = 1 - |R|^2$, it is easy to show that the optimum value of $|R|^2$ is 0.25, which is equivalent to $\kappa = 0.011$.

For a lossless dielectric membrane, it may be shown that the Fresnel formulae for reflection and transmission at each individual interface are

$$r = \frac{\sin(\theta_i - \theta_t)}{\sin(\theta_i + \theta_t)}, \tag{3.108}$$

for TE waves, and

$$r = \frac{\tan(\theta_i - \theta_t)}{\tan(\theta_i + \theta_t)} \tag{3.109}$$

for TM waves.

From the Fresnel formulae, the complex amplitude reflection and transmission coefficients, R and T, respectively, for a thin-film membrane have been calculated by Chim to be[47]

$$R = j\frac{2r\sin(\delta/2)}{1 - r^2 e^{-j\delta}} e^{-j\delta/2}, \tag{3.110}$$

and

$$T = \frac{1 - r^2}{1 - r^2 e^{-j\delta}} e^{-j\delta/2}. \tag{3.111}$$

In Eqs. (3.110) and (3.111) r is the amplitude reflection coefficient from the film surface, δ is the two-way phase change of the light transmitted through the film and back. The phase difference δ is given by

$$\delta = 2k_z nh = \frac{4\pi}{\lambda} nh \cos\theta_t, \tag{3.112}$$

where θ_t is the angle of refraction inside the membrane, λ is the free space wavelength, n is the refractive index of the film, and h is its thickness. It will be observed in the above equations that T and R are $\pi/2$ out of phase, which implies that κ_S and κ_R are π out of phase, as is shown in the experimental curves of Fig. 3.17(a). Thus, phase distortion does not affect the range resolution. However, if the amplitude distortion is severe, the range resolution will change because the system will behave as if the pupil is severely apodized.

Figures 3.19 and 3.20 are theoretical plots of normalized reflected power $I_r = |R|^2$ as a function of incident angle and wavelength, respectively. In the figure two different membrane materials are compared: silicon nitride with a refractive index of $n = 2.2$ and a thickness of 84 nm, and a standard pellicle comprising a plastic film with a refractive index of $n = 1.5$ and a thickness of 2 μm. A further assumption in these plots is that the TE and TM components are equal in amplitude. The latter type of pellicle is commonly used in low numerical aperture commercial objectives. The silicon nitride film forms the basis of the Mirau correlation microscope built by Chim and Kino.[49] From the preceding plots it will be observed that pellicle beamsplitter is best used for low angle of incidence, narrowband measurements. Plastic pellicles are usable for angles up to ~20° (N.A. ~0.3). A thin silicon nitride beamsplitter, on the other hand, works well for values of $N.A. < 0.9$. We note that for the 84-nm-thick film, the value of κ at normal incidence, where $|R|^2 = 0.4$, is 0.0086, about

3.5 Depth and Transverse Resolution of the Interferometric Microscope

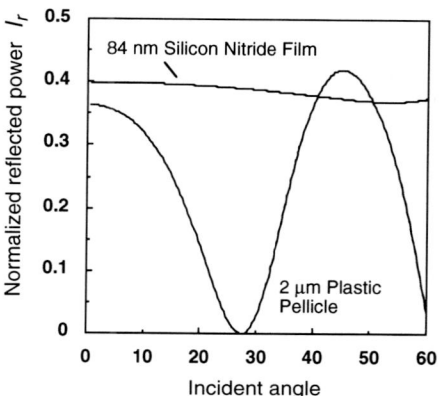

Figure 3.19 Theoretical plot of the normalized reflected power I_r as a function of incident angle. [Courtesy: S. S. C. Chim, "The Mirau correlation microscope—a new tool for optical metrology," Ph.D. Dissertation, Electrical Engineering Department, Stanford University, Stanford, California, USA (June 1991).]

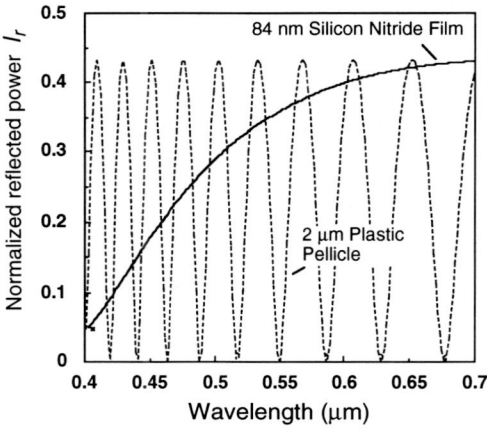

Figure 3.20 Theoretical plot of the normalized reflected power I_r as a function of wavelength. [Courtesy: S. S. C. Chim, "The Mirau correlation microscope—a new tool for optical metrology," Ph.D. Dissertation, Electrical Engineering Department, Stanford University, Stanford, California, USA (June 1991).]

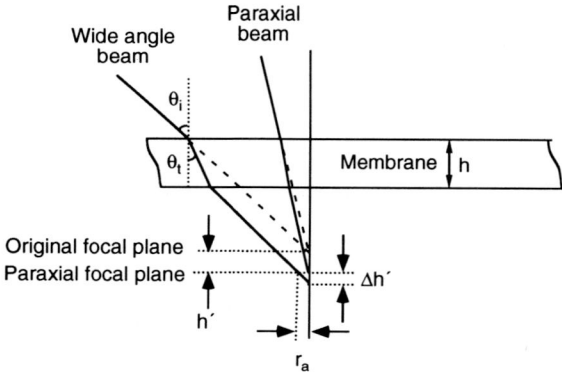

Figure 3.21 Illustration showing the refraction of an optical beam by a membrane.

20% lower than the optimum value of 0.011. To obtain this optimum value, either a thinner film or one of lower refractive index is needed.

Spherical Aberration Due to the Beamsplitter When a thin film, or pellicle, is placed in the path of a focused beam, it tends to introduce spherical aberration into the beam due to the displacement of the off-axis rays as they pass through the material.[50] The thicker the film, the greater the aberrations. We can use ray optics to estimate the aberration, and give a simple physical picture of the effect of the beamsplitter, by determining the radius r_a at which a refracted beam will intercept the focal plane, as illustrated in Fig. 3.21. If this radius, r_a, is greater than the radius of the focused beam estimated from diffraction theory, r_d, then the aberrations are too large and the system will be unsuitable for use in a microscope.

For a beamsplitter with thickness h, it can be shown that in the paraxial approximation the displacement in the z or focus direction of an initially perfectly focused beam is

$$h' = h\left(\frac{n-1}{n}\right). \tag{3.113}$$

In Eq. (3.113) n is the index of refraction of the beamsplitter film. As the ray angles increase, there is an additional displacement $\Delta h'$, given by

$$\Delta h' = \frac{h}{n}\left(1 - \frac{\cos\theta_i}{\cos\theta_t}\right), \tag{3.114}$$

3.5 Depth and Transverse Resolution of the Interferometric Microscope

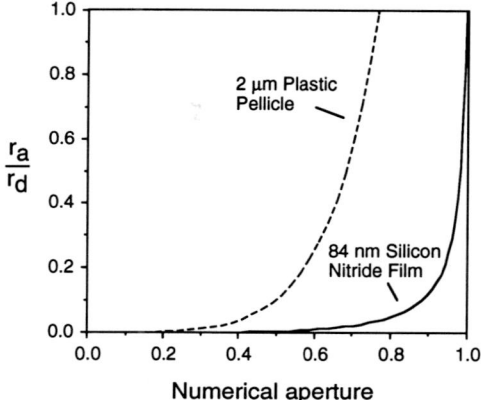

Figure 3.22 A plot of the ratio of the radius of an aberrated beam to the radius of a diffraction-limited beam for both silicon nitride and pellicle membranes. [Courtesy: S. S. C. Chim, "The Mirau correlation microscope—a new tool for optical metrology," Ph.D. Dissertation, Electrical Engineering Department, Stanford University, Stanford, California, USA (June 1991).]

as shown in Fig. 3.21. This extra aberration implies that at the paraxial focal plane, the radius of the focused beam is given by

$$r_a = \frac{h}{n}\left(1 - \frac{\cos\theta_i}{\cos\theta_t}\right)\tan\theta_i, \quad (3.115)$$

where θ_i is the angle of incidence of the outermost ray on the film and θ_t is the angle of the transmitted light within the film calculated from Snell's law. This effect is not important if the calculated radius of the outer ray of the beam is much less than the radius r_d of the beam calculated from diffraction theory,

$$r_d = \frac{0.61\lambda}{\sin\theta_0}. \quad (3.116)$$

In Eq. (3.116) λ is the optical wavelength and $\sin\theta_0$ is the numerical aperture of the beam in air.

Plots of r_a/r_d for both the silicon nitride membrane with $n = 2.2$ and the pellicle membrane with $n = 1.5$ at a wavelength of 540 nm are shown in Fig. 3.22.[49] It will be observed in the figure that a 2-μm-thick plastic membrane can be used with a lens of numerical aperture 0.5 without excessive spherical aberration being introduced into the beam. A small-angle approximation to Eqs. (3.115) and (3.116) with $\theta_i = \theta_0$ and $\sin \theta_i = n \sin \theta_t$, yields the result,

$$\frac{r_a}{r_d} \approx \frac{h(n^2 - 1)\theta_0^4}{1.22 n^3 \lambda}. \tag{3.117}$$

It will be seen that the ratio of the aberrated radius to the diffraction-limited radius increases as the fourth power of the convergence angle, $(N.A.)^4$, and is proportional to the thickness of the film.

The main factor which limits the use of pellicle beamsplitters in a Mirau microscope is the variation in transmission of the pellicle with angle. If we limit the microscope to narrowband illumination, or are not concerned with a uniform illumination spectrum, it is possible to use plastic pellicles with lenses of numerical aperture 0.5 or less. Such instruments are most suited to height and surface roughness measurements where transverse resolution is not a major concern. The 84-nm-thick silicon nitride membrane can be used with a lens of numerical aperture of 0.8, but it is necessary to use a 50-nm-thick membrane for a numerical aperture of 0.9. Therefore in order to accurately measure both height and width with a wide-aperture lens, a thin-film beamsplitter with a thickness of 100 nm or less is needed. The Mirau interferometric microscope designed by Chim uses two membranes, one for the beamsplitter and the second to support a mirror used as the reference. In this case, the values of r_a/r_d shown in the plot must be doubled.

Point Spread Function with a Finite-Thickness Beamsplitter As we have seen, the spherical aberrations due to the beamsplitter cause the spot size of the beam to increase. We can make a direct calculation of the PSF by multiplying the pupil functions associated with the point spread functions h_S and h_R by the parameters $\kappa_S = T^4$ and $\kappa_R = R^2 T^2$, respectively. By employing the paraxial approximation, we may write,

$$h_S(x,y) = h_S(r) = 2\pi \int \kappa_S P(k_r) J_0(k_r r) k_r \, dk_r, \tag{3.118}$$

and

$$h_R(x,y) = h_R(r) = 2\pi \int \kappa_R P(k_r) J_0(k_r r) k_r \, dk_r. \tag{3.119}$$

We use Eqs. (3.106) and (3.107) with $u_S(x_s, y_s) = \delta(x,y)$ to calculate the PSF of the microscope. If the films used in the Mirau interferometer are infinitesimally thin, then the calculations in Eq. (3.118) and (3.119) reduce to the normal Airy pattern of a focused lens.

3.5 Depth and Transverse Resolution of the Interferometric Microscope

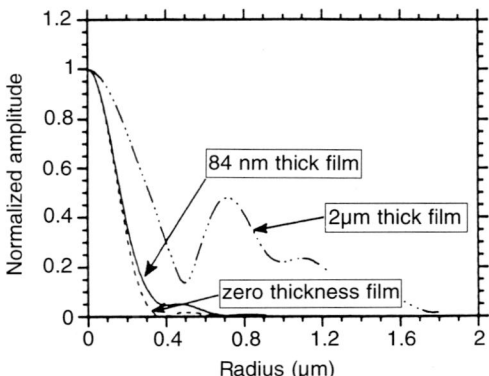

Figure 3.23 Comparisons between three point spread functions of the MCM assuming membrane thicknesses of zero, 84 nm, and 2 μm, respectively. [Courtesy: S. S. C. Chim, "The Mirau correlation microscope—a new tool for optical metrology," Ph.D. Dissertation, Electrical Engineering Department, Stanford University, Stanford, California, USA (June 1991).]

For finite-thickness films, however, very different results can be obtained. Figure 3.23 is a theoretical plot of the PSF as a function of the radial distance r, for three different membrane thicknesses, zero, 84 nm, and 2 μm. The illumination wavelength is assumed to be 500 nm and the numerical aperture is taken to be $N.A. = 0.8$. It will be seen that the effect of the 84-nm-thick films is to increase the width between half-power points (0.707 amplitude) by only about 2%. However, the width of the response at lower levels of normalized signal is considerably worsened by the finite-thickness membranes and thus it has a greater effect on the 20–80% edge response. These ideas are further illustrated in Fig. 3.24. Figure 3.24(a) plots the normalized PSF vs. normalized radius $V = 2\pi r\, N.A./\lambda$ for an MCM using an 84-nm-thick beamsplitter and differential numerical aperture lenses. Figure 3.24(b) is the same plot for an MCM using a 2-μm-thick pellicle beamksplitter. It will be noted in these illustrations that the FWHM resolution deteriorates by about 10% for $N.A. = 0.9$ and is badly distorted for large $N.A.$ with 2-μm-thick pellicles.

As we have discussed for the confocal microscope, it is often more convenient to measure the edge response, rather than the PSF of a microscope, by scanning over a cleaved edge of silicon. We may calculate the intensity edge response as the square of the integral of $h(r)$, taken over the edge using the geometry shown in Fig. 3.25.

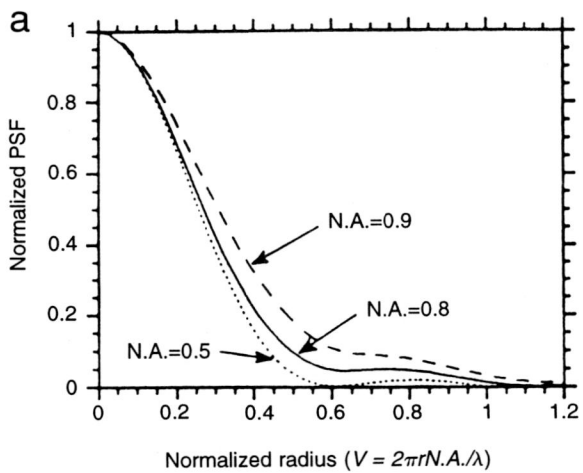

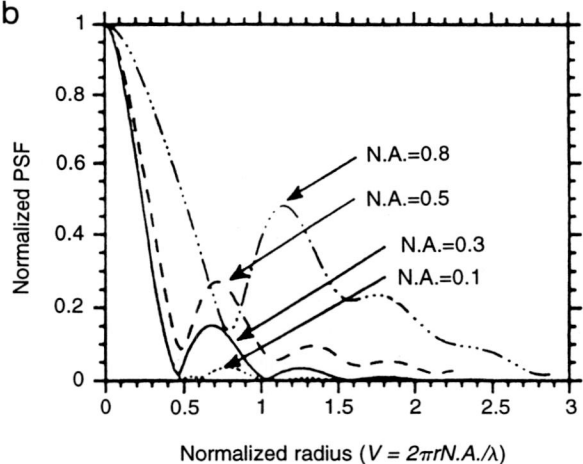

Figure 3.24 Point spread functions of the MCM plotted for several different numerical apertures. (a) 84-nm-thick silicon nitride membranes; (b) 2-μm-thick Mylar pellicles. [Courtesy: S. S. C. Chim, "The Mirau correlation microscope—a new tool for optical metrology," Ph.D. Dissertation, Electrical Engineering Department, Stanford University, Stanford, California, USA (June 1991).]

3.5 Depth and Transverse Resolution of the Interferometric Microscope

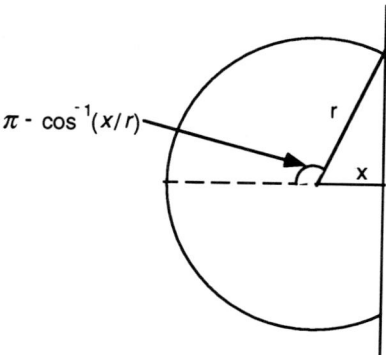

Figure 3.25 The geometry used for calculating the edge response function.

$$Edge(x) = 4\pi^2 \left| \int_0^x h(r)r\,dr + \int_x^\infty h(r)[1 - (1/\pi)\cos^{-1}(x/r)]r\,dr \right|^2. \quad (3.120)$$

Plots of two such theoretical edge responses for the PSF with infinitesimally thin membranes and for 84-nm-thick membranes are shown in Fig. 3.26, along with the experimental edge response for the 84-nm-thick membranes. The theoretical and experimental edge responses for the 0.8 N.A. system are in good agreement. The d(Edge) or 20–80% width has been measured to be 0.23 μm.

In Table 3.3, we compare the theoretical and experimental edge responses of several types of confocal and interferometric microscopes. It will be observed that all the microscopes, when optimized, give about the same performance.

Table 3.3 The Edge Response of Four Different Types of Optical Microscope

Type of microscope	Value of K(Edge) for 20–80% edge resolution, where d_x(Edge) = K(Edge)$\lambda/N.A.$
Ideal theoretical confocal or interferometric microscope	K(Edge) = 0.33
MCM, theory, with 84-nm-thick membranes, $N.A. \leq 0.8$	K(Edge) = 0.37
MCM, experimental, with 84-nm-thick membranes, $N.A. \leq 0.8$	K(Edge) = 0.37
Linnik microscope, experimental, $N.A. \leq 0.9$	K(Edge) = 0.38
RSOM, experimental, $N.A. \leq 0.95$	K(Edge) = 0.39

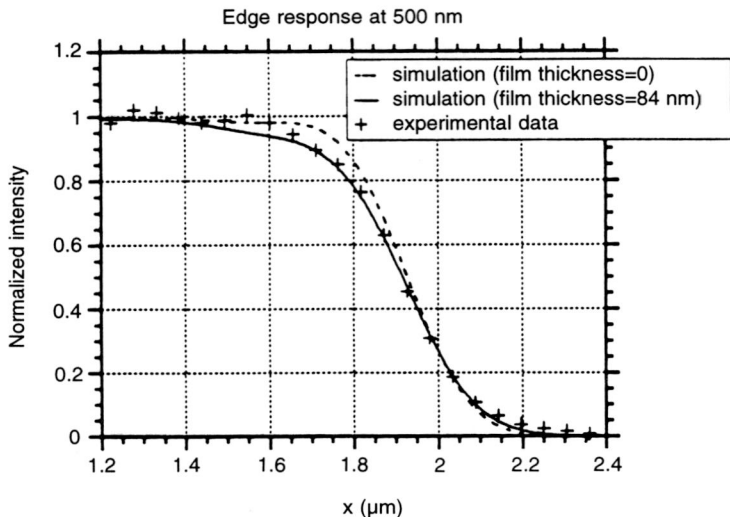

Figure 3.26 Comparison between theoretical and experimental edge responses of the MCM. [Courtesy: S. S. C. Chim, "The Mirau correlation microscope—a new tool for optical metrology," Ph.D. Dissertation, Electrical Engineering Department, Stanford University, Stanford, California, USA (June 1991).]

3.6 The Near-Field Scanning Optical Microscope (NSOM)

In Chapter 2, we described two types of near-field microscopes, those with their definition based on the size of a small pinhole or probe, and those based on the use of the solid immersion lens. We shall first discuss the theory of operation of these microscopes and the limitations on definition and efficiency imposed by the tapering of the fiber. In Section 3.7 we will discuss the parameters of the solid immersion lens.

3.6.1 Attenuation in a Tapered Rod or Fiber

We shall first consider the efficiency and transverse response of a tapered fiber near-field optical probe. In a near-field microscope operating in transmission mode, the light loss is dominated by two factors, the attenuation of the cutoff mode in the fiber and the power radiated from the pinhole. It is assumed that the only signal observed is the radiation field from the pinhole that is relayed through an objective lens to the detector, as illustrated in Fig. 3.27.

3.6 The Near-Field Scanning Optical Microscope (NSOM)

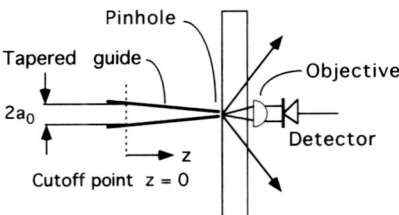

Figure 3.27 Transmission NSOM with objective lens and detector.

The Tapered Fiber We consider the situation for a fiber that is tapered to a diameter that is smaller than the cutoff diameter of the principal mode of the waveguide at the pinhole end. In practice, the fiber is of the order of 125 μm in diameter and is slowly tapered to a few microns in diameter. As the fiber decreases in size, the beam that was originally confined to the 5-μm core leaks into the cladding. With an adiabatic taper, the basic LP_{01} mode of the fiber maintains its original form and becomes a TE_{11} mode of the metal-covered fiber, the one with the least attenuation in the cutoff region. The TE_{11} mode of a circular waveguide of radius a, filled with a medium of refractive index n, has a field E_ϕ that varies as $J_1'(k_c r)$, where $k_c = 2\pi n/\lambda_c$. In this expression J_1 is a first-order Bessel function of the first kind, the sign ' denotes differentiation with respect to the argument, n is the index of refraction of the material and λ_C the free space wavelength at the cutoff frequency.[51] Since the boundary condition on the field E_ϕ implies that $J_1'(k_c a) = 0$, the cutoff frequency is given by the relation $k_c a = 1.84$. For a quartz fiber of refractive index $n = 1/5$, and a free space wavelength $\lambda_C = 0.632$ μm, the cutoff radius is $a_0 = 0.195\lambda = 120$ nm. Thus, for a pinhole and hence a guide 50 nm in diameter, the TE_{11} waveguide mode is well below the cutoff radius.

For a wave in a uniform fiber guide of radius a smaller than the cutoff value, ($ka < 1.84$) the attenuation a per unit length is

$$\alpha = \sqrt{k_c^2 - k^2} = \sqrt{(1.84/a)^2 - k^2} \text{ nepers/unit length.} \quad (3.121)$$

The attenuation is typically dominated by the first term in Eq. (3.121), and is determined essentially by the diameter of the guide.

For a guide with a linear taper from the cutoff point, $a = a_0(1 - z/z_0)$, where $a_0 = 1.84/k$, and the radius is zero at $z = z_0$, as shown in the

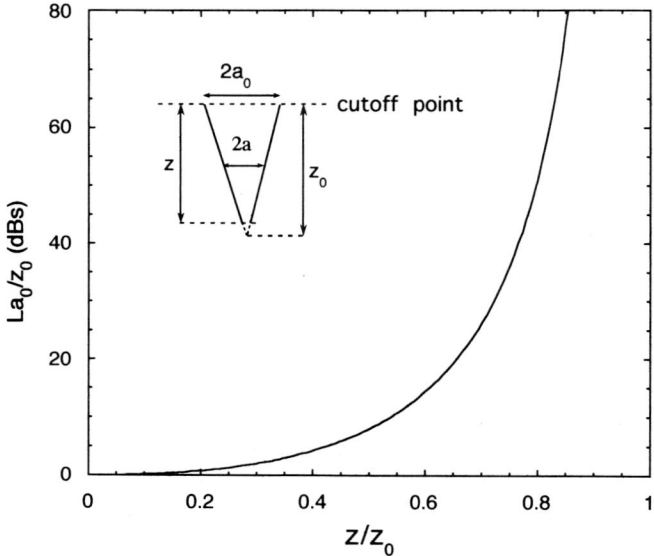

Figure 3.28 Normalized attenuation in the cutoff region of the guide plotted from Eq. (3.122) as a function of normalized distance z/z_0.

inset in Fig. 3.28, the total attenuation L in dBs from $z = 0$ to z can be written in the approximate form

$$L = 8.68 \int_0^z \alpha \, dz = 16 \frac{z_0}{a_0} \int_0^{z/z_0} \left[\frac{1}{(1-s)^2} - 1 \right] ds \text{ dBs/unit length.} \quad (3.122)$$

The attenuation is plotted as a function of z/z_0 in Fig. 3.28.

We now consider a tapered guide with a flat end of radius $a = 30$ nm or 0.25 of the cutoff radius. In this case the value of $z/z_0 = 0.75$. From Fig. 3.28, the attenuation is $La_0/z_0 = 37$ dBs. In order to keep the attenuation L in the cutoff region less than 50 dBs, the angle of the taper must be such that $z_0/a_0 < 1.35$ or about 36°. In practice, the tips look roughly parabolic in shape, after which the shank tapers slowly up to a diameter of ~125 μm. Calculations of this type are crude at best but serve to emphasize the fact that the taper near the tip must be rapid, and the attenuation in the cutoff region can be very high. Estimates of the attenuation of a tapered guide in the collection mode, using an analysis in spherical coordinates, have been carried out by Roberts.[52]

3.6 The Near-Field Scanning Optical Microscope (NSOM)

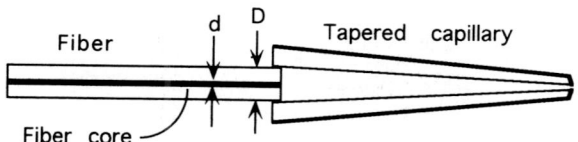

Figure 3.29 Schematic illustration of a tapered capillary covered with metal and excited from a fiber.

The Tapered Pipette The alternative to a tapered fiber is to use a pipette, as illustrated in Fig. 3.29. In this case, the taper beyond the cutoff point attenuates the light in much the same way as the tapered optical fiber, but now the excitation is not as efficient. An estimate of the excitation efficiency can be obtained by considering the collection mode; in this case only a small portion of the beam transmitted from the pinhole strikes the fiber core. Therefore, the excitation efficiency must be approximately the ratio of the area of the core (πd^2) to the area of the hole in the capillary (πD^2). For a 5-μm-diameter core, and a 125-μm-diameter cladding, this estimate yields an extra loss over and above the loss in the cutoff region of 28 dBs. Since the loss in the cutoff region will be comparable to that of tapered fiber, the use of a tapered fiber is the better choice.

A Dielectric Tip Another possibility is to use a rapidly tapered tip of a high refractive index material such as diamond, $n = 2.4$, and not deposit metal on the tip. In this case, the wave is not cut off as the tip tapers down in diameter, so a guided wave is obtained all the way to the tip. However, as the tip diameter is decreased to a small fraction of a wavelength, the wave velocity approaches that of light in air, the outside medium, and there are fringing fields outside the tip that extend out to distance comparable to a half-wavelength or more. This approach, although it is more efficient, does not have the resolution of a metal-covered guide.

3.6.2 The Fields outside the Pinhole

We have determined how the fields are attenuated in the guide. We must now examine how the fields fall off outside the pinhole and how near the sample must be to the pinhole to obtain good transverse resolution.

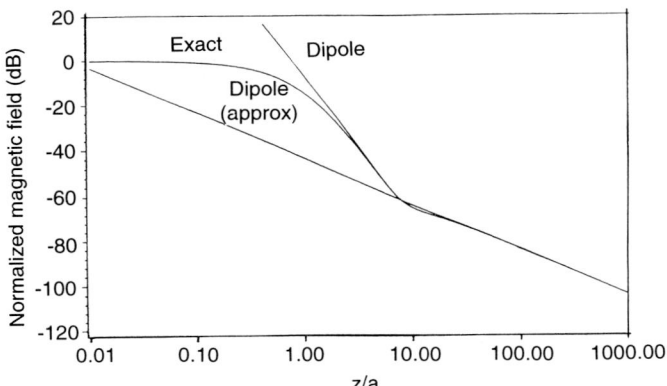

Figure 3.30 A plot of the normalized x polarized magnetic field versus z/a for $a = \lambda/50$. [From Y. Leviatan, "Study of near-zone field of a small aperture," J. Appl. Phys. **60**, 1577–1583 (1986), with permission.]

It was shown by Bethe that for a circular pinhole in a thin sheet, the fields behave as if they are equivalent to those generated by a magnetic dipole lying in the plane of the aperture, and an electric dipole normal to the plane.[53] We would expect from this picture that the fields due to the small hole would fall off as $1/z^3$ near to the hole and give rise to weak radiation field that fall off as $1/z$ far from the hole. Bethe's picture is not entirely adequate, however, because we are interested in the fields very close to the hole, where it behaves as a distributed set of dipoles.

Calculations have been made by Roberts[54] and Leviatan[55] of both the fields within one pinhole diameter of the aperture and the far-field pattern located at some distance from the pinhole. Leviatan used the assumption of uniform plane wave excitation on the opposite side of a thin sheet containing the pinhole. He then worked out the distribution of the equivalent magnetic current sources required and determined the normalized x-polarized magnetic field at the axis as a function of normalized distance z/a from the pinhole. One of these plots is shown in Fig. 3.30 for a pinhole radius $a = \lambda/50$. For $z/a < 1$, near the pinhole, the field is uniform and the various dipole approximations are totally inadequate. Beyond this point, the field falls off like a static dipole, $1/z^3$, and then at a distance of $z/a \approx 25$ or $z \approx \lambda/2$, the field falls off like those of a radiating dipole, $1/z$. The calculations have been made for different hole sizes and confirm that the $1/z$ slope starts at $z \approx \lambda/2$ for all pinhole sizes.

3.6 The Near-Field Scanning Optical Microscope (NSOM)

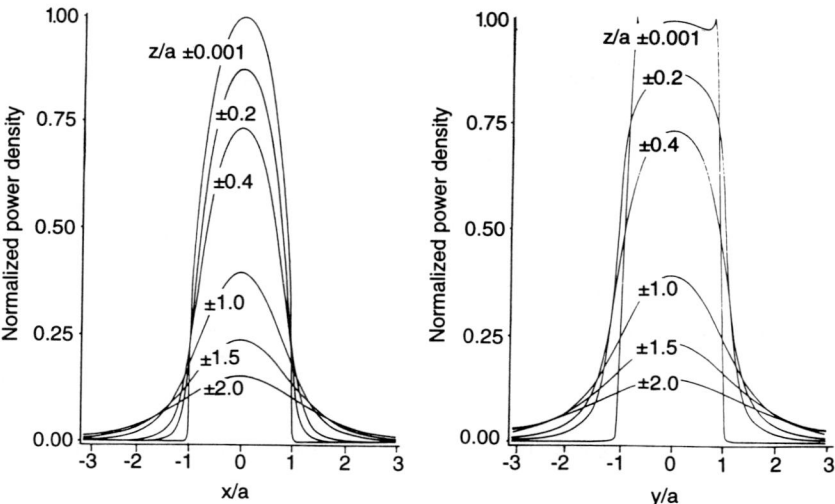

Figure 3.31 Plots of normalized real power density at different values of z/a, as functions of x/a and y/a, respectively. [From Y. Leviatan, "Study of near-zone field of a small aperture," J. Appl. Phys. **60**, 1577–1583 (1986), with permission.]

The form of the fields of a TE_{11} mode passing through the cutoff waveguide are

$$H_z = H_0 J_1(k_c r) \cos \varphi$$

$$\frac{-jk\eta}{\alpha} H_\varphi = E_r = \frac{jk\eta H_0}{k_c} \frac{J_1(k_c r)}{k_c r} \sin \varphi \qquad (3.123)$$

$$\frac{jk\eta}{\alpha} H_r = E_\varphi = \frac{jk\eta}{k_c} J_1'(k_c r) \cos \varphi.$$

In Cartesian coordinates, we can write

$$E_y = E_r \sin \phi + E_\phi \cos \phi. \qquad (3.124)$$

Very near the axis the field E_y is uniform, but it falls off monotonically from the center along the y axis, dropping to about 60% of its axial value at the edge of the guide, as do the H_X fields along the y axis. Leviatan's plots for the x and y variations of the normalized power density at different normalized distances from the pinhole, z/a, are shown in Fig. 3.31. The calculations made by Leviatan and Roberts give H fields that are almost uniform near the hole but increase slightly near the edges. However, the detailed shape of the field makes little difference to the way the fields fall off away from the hole. As will be seen from Fig. 3.31, the fields tend to fall off with x and y at a plane a short distance from the hole.

These results demonstrate that the radiation fields observed at the detector remain fairly well collimated in both principal cross sections up to a distance z/a of at least one aperture radius. At this distance the half-power points are 1.1 pinhole diameters apart in the x direction and 1.4 pinhole diameters apart in the y direction. It should be noted that it is the radiation power that is received by the detector and that the radiation power density is the relevant parameter of interest. However, if the instrument was operated in reflection mode, then the total E field associated with the pinhole would be the field of interest, and we would have to account for the quasistatic terms or evanescent field of the pinhole. The results for transverse resolution would only be slightly different from those shown here.

In conclusion, the power density at a normalized distance $z/a = 1$ is only slightly less than at the hole itself, and the transverse definition is determined by the hole diameter up to this distance. For $z/a \gg 1$, the beam spreads rapidly, the fields fall off as $1/z$ and the definition becomes very poor. We will now examine the alternative approach, the use of the solid immersion lens.

3.7 The Solid Immersion Microscope (SIM)

The solid immersion microscope described in Chapter 2 uses a solid immersion lens as an add-on to a confocal microscope. The problem of finding its transverse and range resolutions is more difficult than for the CSOM because there is an interface between the SIL and the air. Furthermore, the phase and amplitude changes of the waves passing through the interface depend on their polarization. Scalar theory is, therefore, inadequate. Instead, a more rigorous vector field theory based on the work of Richards and Wolf must be employed.[6] The application of this theory to determine depth and transverse response of a CSOM is described in Appendix A. For the depth response, the results obtained are identical to that of the scalar theory.

3.7.1 The Transverse and Longitudinal Magnifications of the SIL

We shall first determine the transverse and longitudinal magnifications of the SIL. We consider the configurations shown in Figs. 3.32(a) and 3.32(b) for a hemispherical and stigmatic lens, respectively. It follows

3.7 The Solid Immersion Microscope (SIM)

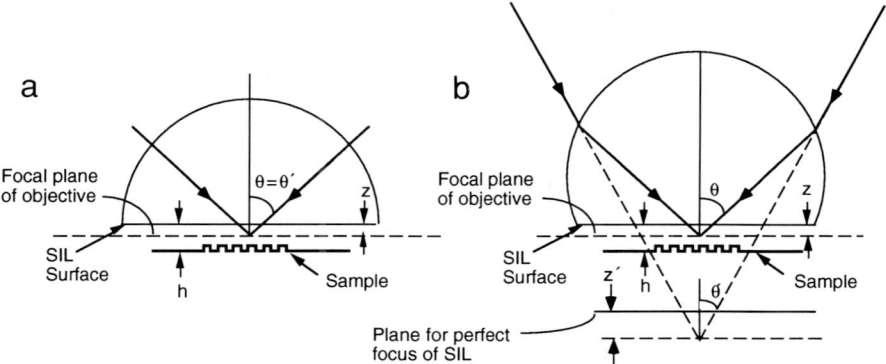

Figure 3.32 A schematic of the configuration considered in the theory for the SIL; (a) a hemispherical lens; (b) a stigmatic lens.

from Eq. (3.1), the sine condition, that the transverse magnification of the system, M_T, is

$$M_T = n \sin \theta / \sin \theta', \qquad (3.125)$$

where n is the refractive index of the SIM.

When a hemispherical SIM is used, the rays pass into the lens along its radii, so that $\sin \theta = \sin \theta'$, and the transverse magnification is $M_T = n$. Consequently if the objective lens is moved a small distance x in the transverse direction, the image point inside the SIM moves a distance $x = x'/n$. Thus, there is an additional demagnification due to the SIM of $1/n$, and the tolerance to transverse movement of the objective is increased by a factor n. Similarly, it follows from ray optics that when the objective is moved a distance z' so that its focal point moves the same distance, the image point inside the SIM moves a distance $z = z'/M_T^2$ or z'/n^2. Hence the longitudinal magnification is $M_L = n^2$.

A similar analysis can be carried out for the stigmatic lens illustrated in Fig. 3.32(b) and described more fully in Chapter 2. In this case, the image point is at a distance a/n from the center of the SIM, and the focus of the objective at a distance na from the center of the SIM, with $\sin \theta = n \sin \theta'$. Consequently, the transverse magnification is $M_T = n^2$, and the longitudinal magnification is $M_L = n^3$. The great advantage of this system is that if an SIM with a refractive index $n \sim 2$ is used, numerical aperture of the objective need only be of the order of 0.4 to obtain an effective numerical aperture for the SIL of 1.6. This makes the design of the objective relatively simple and gives the possibility of using a relatively long working distance objective.

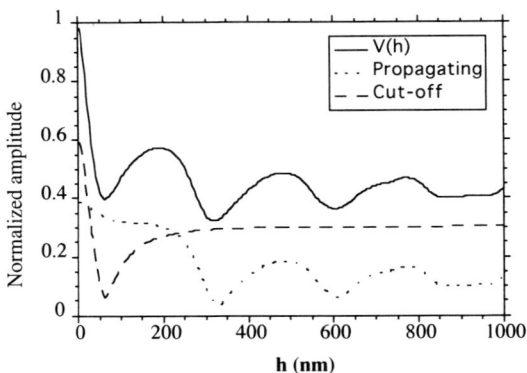

Figure 3.33 Graph of $V(0,h)$ for propagating and cutoff modes ($n = 2.0$). [From G. S. Kino and S. M. Mansfield, "Solid immersion microscope for real-time near field imaging," AIP Conference Proceedings **241**, 61–69, H. Kumar Wickramasinghe, editor (American Institute of Physics, 1992), with permission.]

3.7.2 The Depth Response of the SIM

The depth response calculation for the SIM is complicated by the fact that light is reflected from both the bottom surface of the SIL and the sample. If the signal reflected from the object has amplitude A and that from the surface of the SIL has amplitude B, then the detector will have a component proportional to the product AB rather than just A^2. This coherent detection mechanism has the advantage that it will emphasize weak signals reflected from the object but the disadvantage that there is a background glare. The theoretical depth response from a plane reflector located a distance h from the bottom of the SIL is given by the relation

$$|V(z,h)|^2 = \left| \int_0^{\theta_0} [R_\parallel(\cos\theta, h) + R_\perp(\cos\theta, h)] e^{-2jk_1 z \cos\theta} \cos\theta \sin\theta \, d\theta \right|^2$$

(3.126)

In Eq. (3.126), $R_\parallel$ and $R_\perp$ are the parallel and perpendicular amplitude reflection coefficients. These parameters can be determined in much the same way as is done for thin films in Chapter 5. However, in this case, it is necessary to take account of the fact that k_z may be imaginary in the air gap if the beam is focused to a point below the interface, as is discussed in the analysis below for the depth response.

A theoretical plot of $V(0,h)$ for a perfect mirror reflector and the various terms that contribute to it is shown in Fig. 3.33. There are two

3.7 The Solid Immersion Microscope (SIM)

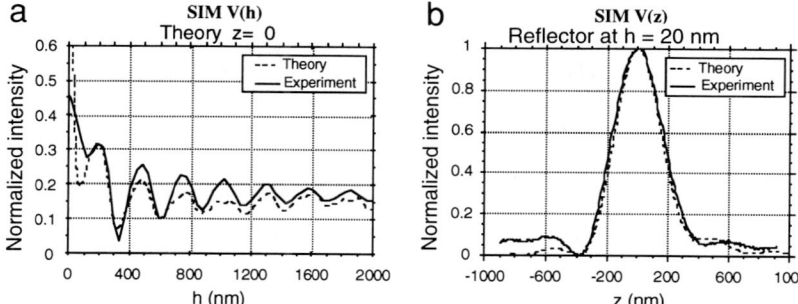

Figure 3.34 Intensity depth response of the SIM with a perfect reflector: (a) with $z = 0$, the position of the reflector, h, is varied; (b) with $h = 20$ nm, the focal position z is varied. [From G. S. Kino and S. M. Mansfield, "Solid immersion lens photon tunneling microscope," *Scanning Microscopy Instrumentation*, G. S. Kino, editor, SPIE **1556**, 2–10 (1992), with permission.]

contributions to the output signal. The first is from the cutoff modes ($\sin \theta \geq 1/n$) or the fields that fall off exponentially from the surface of the SIL. The second is from the propagating modes ($\sin \theta < 1/n$). The signal falls off very rapidly, within 50 nm of the surface of the SIL due to the behavior of the cutoff modes, and then becomes almost periodic with approximately a half-wavelength period. The period oscillations correspond to the interference between the signal reflected from the interface and the propagating small-angle rays reflected from the mirror.

Experimental Results To test the theoretical depth calculations, Kino and Mansfield used as an object a small steel ball bearing with a radius of curvature of 200 μm.[28] They employed a curved sample rather than a plane mirror to allow for better contact with the $n = 2.0$ SIL. Two different experiments were performed. The first was a $V(h)$ experiment where the SIL was placed precisely on the focal plane of the objective and the sample was moved from being in contact with the SIL out to a distance of about 2 μm. The second was a $V(z)$ experiment where the sample was brought as close to contact with the SIL as possible, and the objective lens was scanned in the z direction to scan the focal plane through the SIL/object interface. The results of the first experiment are shown with the theoretical curves in Fig. 3.34(a) and those for the second experiment in Fig. 3.34(b). Dashed lines denote the theory, and full lines the experimental results.

It was evidently not possible to make good contact between the SIL and the ball. This fact explains why the experimental curves do not match the theoretical ones for separations less than 100 nm. There are several possible reasons for the less than perfect contact. The most likely one is that the SIL had small contaminants on it that were on the scale of 0.1 μm and that kept the surfaces separated by a small distance. Nevertheless, these curves show a good match with the theoretical predictions.

The theoretical calculations of the depth response for an $n = 2.0$ SIL and a perfectly reflecting sample suggest that the signal should drop from a maximum with the sample and SIL in contact to a minimum when the sample is moved a distance $h = 60$ nm from the surface of the SIL. This intensity curve, called $|V(h)|^2$, compares well to the experimental results, as shown in Fig. 3.34(a). If the separation between the SIL and the object is fixed and the focus is scanned, a curve more reminiscent of a typical confocal depth response is achieved. This $|V(z)|^2$ curve is shown in Fig. 3.34(b). When the sample is in contact with the SIL, the $|V(z)|^2$ curve for a confocal microscope operating at an effective wavelength λ/n is obtained.

3.7.3 The Transverse Response of the SIM

The theoretical transverse response of the SIM has been calculated by Kino and Mansfield using the vector field theory of Richards and Wolf to find the electromagnetic fields close to the focal plane of the objective, with the expressions modified to account for the SIL.[27,28] In general, the amplitude of the field at any point depends on both the position of the bottom surface of the SIL relative to the focal plane, z, and the position of the sample relative to the bottom surface of the SIL, h, as shown in Fig. 3.32. The field distribution also depends on the wavelength of the illumination, the numerical aperture of the objective, and the refractive index of the SIL. Following Richards and Wolf's theory we decompose the focused beam inside the SIL into plane wave components. These plane wave components are transmitted through the glass-air interface of the SIL with transmission coefficients $\tau_\parallel$ and $\tau_\perp$, for waves with the E field polarized parallel and perpendicular to the interface, respectively. The point spread function for a confocal microscope $h(r)$ is derived from vector theory for a confocal microscope in Appendix A. For the SIM, the derivation is somewhat more complicated, and yields the following result:

$$h(r) = I_0^2 - I_2^2, \tag{3.127}$$

3.7 The Solid Immersion Microscope (SIM)

where

$$I_0 = \int_0^{\theta_0} d\theta \cos^{1/2}\theta \sin\theta (\tau_\parallel \cos\theta + \tau_\perp) J_0(kr\sin\theta) e^{-j(k_{1z}z - k_{0z}h)}, \quad (3.128)$$

and

$$I_2 = \int_0^{\theta_0} d\theta \cos^{1/2}\theta \sin\theta (\tau_\perp - \tau_\parallel \cos\theta) J_2(kr\sin\theta) e^{-j(k_{1z}z - k_{0z}h)}. \quad (3.129)$$

The z components of the propagation constants in air and in the SIL are

$$k_{0z} = k\sqrt{1 - n^2 \sin^2\theta}, \quad (3.130)$$

and

$$k_{1z} = n k \cos\theta, \quad (3.131)$$

respectively, with $k = 2\pi/\lambda$.

The transmission coefficients $\tau_\parallel$ and $\tau_\perp$, are functions of θ given by the relation,

$$\tau_{\parallel/\perp} = \frac{2Z_{0\parallel/\perp}}{Z_{0\parallel/\perp} + Z_{1\parallel/\perp}} \quad (3.132)$$

where

$$Z_{0\parallel} = \eta\sqrt{1 - n^2\sin^2\theta}; \quad Z_{1\parallel} = \frac{\eta\cos\theta}{n}, \quad (3.133)$$

$$Z_{0\perp} = \frac{\eta}{\sqrt{1 - n^2\sin^2\theta}}; \quad Z_{1\perp} = \frac{\eta}{n\cos\theta}, \quad (3.134)$$

and $\eta = \sqrt{\mu_0/\varepsilon_0}$ is the wave impedance of free space.

For rays that make angles with the SIL surface normal that are larger than the critical angle, $\sin\theta > 1/n$, the propagation constant, k_{0z}, in air becomes imaginary and the transmission coefficients become complex. Thus, the large-angle rays in the SIL are totally internally reflected. As the corresponding values of k_{0z} in the air gap are imaginary, the fields associated with these rays fall off exponentially in the air gap. However the small-angle rays in the SIL still propagate in the air gap. In order to get good range resolution and to eliminate glare from reflecting objects at different depths, it is useful to incorporate the SIL in a confocal microscope. In addition, the confocal focusing techniques help to improve the contrast in the image.

If the amplitude point spread function (PSF) of the lens is $h(r)$, the amplitude PSF of the SIM, when used in a CSOM is $h^2(r)$, provided that

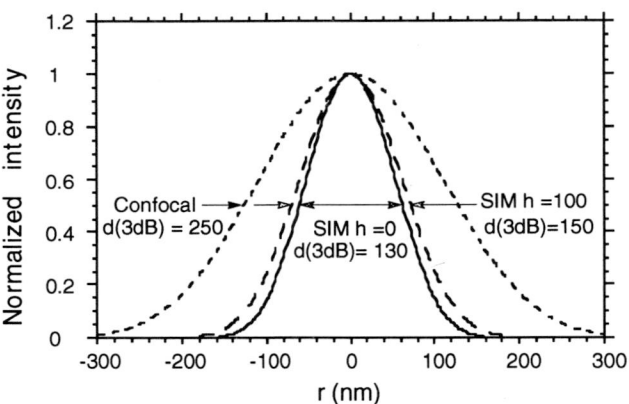

Figure 3.35 Point spread function for the confocal microscope and SIM with $h = 0$ and $h = 100$ nm. [From G. S. Kino and S. M. Mansfield, "Solid immersion lens photon tunneling microscope," *Scanning Microscopy Instrumentation*, G. S. Kino, editor, SPIE **1556**, 2–10 (1992), with permission.]

the signal from the reflecting object is much larger than the reflected signal from the interface. Solving the integrals numerically, the intensity PSF $|h^2(r)|^2$ at the detector is found to have the form shown in Fig. 3.35. The PSF is a function of both the position z of the flat surface of the SIL relative to the focal plane of the objective and the distance of the object h from the bottom surface of the SIL. For a perfectly focused SIL with the object in contact with the SIL, the PSF matches closely the PSF for a confocal microscope operating at a wavelength λ/n.

Figure 3.35 shows calculations made for an $n = 2.0$ hemispherical SIL operated at a wavelength of 536 nm with an $N.A. = 0.8$ objective lens. The PSF for the SIM is half as wide as for the confocal microscope. If the object is moved 100 nm away from the SIL, which is approximately the spot size, the width of the PSF increases by 15% and the sidelobes stay well suppressed. Thus it is possible to have a finite distance between the SIM and the object; the reason is that the lens behaves as if the fields are apodized. Apodization smoothes out the Fourier transform but does not significantly increase the width of the main lobe.

Further calculations have been made by Mansfield for larger gap spacing.[27] His results indicate that as the air gap increases, the response is dominated by the interference term between the reflected signals from the object and the air-glass interface. The transverse intensity response then becomes more like $|h(r)|^2$ rather than $|h^2(r)|^2$.

3.7 The Solid Immersion Microscope (SIM)

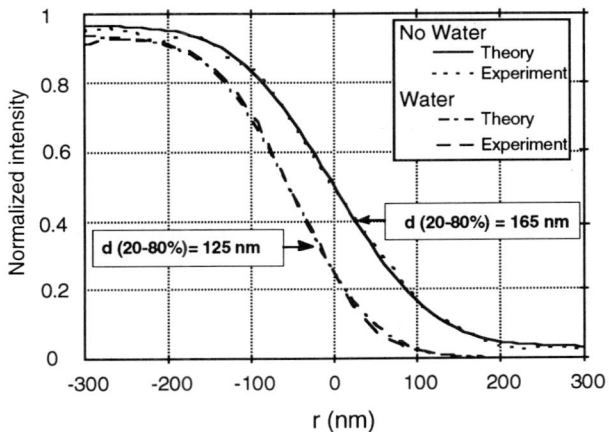

Figure 3.36 Experimental and theoretical edge responses for the near-field microscope ($n = 1.65$). [From G. S. Kino and S. M. Mansfield, "Solid immersion lens photon tunneling microscope," *Scanning Microscopy Instrumentation*, G. S. Kino, editor, SPIE **1556**, 2–10 (1992), with permission.]

Experimental Measurements of the Transverse Response An image of a grating with 200-nm period, shown in Chapter 2, gives a good check on the resolution of the SIM. Kino and Mansfield have also measured the edge response of the corner of a cleaved silicon sample with the SIM using a mercury vapor arc source and a blue filter with the microscope in different situations. They tested an $n = 1.65$ SIL with an effective $N.A.$ of 1.32 at a wavelength of 450 nm.[56] With air between the sample and the surface of the SIL, as discussed already, the amplitude of the signal reflected from the sample is smaller than that from the interface between the SIL and air. Hence the output from the detector is proportional to the amplitude edge response function. The experimental plots shown in Fig. 3.36 compare favorably with a theory based on this assumption. On the other hand, when water is placed between the SIL and the sample, the reflected signal from the surface of the SIL is very small, and the intensity edge response function is observed; again the experiments and theory in Fig. 3.36 compare favorably with each other.

The theory has been extended to deal with the use of the SIL on a magneto-optical storage disk, and the experimental results obtained agree well with the theory.[57] In this application the SIM has far better efficiency and a faster image frame time than the NSOM, but its relatively large area makes the problem of bringing the lens near the sample more difficult. The NSOM, on the other hand, has better resolution. In applications where speed and improved in resolution are needed, the SIL is a useful add-on unit to the RSOM.

3.8 Conclusion

We have examined both the depth response and transverse resolution of the CSOM and interference optical microscope. Approximate and exact formulae were presented for the CSOM showing the tradeoff between resolution and efficiency on the one hand and pinhole size on the other. The role of lens aberrations and their influence on the depth response were also discussed. The equations for the depth and transverse resolution of the optical interference microscope in the paraxial approximation were also derived and were found to be identical to those for the CSOM in the same limits. The results presented in this chapter indicate that the improvement in transverse resolution of a CSOM and optical interference microscope when compared with a standard optical microscope is small. The strength of the CSOM and optical interference microscope remains their ability to increase the contrast of the detected image and remove glare from out-of-focus planes from the image. In addition, the optical interference microscope has the advantage that it can give a direct measurement of phase.

An analysis of the basic limitations on efficiency and resolution of the NSOM has been given, and the confocal microscope theory has been extended to deal with the range and transverse resolution of the SIM.

References

1. A. Atalar, "An Angular-spectrum approach to contrast reflection acoustic microscopy," J. Appl. Phys. **49,** 5130–5139 (1978).
2. H. K. Wickramasinghe, "Contrast and imaging performance in the scanning acoustic microscope," J. Appl. Phys. **50,** 664–672 (1979).
3. C.-H. Chou and G. S. Kino, "The evaluation of V(z) in a type II reflection microscope," IEEE Trans. on Ultrasonics, Ferroelectrics and Freq. Control **UFFC-34** (3), 341–345 (1987).
4. I. J. Cox, D. K. Hamilton, and C. J. R. Sheppard, "Observation of optical signatures of materials," Appl. Phys. Lett. **41,** 604–606 (1982).
5. T. R. Corle, C.-H. Chou, and G. S. Kino, "Depth response of confocal optical microscopes," Opt. Lett. **11,** 770–772 (1986).
6. B. Richards and E. Wolf, "Electromagnetic diffraction in optical systems II. Structure of the image field in an aplanatic system," Proc. Roy. Soc. London Ser. **A 253,** 358 (1959).
7. G. S. Kino and G. Q. Xiao, "Real-time scanning optical microscopes." *Confocal Microscopy,* T. Wilson, editor (Academic Press, 1990).
8. M. Born and E. Wolf, *Principles of Optics,* 6th ed. (Pergamon, 1983).

9. C. J. R. Sheppard and T. Wilson, "Effects of high angles of convergence on V(z) in the scanning acoustic microscope," Appl. Phys Lett. **38**, 858–859 (1981).
10. G. S. Kino and G. Q. Xiao, "Real-time scanning optical microscopes." *Confocal Microscopy*, T. Wilson, editor (Academic Press, 1990).
11. C. J. R. Sheppard, "Depth of field in optical microscopy," Journal of Microscopy **149** Pt. 1, 73–75 (1988).
12. S. Kimura, and C. Munakata, "Calculation of three-dimensional optical transfer function for a confocal scanning fluorescent microscope," JOSA A **6**, 1015–1019 (1989).
13. T. Wilson, "The role of the pinhole in confocal imaging systems," *Handbook of Biological Confocal Microscopy*, J. B. Pawley, editor (Plenum Press, 1990).
14. J.-A. Conchello and J. W. Lichtman, "Theoretical analysis of a rotating-disk partially confocal scanning microscope," Appl. Opt. **33**, 585596 (1994).
15. T. Wilson, "Optical sectioning in confocal fluorescent microscopes," Journal of Microscopy **154** Pt. 2, 143–156 (1989).
16. P. A. Benedetti, V. Evangelista, D. Guidarini, and S. Vestri, "Confocal-line microscopy," Journal of Microscopy **165** Pt. 1, 119–129 (1992).
17. T. Wilson and S. J. Hewlett, "Imaging in scanning microscopes with slit-shaped detectors," Journal of Microscopy **160** Pt. 2, 115–139 (1990).
18. T. Wilson, "Optical aspects of confocal microscopy," *Confocal Microscopy*, T. Wilson, editor (Academic Press, 1990).
19. D. K. Hamilton, T. Wilson, and C. J. R. Sheppard "Experimental observations of the depth-discrimination properties of scanning microscopes," Opt. Lett. **6**, 625–626 (1981).
20. D. K. Hamilton and C. J. R. Sheppard, "Interferometric measurements of the complex amplitude of the defocus signal V(z) in the confocal scanning optical microscope," J. Appl. Phys. **60**, 2708–2712 (1986).
21. T. Wilson and A. R. Carlini, "The effect of aberrations on the axial response of confocal imaging systems," Journal of Microscopy **154** Pt. 3, 243–256 (1989).
22. C. J. R. Sheppard, "Influence of spherical aberration on axial imaging of confocal reflection microscopy," Appl. Opt. **33**, 616–624 (1994).
23. H. E. Keller, "Objective lenses for confocal microscopy," *Handbook of Biological Confocal Microscopy*, J. B. Pawley, editor (Plenum Press, 1990).
24. T. Wilson and A. R. Carlini, "Effect of detector displacement in confocal imaging systems," Appl. Opt. **27**, 3791–3799 (1988).
25. A. Boyd, "Colour coded stereo images from the tandem scanning reflected light microscope (TSRLM)," Journal of Microscopy **146** Pt. 1, 137–142 (1987).
26. C. J. Cogswell, D. K. Hamilton, and C. J. R. Sheppard, "Colour confocal reflection microscopy using red, green and blue lasers," Journal of Microscopy **165** Pt. 1, 103–117 (1992).
27. S. M. Mansfield, "Solid immersion microscopy," Ph.D. Dissertation, Department of Applied Physics, Stanford University, Stanford California, USA (March 1992).
28. G. S. Kino and S. M. Mansfield, "Solid immersion lens photon tunneling microscope," *Scanning Microscopy Instrumentation*, SPIE **1556**, 2–10 (1992).
29. G. S. Kino, T. R. Corle, P. C. D. Hobbs, and G. Q. Xiao, "Optical sensors

for range and depth measurements," *The Changing Frontiers of Optical Techniques for Industrial Measurement and Control,* Proc. 5th International Congress on Applications of Lasers and Electro-Optics ICALEO '86, C. M. Penney and H. J. Caulfield, editors, 93–102 (Springer-Verlag, 1987).
30. G. Q. Xiao, "Confocal optical imaging systems and their applications in microscpy and range sensing," Ph.D. Dissertation, Physics Department, Stanford University, Stanford California, USA (December 1989).
31. D. R. Sandison, D. W. Piston, R. M. Williams, and W. W. Webb, "Quantitative comparison of background rejection, signal-to-noise ratio, and resolution in confocal and full-field laser scanning microscopes," App. Opt. **34,** 3576–3588 (1995).
32. T. Wilson and A. R. Carlini, "Size of the detector in confocal imaging systems," Opt. Lett. **12,** 227–229 (1987).
33. T. Wilson and A. R. Carlini, "Three-dimensional imaging in confocal imaging systems with finite sized detectors," Journal of Microscopy **149** Pt. 1, 51–66 (1988).
34. S. Kimura and T. Wilson, "Effect of axial pinhole displacement in confocal microscopes," Appl. Opt. **32,** 2257–2261 (1993).
35. H. Shuman, "Contrast in confocal scanning microscopy with a finite detector," Journal of Microscopy **149** Pt. 1, 67–77 (1988).
36. T. Wilson and S. J. Hewlett, "Optical sectioning strength of the direct-view microscope employing finite-sized pinhole arrays," Journal of Microscopy **163** Pt. 2, 131–150 (1991).
37. J. W. Goodman, *Introduction to Fourier Optics* (McGraw-Hill, 1968).
38. H. H. Hopkins, "On the diffraction theory of optical images," Proc. Roy. Soc. **A217,** 408–432 (1953).
39. H. H. Hopkins, "Applications of coherence theory in microscopy and interferometry," JOSA **47,** 508–526 (1957).
40. E. C. Kintner, "Method for the calculation of partially coherent imagery," Appl. Opt. **17,** 2747–2753 (1978).
41. T. Wilson and C. J. R. Sheppard, *Theory and Practice of Scanning Optical Microscopy* (Academic Press, 1984).
42. E. Hecht, and A. Zajac, *Optics* (Addison Wesley, 1979).
43. G. S. Kino, "Fundamentals of scanning systems" in *Scanned Image Microscopy,* E. A. Ash, editor (Academic Press, 1980).
44. C. J. R. Sheppard and M. Gu, "Edge setting criterion in confocal microscopy," Appl. Opt. **31,** 4575–4577 (1992).
45. T. R. Corle, "Studies in confocal scanning optical microscopy," Ph.D. Dissertation, Applied Physics Department, Stanford University, Stanford, California, USA (June 1989).
46. A. R. Carlini and T. Wilson, "The role of pinhole size and position in confocal imaging systems." *Scanning Imaging Technology* T. Wilson and L. Balk, editors, SPIE **809,** 97–100 (1987).
47. S. S. C. Chim, "The Mirau correlation microscope—a new tool for optical metrology," Ph.D. Dissertation, Electrical Engineering Department, Stanford University, Stanford, California, USA (June 1991).

48. M. Davidson, K. Kaufman, I. Mazor, and F. Cohen, "An application of interference microscopy to integrated circuit inspection and metrology," *Integrated Circuit Metrology, Inspection, and Process Control,* Kevin M. Monahan, editor, SPIE **775,** 233–247 (1987).
49. S. S. C. Chim and G. S. Kino, "The correlation microscope," Optics Letters **15,** 579–581 (1990).
50. D.-X. Shi and M. L. Wolbarsht, "Microscope objectives, cover slips, and spherical aberration: comments," Appl. Opt. **27,** 2106 (1988).
51. S. Ramo, J. R. Whinnery, and T. Van Duzer, *Fields and Waves in Communication Electronics,* Chapter 8, 3rd edition (John Wiley & Sons, 1994).
52. A. Roberts, "Small-hole coupling of radiation into a near-field probe," J. Appl. Phys. **70,** 4045–4049 (1991).
53. H. A. Bethe, "Theory of diffraction by small holes," Phys. Rev. **66,** 163–182 (1944).
54. A. Roberts, "Near-zone field behind circular apertures in thick, perfectly conducting screens," J. Appl. Phys. **65,** 2896–2899 (1989).
55. Y. Leviatan, "Study of near-zone field of a small aperture," J. Appl. Phys. **60,** 1577–1583 (1986).
56. G. S. Kino and S. M. Mansfield, "Solid immersion microscope for real-time near field imaging," AIP Conference Proceedings **241,** H. Kumar Wickramasinghe, editor, 61–69 (American Institute of Physics, 1992).
57. S. Hayashi, G. S. Kino, and I. Ichimura "Solid immersion lens for optical storage," *Three Dimensional Microscopy: Image Acquisition and Processing II,* T. Wilson and C. J. Cogswell, editors, SPIE **2412,** 80–87 (1995).

CHAPTER 4

Phase Imaging

4.1 Introduction

Conventional optical microscopes measure only the amplitude component of an optical wave by detecting the reflected or transmitted intensity. However, a great deal of additional information can be obtained by measuring its phase. The extremely high frequency of an optical wave precludes making a direct measurement of its phase with a single light beam. Instead phase images are formed by converting the phase information into amplitude variations, or by heterodyning the signal so that the phase appears on a carrier at a lower frequency.

The development of phase imaging in optical microscopy began in 1935 when Zernike introduced his method of phase-contrast imaging for the standard optical microscope. In phase-contrast imaging both the amplitude and phase changes of the sample appear as intensity (or contrast) changes in the image. This technique made it possible to observe, for the first time, small changes in thickness and refractive index of biological samples. Since Zernike's pioneering work, numerous other techniques for phase imaging have been proposed.

This chapter will concentrate on phase imaging techniques developed for the CSOM and interference optical microscope. It is divided into four sections after the Introduction. Section 4.2, Phase-Contrast Imaging in Conventional Microscopes, describes the Zernike phase-contrast imaging technique and how it is applied in standard microscopes.

Section 4.3, Phase-Contrast Imaging in the CSOM, describes phase imaging in the CSOM using an interferometer and then goes on to describe techniques using two detectors, electro-optic modulation, and an ac Zernike method. Interferometric methods combine the light from separate reference and sample beams at the same frequency to produce an interferogram of the sample. The phase information can be extracted from the fringe structure of the interferogram. This section concludes with a brief section on heterodyne phase imaging techniques. Acousto-optic modulators are used to heterodyne a sample and reference beam, which are commonly at two different frequencies. In a heterodyne system the phase of the low-frequency output provides a direct measurement of the phase difference between the reference and sample beams. Phase images can also be obtained by using an electro-optic cell to periodically phase shift a portion of the beam. The phase data can be extracted from the amplitude of the signal at the modulation frequency. The latter two methods rely on the fact that the data from a single-pinhole CSOM is obtained point by point, and thus electronic processing can be used to measure the phase and amplitude of the signal reflected from each point separately.

In an RSOM, other techniques must be used. Phase-contrast imaging has been demonstrated in these microscopes by periodically defocusing the sample to generate a differential depth response. This method, like Zernike's, converts the phase to an amplitude or intensity change and thus provides an indirect measurement of phase.

Section 4.4, *Differential interference contrast* (DIC) *imaging* describes the basic ideas behind another class of phase imaging techniques introduced by Nomarski in 1955. As the name implies, DIC imaging generates a differential image of the sample in which the intensity variations are proportional to the phase and amplitude changes at an edge. Most of the research in this area has concentrated on scanning microscopes with a large depth of focus, although there have been adaptations of Nomarski's ideas to the CSOM and RSOM.

To conclude the chapter, in Section 4.5, Phase Imaging with an Interference Microscope, we will discuss phase measurements with an interference optical microscope. Interference microscopes provide a direct measurement of phase from the processed data. Furthermore, they are more sensitive than the CSOM to phase changes on the sample. The increased phase sensitivity can be used for image processing or other phase-sensitive applications.

4.2 Phase-Contrast Imaging in Conventional Microscopes

The technique most commonly used for phase-contrast imaging in a standard microscope was first published by Zernike in 1935.[1,2,3] Zernike's method converts phase information on the sample to amplitude information or contrast in the image. By producing amplitude images from phase specimens, phase-contrast imaging made it possible to observe details of transparent unstained objects with a clarity and resolution never before achieved.[4] The method was so successful that it was hailed as the greatest

4.2 Phase-Contrast Imaging in Conventional Microscopes

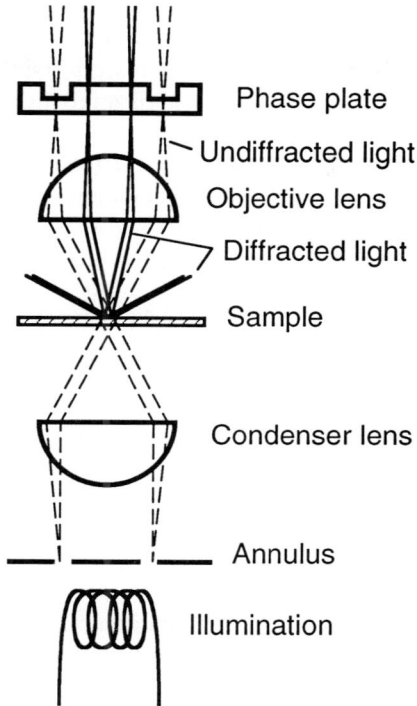

Figure 4.1 Simplified schematic of a Zernike phase-contrast microscope.

advance in microscopy in a century and garnered the Nobel Prize in Physics for Zernike in 1953.

A simplified schematic of a Zernike phase microscope is shown in Fig. 4.1.[5,6] The sample is illuminated through an annular ring located at the back focal plane of the condenser lens. When a transparent sample is put under the microscope, most of the light passes through it without being absorbed; therefore, it is difficult to observe with a standard microscope any contrast due to the sample's internal structure. However, a transparent sample will change the phase of the light passing through it, and the structure of the sample will give rise to higher-order spatial frequency components in the phase-shifted light. This light is collected by an objective lens which contains an annular phase filter in its back focal plane. The phase filter is designed to pass the undiffracted illumination through a region with a shorter optical path length than the diffracted light. The phase filter retards this light by either $\pi/2$ radians or $3\pi/2$ radians. The diffracted light does not undergo an additional phase shift. An image is formed by the interference of the diffracted and undiffracted light at the detector. In practice the phase annulus is also darkened slightly so that the intensities of the diffracted and undiffracted light components are roughly equal.

Mathematically, this process can be described by writing the amplitude transmission of a transparent phase specimen, such as a living cell, as

$$t(x,y) = e^{j[\phi(x,y) - \bar{\phi}]} \approx 1 + j[\phi(x,y) - \bar{\phi}]. \tag{4.1}$$

In Eq. (4.1) $\bar{\phi}$ is the average phase of the light transmitted by the object and $\phi(x,y)$ is the phase variation at a point (x,y) on the sample. Furthermore, it is assumed that the phase shift $\phi(x,y) - \bar{\phi}$ is small so that higher order terms can be dropped in the expansion.[7] In the series expansion, the first term represents the undiffracted light while the second term is the light diffracted by the sample. To this approximation, the image produced by a conventional microscope is of the form

$$I(x,y) = |t(x,y)|^2 = |1 + j[\phi(x,y) - \bar{\phi}]|^2 = 1. \tag{4.2}$$

If a phase ring is used to retard the undiffracted illumination by $\pi/2$ radians, the transmission of the sample and the phase ring, to first order in $[\phi(x,y) - \bar{\phi}]$, is

$$t(x,y) = e^{-j\frac{\pi}{2}} + j[\phi(x,y) - \bar{\phi}] = j[\phi(x,y) - \bar{\phi} - 1]. \tag{4.3}$$

The image intensity is then given by

$$I(x,y) = |j[\phi(x,y) - \bar{\phi} - 1]|^2 \approx (1 + 2\bar{\phi}) - 2\phi(x,y). \tag{4.4}$$

Equation (4.4) illustrates that, in a Zernike phase-contrast microscope, when the phase change is small, the image intensity at each point is linearly related to the phase change of the light traveling through the sample at that point.

The Zernike system just described is often called a positive phase-contrast system. The minus sign in Eq. (4.4) means that thin phase objects look darker than the background illumination. If the phase ring contains a $3\pi/2$ phase delay, a negative phase-contrast image is generated. Negative phase contrast causes thin phase objects to appear brighter than the surrounding background. By using a color dispersive material in the phase filter, it is also possible to add color to the image. The phase object may appear, for example, as red on a green background.[8]

Halos in the Image One disadvantage of the Zernike phase-contrast microscope just described is that structures in the image have a halo around them. The halo occurs because the *point spread function* (PSF) of the light traveling through the phase ring has sidelobes with a larger amplitude than the PSF of the light passing through the center region. For example, if the ring is infinitesimally thin, the PSF of the light passing through it will be a zeroth order Bessel function, which has a first sidelobe with 0.4 of the amplitude of the main lobe. Some of the light scattered by the sample passes through the phase ring and suffers an unwanted phase shift. The sidelobes in the phase response form a weak image with the opposite contrast to the primary image and creates a halo.[9,10]

4.3 Phase-Contrast Imaging in the CSOM
4.3.1 Phase Imaging with an Interferometer

In the CSOM, rather than adopting the Zernike phase-contrast method, phase measurements were initially made by adding a reference arm to the microscope to form an interferometer. The first such systems were reported in the literature by Leiner and Moore[11] and by Brakenhoff.[12] Leiner and Moore worked on a system employing a spatially-coherent source and a point detector, but the optical system was not confocal. Brakenhoff added a reference path to a transmission CLSM to form a Mach-Zehnder type of interferometer. Shortly thereafter, Hamilton and Sheppard at Oxford University also reported a reflection CSOM based on a Michelson interferometer.[13] The latter two instruments are described in Chapter 2.

Interferometric CSOMs combine the shallow depth response of a CSOM with phase imaging capability. The shallow depth response removes the 2π phase ambiguity for thicker samples. In addition, because the point detector responds to light only on the optical axis of the lens, the sidelobes are much smaller and the halo effect is not present. The influence of lens aberrations can change the phase front of the reference beam across its diameter and hence the measured phase at different points in the field.[14]

Vibration and Thermal Drift Phase images are produced in interferometric CSOMs by extracting the phase information from the interference term in the depth response function of the microscope. The technique is more sensitive than Zernike phase-contrast imaging, because the difference between the phases of signals reflected from a fixed reference and from the sample is taken, while the Zernike technique measures the difference between the phases of wide-angle and narrow-angle rays. On the other hand, the Zernike method has the major advantage that it is not as sensitive to vibrations or thermal drift because the two path lengths involved are almost identical and coincident. In interferometric confocal microscopes, the path length of the reference arm may be very different from that in the sample arm, and the system must be made very rigid to obtain good images. The vibration and thermal drift problems must be considered when designing an interferometric microscope. They are practical only if these effects can be minimized.

The theory of phase imaging for the confocal interference microscope has been published by Hamilton and Sheppard.[13,15] We will rederive their depth response result in order to illustrate the origin of the various terms in the interference equation and to compare different types of phase-contrast systems on the same basis. The following section will build upon the results of Chapter 3 for calculating the depth response of a CSOM.

Signal from a Plane Reflector For a confocal interference microscope, the total field at the pinhole detector is the sum of the fields reflected by the sample, S, and the reference arm, R, of the interferometer, respectively. To the paraxial approximation, we find that

$$E_{\text{det}} = \int_0^{\theta_0} (S + R) \sin \theta \, d\theta. \tag{4.5}$$

The normalized intensity at the detector is obtained by squaring Eq. (4.5) and introducing the normalization constant $1 - \cos \theta_0$, which gives a unit output when $S + R = 1$,

$$I_{CSOM} = |E_{\text{det}}|^2 = \left| \frac{1}{1 - \cos \theta_0} \int_0^{\theta_0} (S + R) \sin \theta \, d\theta \right|^2. \tag{4.6}$$

It is important to note that Eq. (4.6) is different from the depth response of an interference microscope, such as the Mirau or Linnik microscope, for which the amplitudes of each component of the angular spectrum at the detector are added to yield a response of the form

$$I_{Interference} = \frac{1}{1 - \cos \theta_0} \int_0^{\theta_0} |(S + R)|^2 \sin \theta \, d\theta. \tag{4.7}$$

This result should also be contrasted with that of the Zernike microscope for which the response to a plane reflector, normalized to the signal for no phase shift, is

$$I_{Zernike} = \left| \frac{1}{1 - \cos \theta_0} \left(\int_0^{\theta_1} S \sin \theta \, d\theta + e^{-j\pi/2} \int_{\theta_1}^{\theta_0} S \sin \theta \, d\theta \right) \right|^2, \tag{4.8}$$

4.3 Phase-Contrast Imaging in the CSOM

where θ_1 is the angle of the aperture of the central region of the Zernike system.

In the confocal interference microscope, the reflected amplitude from a plane sample located a distance z from the focus can be written as $S = S_0 \exp - 2jkz \cos\theta$. Similarly, the reflected amplitude from the reference mirror, located at $z = 0$, is $R = R_0 \exp - j\phi_R$, where S_0 and R_0 are proportional to the Fresnel reflection coefficients of the sample and reference mirror, which are assumed to be constant with angle. The variable ϕ_R is a static phase shift acquired by the reference beam when it reflects from the reference surface. Substituting these values into Eq. (4.6) we obtain

$$I_{\text{det}} = \left| \frac{1}{1 - \cos\theta_0} \int_0^{\theta_0} (S_0 e^{-2jkz\cos\theta} + R_0 e^{-j\phi_R}) \sin\theta \, d\theta \right|^2. \quad (4.9)$$

Equation (4.9) can be directly integrated using the paraxial approximation to obtain the result

$$I_{\text{det}} = \left| \frac{1}{1 - \cos\theta_0} \left[S_0 \frac{e^{-2jkz\cos\theta}}{2jkz} \bigg|_0^{\theta_0} + \frac{R_0}{2} e^{-j\phi_R} \cos\theta \bigg|_0^{\theta_0} \right] \right|^2. \quad (4.10)$$

The limits of Eq. (4.10) can be evaluated to give

$$I_{\text{det}} = S_0^2 \left[\frac{\sin kz(1 - \cos\theta_0)}{kz(1 - \cos\theta_0)} \right]^2 + R_0^2 \\ + 2 S_0 R_0 \left[\frac{\sin kz(1 - \cos\theta_0)}{kz(1 - \cos\theta_0)} \right] \cos [kz(1 + \cos\theta_0) - \phi_R]. \quad (4.11)$$

The first two terms in the numerator of Eq. (4.11) are the reflected intensity from the sample and the reference surfaces. If $R_0 = 0$, Eq. (4.11) reduces to the standard depth response of a CSOM. The third term in Eq. (4.11) is the interference term. It contains the phase information multiplied by the correlation function $\sin X/X$ where $X = kz(1 - \cos\theta_0)$. This expression is identical in form to the interference term obtained in a Mirau or Linnik interference microscope. In both instruments the correlation function represents the natural reduction of the fringe amplitude when a high-numerical-aperture lens is defocused. The expressions are identical because, in the interference CSOM, a coherent source is used, which is equivalent to illuminating through a single pinhole. In the Mirau or Linnik microscope, the spatially-incoherent source behaves as if each point on it is a separate source.

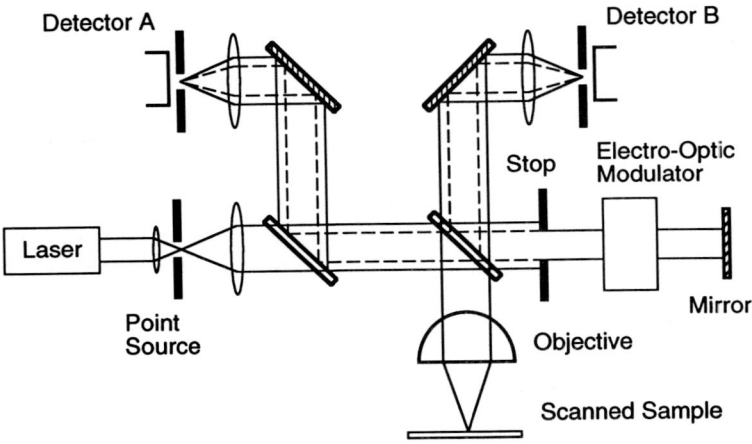

Figure 4.2 Schematic of the confocal interference microscope of Hamilton and Sheppard with an electro-optic cell added to the reference arm. [From H. J. Matthews, D. K. Hamilton, and C. J. R. Sheppard, "Surface profiling by phase-locked interferometry," Appl. Opt. **25**, 2372–2374 (1986), with permission.]

We will find it convenient to write Eq. (4.11) in the form

$$I = I_1 + I_2 \cos(\phi_S - \phi_R), \tag{4.12}$$

where I_1 is the reflected intensity term, I_2 denotes the amplitude of the interference term, and $\phi_S = kz(1 + \cos\theta_0)$. It will be observed that the term containing the phase information varies with ϕ_S and thus depends on the average phase change over all rays impinging on the sample rather than a difference term between inner and outer rays, as in the Zernike device.

Dual Detectors To extract the phase information from Eq. (4.12), Matthews, Hamilton, and Sheppard used two detectors in their microscope, as discussed in Chapter 2. They exploited the fact that, when a beam is deflected by 90° in a beamsplitter, it suffers a $\pi/2$ phase change compared to the beam passing straight through the beamsplitter. Thus, when two beams are combined in a beamsplitter, as shown in Fig. 4.2, the outputs from the two different faces of beamsplitter correspond to the sum and difference of the original beams. It follows that the two detectors shown in Fig. 4.2 produce a signal of the form,

$$I_{A,B} = I_1 \pm I_2 \cos(\phi_S - \phi_R). \tag{4.13}$$

4.3 Phase-Contrast Imaging in the CSOM

The sum and difference of the detector signals result in images proportional to the phase and amplitude of the reflected light,

$$I_{Sum} = I_A + I_B = 2I_1, \quad (4.14)$$

and

$$I_{Diff} = I_A - I_B = 2I_2 \cos(\phi_S - \phi_R). \quad (4.15)$$

The sum signal, I_{Sum}, is a normal CSOM intensity image, while the difference signal, I_{Diff}, is an interferogram of the sample modulated by its reflectivity and the correlation function of the two beams. The interferogram can be analyzed to extract the phase information by using one of the techniques developed for the Mirau or Linnik interference microscopes, described in Section 4.5. Alternatively an electro-optic cell placed in the reference arm of Fig. 4.2 can provide a direct measurement of the phase difference between the two beams, as described below.

4.3.2 Electro-optic Phase Imaging

To directly measure the phase and amplitude in an interference CSOM, two different techniques have been developed by the Oxford group.[16] One method is to use an electro-optic phase modulator in the reference path, as shown in Fig. 4.2, and record the difference signal, Eq. (4.14), when the EO cell is set to zero phase shift,

$$I_0 = 2I_2 \cos(\phi_S - \phi_R). \quad (4.16)$$

The difference signal is then measured with the EO cell set to add a $\pi/2$ phase shift to the reference arm,

$$\begin{aligned} I_{\pi/2} &= 2I_2 \cos(\phi_S - \phi_R - \pi/2) \\ &= 2I_2 \sin(\phi_S - \phi_R). \end{aligned} \quad (4.17)$$

Dividing the two signals gives the phase,

$$\frac{I_{\pi/2}}{I_0} = \tan(\phi_S - \phi_R). \quad (4.18)$$

Quadrature Phase Imaging An alternative method of directly obtaining the phase of the signal is to adjust the phase delay through the EO modulator so that the difference signal at the detector is zero. This condition is met when the phase term $\phi_S - \phi_R = \pi/2$. The phase delay of the light through the EO cell is continuously adjusted as the beam scans to maintain zero signal. The two beams are thus locked in quadrature and the phase information is obtained from the driving signal for the EO cell.

The defocus distance, or the height change on the sample, is given by the phase term in Eq. (4.11),

$$kz(1 + \cos \theta_0) + \phi_R = \pi/2. \tag{4.19}$$

Solving for z, we find that

$$z = \frac{\frac{\pi}{2} - \phi_R}{k(1 + \cos \theta_0)}. \tag{4.20}$$

4.3.3 The ac Zernike Technique

The Zernike technique uses the phase difference between inner and outer rays of a beam to produce a phase-contrast image. To obtain good contrast, the amplitude of the beam passing through the annulus must be decreased, so as to make the amplitudes of the diffracted and undiffracted parts of the beam roughly equal; this stratagem reduces the sensitivity of the microscope.

A different method of phase imaging, demonstrated by Corle and Kino, develops an ac phase-contrast signal. Their *ac Zernike* technique uses an EO cell placed in the illumination path of the microscope to periodically adjust the focus of the beam on the sample.[17,18] The periodic focus adjustment produces a differential of the depth response curve as an ac signal which passes through zero at the focal point, as shown in Fig. 4.3. The change of intensity near the zero crossing is used to measure height changes on the sample or to continually focus on a moving sample such as an optical disk by using the positive and negative excursions in the differential depth response to indicate if the reflector is located above or below the focal point. This instrument, however, does not provide a direct measurement of phase because the image intensity varies with both the reflected amplitude and phase of the light. In that sense, it is closer to the classical Zernike phase-contrast system than the techniques developed at Oxford.

The EO cell, used by Corle and Kino in their experiments, is shown in Fig. 4.4. This PLZT cell consists of an inner and outer region, much like the mask used by Zernike, on which transparent indium tin oxide electrodes are deposited. The cell is placed in front of the beamsplitter of a CSLM and the outer electrodes are modulated by a voltage $V \cos \Omega t$, as shown in Fig. 4.5. The principle of operation can be described quite simply. If the outer region of the cell covers half the area of the beam and introduces a π phase shift in the illumination when a voltage is applied, the light that passes through the two sections of the cell will add

4.3 Phase-Contrast Imaging in the CSOM

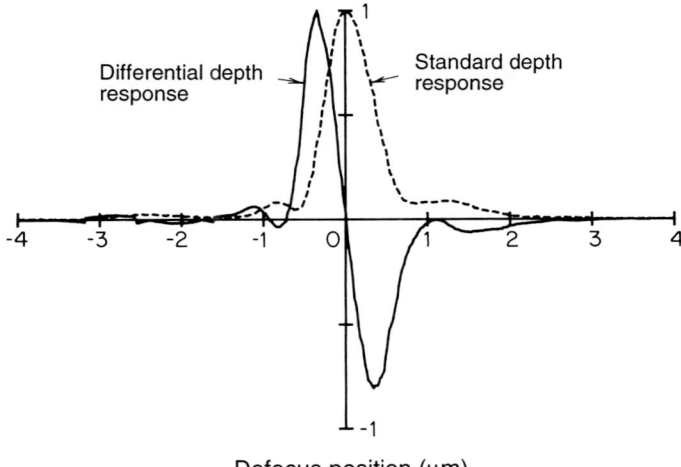

Figure 4.3 The differential depth response obtained using an EO cell in the CSOM.

out of phase at the focal plane and therefore cancel. The two components of the light, however, will not be completely out of phase at a different z position, and the output signal will be finite. Looked at another way, changing the voltage on the EO cell has effectively moved the focal point of the system. In practice, the phase shift introduced into the outer portion

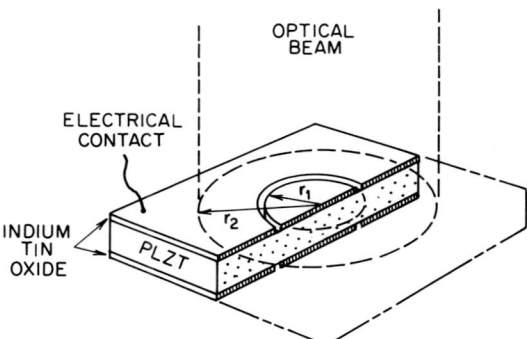

Figure 4.4 Illustration of the EO cell used to dither the focus position for phase-contrast imaging. [From G. S. Kino, T. R. Corle, and G. Q. Xiao, "New types of scanning optical microscopes," *Integrated Circuit Metrology, Inspection and Process Control II,* Kevin M. Monahan, editor, SPIE **921**, 116–122 (1988), with permission.]

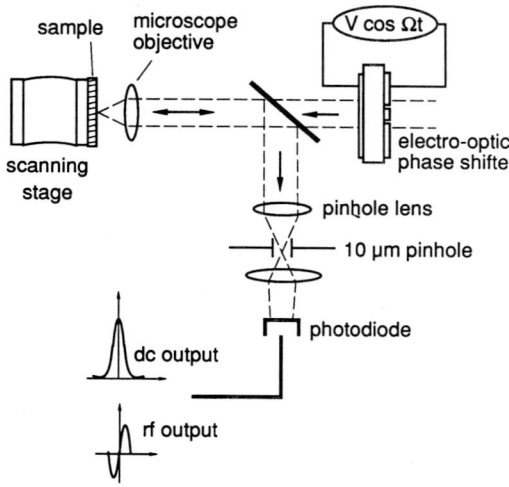

Figure 4.5 Schematic illustration of a CSOM containing an EO cell for phase measurements.

of the beam is not π but approximately 0.2π radians. The voltage is periodically modulated at a frequency Ω, causing a periodic focal shift.

The periodic defocus produces a differential depth response at the modulation frequency Ω and a conventional depth response at the dc and the second harmonic. The zero crossing in the ac response arises because, as the voltage is changed, the focal position changes monotonically with the applied voltage. There are two passes through the focal plane in each cycle, Ω, resulting in two maxima of the detected intensity. Thus, when the sample is located at the focal plane of the objective lens, the electro-optic modulation converts the intensity signal into its second harmonic (2Ω) and a zero output is obtained at the driving frequency Ω.

It should be noted that a second method of obtaining the same result is to periodically vibrate the sample, objective lens, or pinhole.[19] However, mechanical vibration, while it is simpler, will limit the scanning frequency. Furthermore, if the system is not well damped, it is a possible source of noise for precision measurements.

Theory for ac Phase Contrast with a Plane Reflector We can derive a simple formula for the depth response of the ac Zernike system in terms of the phase shift through the EO cell. Following the treatment given in Chapter 3, the normalized amplitude depth response, $V(z)$, of this system

4.3 Phase-Contrast Imaging in the CSOM

to a plane reflector located at a distance z from the focus can be written in the paraxial form as

$$V(z) = \frac{\int_0^{\theta_1} e^{-2jkz\cos\theta} \sin\theta \, d\theta + e^{-j\phi} \int_{\theta_1}^{\theta_0} e^{-2jkz\cos\theta} \sin\theta \, d\theta}{1 - \cos\theta_0}, \quad (4.21)$$

where the rays passing through the inner and outer electrodes subtend angles θ_1 and θ_0 to the axis, respectively, and ϕ is the phase delay through the outer part of the electro-optic cell.

The amplitude of the field at the detector is, therefore, of the form

$$V(z) = \frac{(e^{-2jkz\cos\theta_1} - e^{-2jkz}) + e^{-j\phi}(e^{-2jkz\cos\theta_0} - e^{-2jkz\cos\theta_1})}{2j(kz)^2(1 - \cos\theta_0)} \quad (4.22)$$

We are concerned with the output signal from the detector or $I_{det} = |V(z)|^2$ and, in particular, the term that varies with ϕ, $I_{det,\phi}$, which, from Eq. (4.22), can be written as

$$I_{det,\phi} = 2\cos[kz(1 - \cos\theta_0) + \phi] \frac{\sin kz(1 - \cos\theta_1) \sin kz(\cos\theta_1 - \cos\theta_0)}{[kz(1 - \cos\theta_0)]^2}. \quad (4.23)$$

The first term of Eq. (4.23) represents the variation with ϕ of the output signal, and the last term represents the envelope of the z variation, i.e., the modified characteristics of the confocal microscope. It will be noted that at $z = 0$, this term is maximum when

$$1 - \cos\theta_0 = 2\cos\theta_1, \quad (4.24)$$

which corresponds, in the paraxial approximation, to the two regions of the EO cell having equal area.

If a small ac signal of frequency Ω is applied to the EO cell, the resulting phase shift can be written $\phi = \phi_0 \sin\Omega t$. Expanding the cosine term of Eq.(4.23) using the trigonometric angle-sum relations to first order in ϕ or z, i.e., assuming that one or the other quantity is small, it follows that

$$I_{det,\phi} \approx A - B\phi_0 \sin\Omega t \sin kz(1 - \cos\theta_0), \quad (4.25)$$

where A and B are constants in this approximation.

It will be observed in Eq. (4.25) that, for small changes in z, the ac output signal passes through zero as $z \to 0$ and that the sign of the output signal changes as the reflector is moved through the focus. It will also be noted that the signal obtained depends on the term $kz(1 - \cos\theta_0)$, i.e., on the difference in phase change between the inner and outer rays, just as with the original Zernike system. As we have pointed out before, this result implies that the ac Zernike system and the conventional Zernike system are less sensitive to phase changes than the interferometric microscopes. However, because the two paths being compared are close to each other physically, the system is not as sensitive to vibration as are many types of interferometric microscopes.

Experimental Results with the ac Zernike System To illustrate the capabilities of this instrument, Corle and Kino imaged a 102-nm-tall 2.4-μm-wide grating of aluminum on aluminum. Figure 4.6(a) is a conventional CSOM image of the sample using a single-pinhole microscope at $\lambda = 633$ nm with a $N.A. = 0.95$ lens. Figure 4.6(b) is a line scan through the image. Figure 4.6(c) is a phase-contrast image of the same sample made with the electro-optic cell running at 60 kHz, and Fig. 4.6(d) is a line scan through the image. The grating pattern is clearly visible in the phase-contrast image, whereas it is not in the conventional CSOM image. By measuring the intensity variations in the line scan and comparing them with the slope of the depth response curve (Fig. 4.3), the thickness of the film was measured to be 101 nm.[20]

Electro-optic Focus Tracking At the present time, an effort is under way to increase the storage density of optical disks by using larger aperture lenses and moving to shorter wavelengths. Because of the decrease in focal depth associated with both these actions, a high-performance focus servo is required to track the disk as it spins. Small variations in the focus position can be caused by perturbations in the thickness of the plastic layer (nominally 1.2 mm thick) covering the disk and lack of flatness of the disk itself. Large-scale changes can be taken up by the usual electromagnetic actuator which moves the objective lens up and down. However, because of its weight, it is difficult to make the response fast enough to cover the rapid but small variations required in the focus position.

The EO cell in the ac Zernike system can also be used to move the focus by electronic means. It was shown by Osato et al. that this form of focus tracking does not deteriorate the point spread function, although there is some decrease in amplitude as the focus is moved from its normal position.[21] To illustrate this point mathematically, it is convenient to use the normalized coordinates defined in Chapter 1 for the axial distance z

4.3 Phase-Contrast Imaging in the CSOM

and radial distance r: $u = 4kz \sin^2(\theta_0/2)$ and $v = kr \sin \theta_0$. Using the normalized intensity distribution in the neighborhood of the focus,[22]

$$I(u,v) = 4 \left| (e^{-j\phi} - 1) \int_0^{b/a} J_0(v\rho) e^{-ju\rho^2/2} \rho \, d\rho + \int_0^1 J_0(v\rho) e^{-ju\rho^2/2} \rho \, d\rho \right|^2, \quad (4.26)$$

where ρ is the radial coordinate normalized to the radius a of the objective lens aperture, b is the radius of the inner disk of the EO cell, and J_0 is a zero-order Bessel function of the first kind.

For the calculations from Eq. (4.26), the value of b has been chosen to be $b = a/\sqrt{2}$, so the inner and outer areas of the beam are identical. This choice gives the best linearity, equal integrals in Eq. (4.26), and maximum focal plane movement while causing the least degradation to the PSF. The intensity variation along the z-axis, $I(u,0)$, as a function of the change in phase shift ϕ calculated from Eq. (4.26) is shown in Fig. 4.7. Each intensity peak moves linearly with ϕ, while the peak intensity decreases as the phase shift increases. The experimental results, shown in Fig. 4.8, confirm the theory.[23] It will also be noted that the shape of the point spread function (PSF) does not change much with ϕ, as is shown in Fig. 4.9, although the maximum amplitude of the signal decreases with $|z|$.

A system of this type has been tested on a compact disk, with the EO cell placed in a servo loop using a confocal detector to allow the focus position to follow the disk. The system gave satisfactory results. However, a disadvantage which prevented commercialization is that the electro-optic cell which was made out of PLZT, needed over 100 V to move the focus a sufficient distance (1–2 μm).

4.3.4 Acousto-optic Phase Imaging

An alternative to electro-optic phase imaging in a CSOM is to use an acousto-optic (AO) cell as the basis of a heterodyne interferometer. Heterodyning has the advantage of providing a simultaneous direct measurement of both amplitude and phase, but the disadvantage of introducing additional complexity. One such instrument, shown schematically in Fig. 4.10, has been built by Jungerman, Hobbs, and Kino.[23] Since this instrument is typical of the heterodyne phase microscopes described in the literature, we will discuss it here in some detail.[24,25]

Light from a laser beam is used to illuminate an AO deflector. The acoustic wave passing along an AO deflector acts as a moving diffraction grating with a period λ_A, equal to the acoustic wavelength. The light of wavelength λ_0, which passes through the AO cell, is split into two beams;

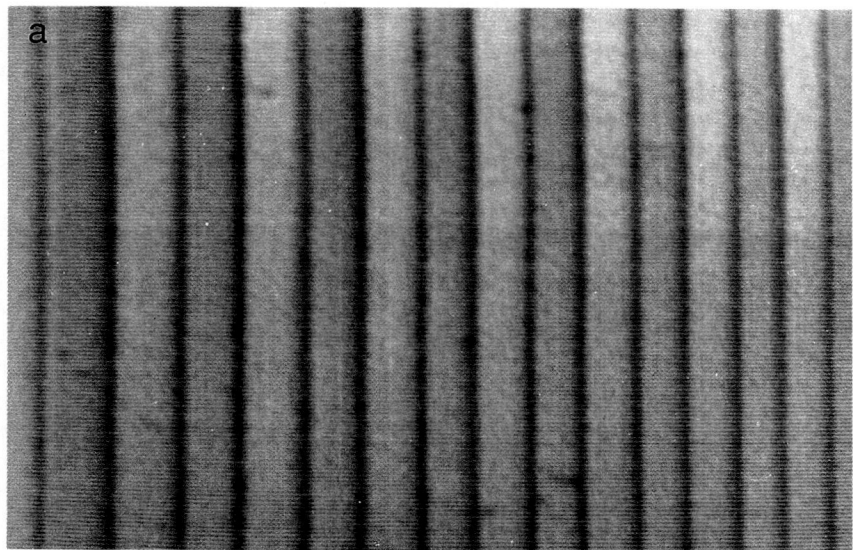

Figure 4.6 Conventional CSOM and EO Zernike phase-contrast image of a 102-nm-tall, 2.4-μm-wide grating of aluminum on aluminum: (a) conventional CSOM image, (b) a line scan through the image, (c) phase-contrast image, and (d) line scan through the image. [From G. S. Kino, T. R. Corle, and G. Q. Xiao, "New types of scanning optical microscopes," *Integrated Circuit Metrology, Inspection and Process Control II,* Kevin M. Monahan, editor, SPIE **921**, 116–122 (1988), with permission.]

4.3 Phase-Contrast Imaging in the CSOM

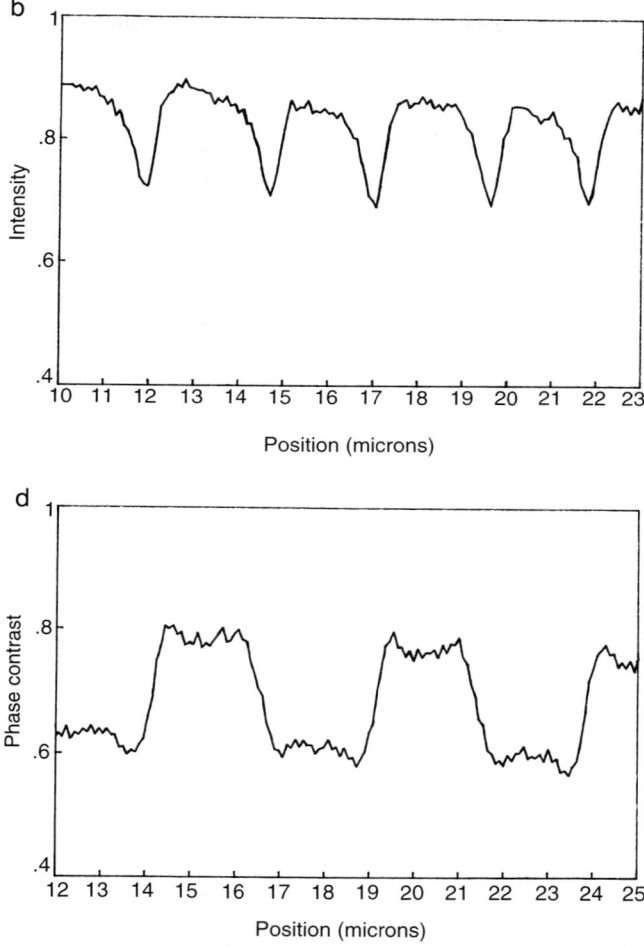

Figure 4.6—*Continued*

one is in the direction of the incident beam and the other is deflected by an angle $\Delta\theta = \lambda_0/\lambda_A$. In addition, the optical frequency ω_0 of the deflected beam is shifted by ω_A, the acoustic frequency.

The undiffracted light provides a stationary reference on the sample. By varying the acoustic frequency, the diffracted beam is scanned across the sample and serves as the probe beam. Upon reflection from the sample, both the reference and sample beams undergo a phase and amplitude change so that $R = R_0 \cos(\omega_O t + \phi_R)$ and $S = S_0 \cos[(\omega_O + \omega_A)t + \phi_S]$.

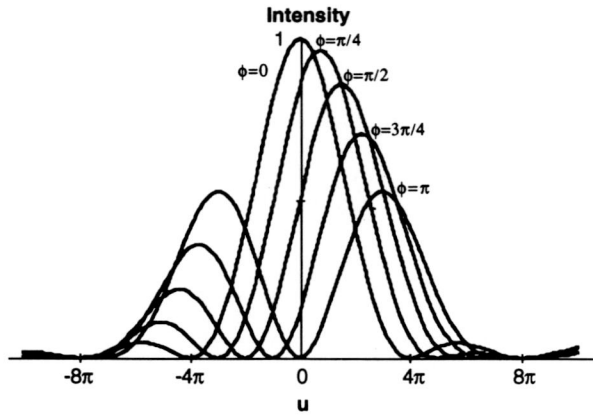

Figure 4.7 The intensity variation along the axis $I(u,0)$ of the ac Zernike phase-modulated microscope for different values of phase shift. [From K. Osato, G. S. Kino, and S. Kubota, "An electro-optic focusing system for a high-density optical disk," Opt. Lett. **18**, 1244–1246 (1993), with permission.]

The beams then retrace their input paths and re-enter the AO cell. A portion of the previously-undiffracted reference beam is deflected to the detector and downshifted by the acoustic frequency ω_A so that $R = R_0 \cos[(\omega_O - \omega_A)t + \phi_R]$, while a portion of the previously-deflected probe

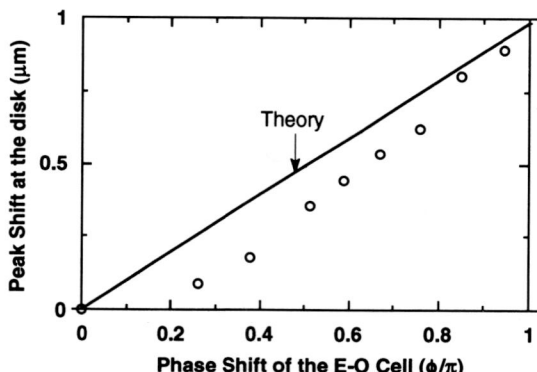

Figure 4.8 Point spread functions at each peak $I(u_P, v)$ (normalized by each of the peak intensities) for several different values of phase shift. [From K. Osato, G. S. Kino, and S. Kubota, "An electro-optic focusing system for a high-density optical disk," Opt. Lett. **18**, 1244–1246 (1993), with permission.]

4.3 Phase-Contrast Imaging in the CSOM

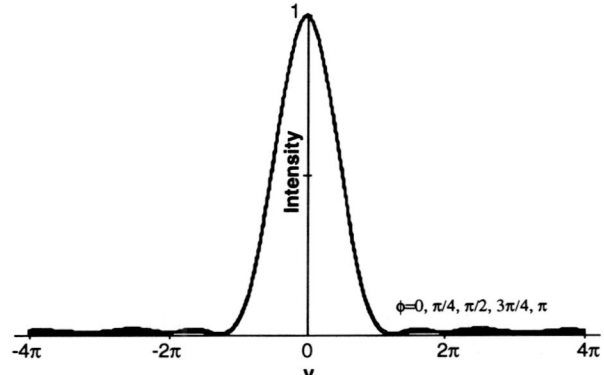

Figure 4.9 Theoretical and experimental phase shift of the EO cell vs. peak shift of the depth response. [From K. Osato, G. S. Kino, and S. Kubota, "An electro-optic focusing system for a high-density optical disk," Opt. Lett. **18**, 1244–1246 (1993), with permission.]

beam passes through the AO cell with no further frequency shifts. The two beams are multiplied at the photodiode, producing an output signal

$$I_{det} = R_0 S_0 \cos[(\omega_O + \omega_A)t + \phi_S]\cos[(\omega_O - \omega_A)t + \phi_R]$$
$$= R_0 S_0 \cos[2\omega_O t + (\phi_S + \phi_R)] \qquad (4.27)$$
$$+ R_0 S_0 \cos[2\omega_A t + (\phi_S - \phi_R)].$$

The second term in Eq. (4.27) is an ac signal at frequency $2\omega_A$ whose phase is the optical phase difference between the fixed reference spot and the scanned spot. Since the amplitude of the light reflected from the fixed spot is constant, the amplitude of the output signal is proportional to the reflected amplitude of the scanned spot.

Extracting the phase from Eq. (4.27) is convenient because the signal obtained is at a frequency where phase can be measured directly by using a lock-in amplifier or other ac detection technique.[26] In the microscope described above, Hobbs was able to extract the phase of the signal electronically by using a successive approximation phase digitizer with a nulling design. The electronic circuit mixed a separately-generated trial signal with the signal from the microscope and adjusted the phase of the trial signal until a null, or zero voltage, was obtained at the output of the phase detector.[27] Hobbs was able to read the phase of the signal at 50,000 points per second with a phase uncertainty of 0.05°.

One issue that arises in scanning microscopes which split the light into separate sample and reference beam paths is that each point in the image will have a different static phase shift added to it because of the

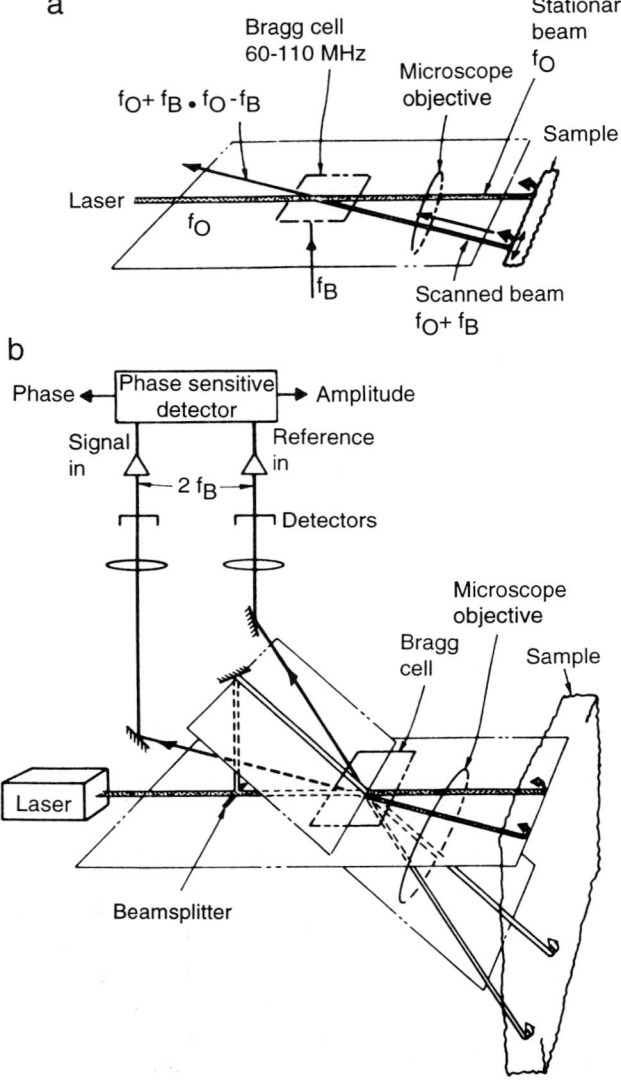

Figure 4.10 Schematic of a scanning interference microscope based on a heterodyne interferometer built by Jungerman et al. [From R. L. Jungerman, P. C. D. Hobbs, and G. S. Kino, "Phase sensitive scanning optical microscope," Appl. Phys. Lett. **45**, 846–848 (1984), with permission.]

4.3 Phase-Contrast Imaging in the CSOM

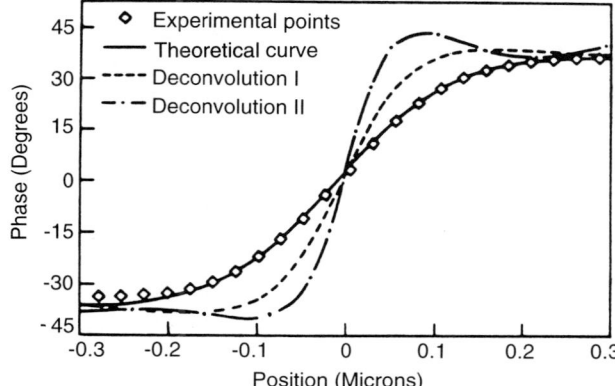

Figure 4.11 Illustration of the application of digital phase filters to improve the edge response in a phase-sensitive optical microscope. [Courtesy: P. C. D. Hobbs, "Heterodyne interferometry with a scanning optical microscope," Ph.D. Dissertation, Department of Applied Physics, Stanford University, Stanford, California, USA (August 1987).]

path length change as the light beam is scanned across the field of view. In addition, a random but repeatable phase change may be introduced as the beam scans through different portions of the optical system, and nonrepeatable phase shifts can arise due to temperature fluctuations, random path length variations, etc. To overcome these difficulties Jungerman's microscope used four beams on the sample rather than two, as shown in Fig. 4.10(b). The two additional beams compensated for the phase errors caused by the scanning of the probe beam, lens aberrations, temperature, and pressure fluctuations in the air path.

Linear Deconvolution An advantage of measuring both the amplitude and phase of the signal is that the resolution of the microscope can be improved using linear deconvolution. In this technique, a digital filter is convolved with the data to increase the amplitude of the high spatial frequencies in the transfer function. By choosing various filters, the lateral resolution can be traded off against the ringing or sidelobe response. The principle is illustrated in Fig. 4.11. In the figure, the original data (diamonds) from a phase imaging microscope is plotted along with the theoretical curve (solid) and two deconvolved experimental curves. The raw data show a 10–90% edge response of 0.23 μm. Deconvolution I uses a relatively gentle filter to achieve a 0.13-μm edge response, with little added ringing in the sidelobes. Deconvolution II uses a different filter to achieve a 0.10-μm edge response, but with extra ringing.

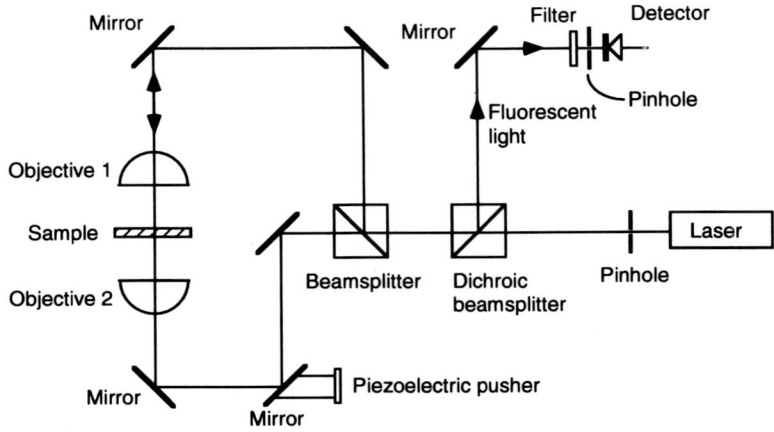

Figure 4.12 Schematic illustration of the 4Pi microscope for generating fluorescent phase images.

The sensitivity of the microscopes described above is ultimately limited by the fact that the sample and reference paths are separated in space. With the exception of the ac Zernike microscope, which is self-referencing, any path length differences or environmental changes that affect the sample and reference channels differently will be imaged as a phase change of the sample. In addition, the microscope has the major disadvantage that it has two reference beams on the sample. Thus, when the sample is moved in the z direction, the reference moves too. Consequently, the depth of focus tends to be increased, but the advantage of the confocal microscope is lost. In addition, special samples are needed which have a flat area over which the second reference beam can be scanned. This approach would be useful only if the references could be placed on a fixed image plane in the body of the microscope.

Fluorescence Phase Images Phase imaging principles are often well suited for giving direct images of transparent biological samples. However, it should be realized that fluorescence phase images cannot be obtained in this manner because the fluorescent light emitted from a sample is incoherent. One device that has been suggested for obtaining fluorescence phase images in a transmission confocal microscope is the 4Pi microscope illustrated in Fig. 4.12. Light enters the sample from two different sides through two identical objective lenses.[28] The sample fluoresces, and the light emitted in each direction travels back through the system in opposite directions along almost equal path lengths to interfere at the detector. As the generation point for the fluorescence is moved away from the focal plane, the two path lengths to the detector vary and fringes appear in the image. In this manner, phase-contrast images may be observed. The system behaves like a confocal microscope if pinholes are placed in front of the detector and the laser source, so that the image disappears if the region being observed is badly defocused. This approach to phase imaging has limited use, however, because of the requirement for two matched objective lenses and the aberrations that would be expected in thick samples.

In the next type of microscope we will discuss, the differential interference contrast microscope, these limitations are overcome by placing the sample and reference beam side by side and scanning them together. As a result of this arrangement, however, it is no longer possible to measure the absolute phase of the reflected light relative to an external reference; only the rate of change of the phase as the beams are scanned over the sample can be measured.

4.4 Differential Interference Contrast Imaging

Since differential interference contrast imaging was first proposed by Nomarski in 1955,[29] the technique has come into wide use because of its simplicity and utility. The ability of Nomarski DIC imaging to highlight edges in an image is unparalleled. It is often possible to see structure in a sample when viewed in a DIC microscope that would not be readily apparent in a normal reflection or transmission image. In addition, the images can have a three-dimensional appearance as if they are being lit from the side, making them visually appealing. The instruments to which the technique has been adapted cover the entire range of optical microscopy, from the standard microscope to the CSOM. Nomarski DIC microscope attachments are available from every major microscope manufacturer.

To add DIC imaging capability to an optical microscope, it is only necessary to add a Wollaston prism and polarizers in the optical path. These components are relatively insensitive to small alignment errors, so they can be inserted and removed in the course of normal operation by the user.

In this section we will discuss Nomarski's method as it has been applied to both the standard optical microscope and the CSOM. This discussion will be followed by a section on a variety of alternative DIC imaging techniques which have been used on scanning optical microscopes.

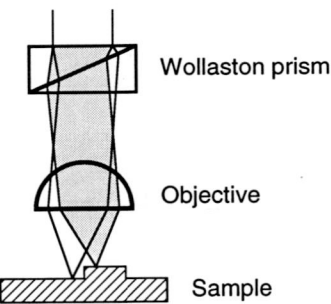

Figure 4.13 Schematic of a Nomarski DIC microscope showing the Wollaston prism and the objective lenses.

4.4.1 The Basic Theory of Nomarski Imaging

The method of DIC imaging originally proposed by Nomarski uses a modified Wollaston prism, made of a birefringent optical material, to produce two closely-spaced orthogonally-polarized spots on the sample, as shown in Fig. 4.13. The prism is placed in the optical path of the microscope so that the two beam paths cross in the back focal plane of the objective. Thus, the microscope is telecentric, as described in Chapter 2. The separation of the two spots is typically a small fraction of a spot diameter. After reflecting from the sample, the light passes back through the objective lens and Wollaston prism and then interferes at the analyzer. If the two beams are reflected from a flat region of the sample, they both have the same path length and there is no interference signal. If, however, the two beams reflect from the top and bottom surfaces of a step, there will be a phase difference between them. The phase difference causes the edge to appear either brighter or darker than the surrounding background. The name DIC imaging arises from the fact that the two spots on the sample *interfere* to produce an image which looks like the *spatial derivative* of the intensity image.

Nomarski's technique for DIC imaging works well on the standard optical microscope, the CSOM, or the RSOM. All that is required is a polarizer in the illumination path, an analyzer in front of the detector, and a Wollaston prism in the back focal plane of the objective. A schematic of an RSOM modified for DIC imaging is shown in Fig. 4.14.

It is instructive to write down, mathematically, how images are formed in a DIC microscope in order to illustrate the difference between the various imaging modes.[30] We begin by expressing the amplitude and phase

4.4 Differential Interference Contrast Imaging

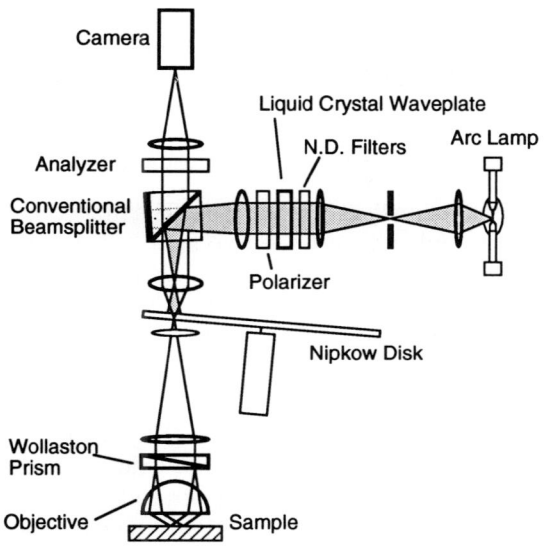

Figure 4.14 Schematic of an RSOM modified for DIC imaging.

of the two spots after they have reflected from the sample as

$$S_1 = R_1 e^{j(\omega_o t + \phi_1)}$$
$$S_2 = R_2 e^{j(\omega_o t + \phi_{wp} + \phi_2)}, \quad (4.28)$$

where ω_0 is the optical frequency, R_1, R_2 are the amplitude reflection coefficients, ϕ_1, ϕ_2 are the phases of the two spots, and the term ϕ_{wp} is a constant phase delay that may be added to one of the polarized beams by the Wollaston prism.

Light travels through the optical system until it reaches the analyzer, where the signals from the two polarizations add to give the result $S_{\text{Analyzer}} = S_1 \cos \theta_1 + S_2 \cos \theta_2$. In the preceding expression, it is assumed that the illuminating beam and the analyzer are polarized in the same direction and that θ_1 and θ_2 are the angles of the polarization direction of the illumination and the analyzer to the optical axis of the Wollaston prism, respectively. After passing through the analyzer, the intensity of the signal at the detector is $I_{\text{det}} = |S_{\text{Analyzer}}|^2$, where

$$I_{\text{det}} = R_1^2 \cos^2 \theta_1 + R_2^2 \cos^2 \theta_2 + 2R_1 R_2 \cos \theta_1 \cos \theta_2 \cos(\phi_{wp} + \Delta\phi). \quad (4.29)$$

Equation (4.29) has been simplified by the substitution $\Delta\phi = \phi_2 - \phi_1$.

Typically the illumination is polarized at 45° with respect to the optical axis of the prism, and the analyzer is oriented perpendicular to the illumination polarization so that $\theta_1 = 45°$ and $\theta_2 = 135°$. With these substitutions Eq. (4.29) reduces to

$$I_{\text{det}} = \tfrac{1}{2}(R_1^2 + R_2^2) - R_1 R_2 \cos(\phi_{wp} + \Delta\phi). \tag{4.30}$$

The microscope image consists of two terms. The first contains the average reflectivity of the sample beneath the two spots. The second depends on the optical phase difference at the edge; it also contains some reflectivity information.

4.4.2 Imaging Modes of a DIC Microscope

When applying the classical Nomarski DIC imaging techniques, the choice of the instrument does not affect the characteristic of the image, since the phase changes of interest normally occur within the depth of focus of the microscope. The following examples were made on an RSOM at Stanford University using an 60×/0.8 $N.A.$ objective lens and accompanying Wollaston prism. They could just as easily have been produced on a standard optical microscope, or CSLM.[31]

Square Law Imaging In a DIC microscope several imaging modes are possible depending on the phase delay through the prism ϕ_{wp}. The phase image can be emphasized and the average reflectivity image eliminated by using a prism with $\phi_{wp} = 0$. In such an instrument the image intensity is given by

$$I_{\text{det}} = \tfrac{1}{2}(R_1^2 + R_2^2) - R_1 R_2 \cos(\Delta\phi). \tag{4.31}$$

On flat portions of the sample, $\Delta\phi = 0$ and $R_1 = R_2$ so that $I_{\text{det}} = 0$ and the average reflectivity is zero. When the two beams pass over an edge, however, $\cos \Delta\phi \neq 1$ and so the image of the edge appears as a bright line on a dark background, as shown in Fig. 4.15(a). In this system the light is linearly polarized when it strikes the sample. Light reflected from flat portions of the sample does not have its polarization changed and so is absorbed by the analyzer. A line scan through the image appears in Fig. 4.15(b).

This imaging strategy is useful for samples such as integrated circuits where large changes in the reflectivity of the sample can obscure the DIC image. However, since the intensity is proportional to the square of the phase change at the edge, $(\Delta\phi)^2$, up edges cannot be distinguished from down edges and small phase changes cannot be easily detected. This

4.4 Differential Interference Contrast Imaging

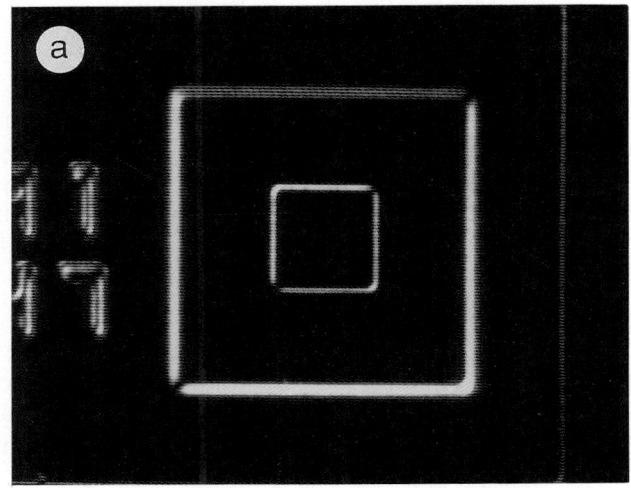

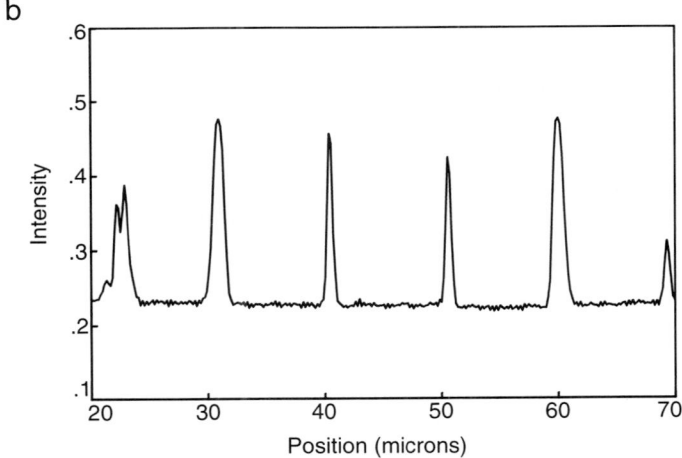

Figure 4.15 Photograph of an integrated circuit in the RSOM operating in DIC imaging mode with $\phi_{wp} = 0$: (a) top image, (b) line scan through the image. [From T. R. Corle and G. S. Kino, "Differential interference contrast imaging on a real time confocal scanning optical microscope," Appl. Opt. **29**, 3769–3774 (1990), with permission.]

disadvantage is largely compensated for by the shallow depth response of the RSOM. The ability of the RSOM to independently image the sample one layer at a time allows the heights of surface features to be determined and up and down edges to be identified. By combining the height and width information from the DIC images, quantitative measurements of slope can also be made.[32]

Linear Imaging To increase the sensitivity of the microscope to small changes in height, it is desirable for the output to be linearly proportional to the phase $\Delta\phi$. This goal can be accomplished by adding a phase shift to one of the spots $\phi_{wp} = \pi/2$. The image intensity in this case is given by

$$I_{\text{det}} = \tfrac{1}{2}(R_1^2 + R_2^2) - R_1 R_2 \sin(\Delta\phi). \tag{4.32}$$

As can be seen from Fig. 4.16(a), the average reflectance changes cannot be eliminated from these images because the light is circularly polarized when it reaches the sample. The reflected light from a flat portion of the sample is rotated into the orthogonal polarization as it travels back through the Wollaston prism and so passes through the analyzer. However, up and down edges can be distinguished from one another because they appear with the opposite contrast in the image, as can be seen in Fig. 4.16(b). This imaging mode is useful for weakly-reflecting samples with small phase changes where changes in reflectivity do not obscure the details of the image.

4.4.3 Polarization-Shifted DIC Imaging

Corle and Kino have developed an imaging mode in which the image intensity is linearly proportional to the phase, $\Delta\phi$, and the background reflectivity changes have been eliminated. This technique uses the *ferroelectric liquid crystal* (FLC) half-wave plate, shown in Fig. 4.14, which is added to a DIC microscope in order to shift the illumination polarization between measurements. The technique is similar to an idea first developed by Vaez-Iravani and See, who used a Pockels cell on a single-pinhole scanning optical microscope to generate DIC images.[33,34]

To obtain a DIC image, the phase delay through the Wollaston prism is set to $\phi_{wp} = \pi/2$, and the polarizer in front of the FLC is oriented so that the two spots interfere constructively at a chosen edge. The image is digitized and a constant offset added. Next, the voltage on the FLC is reversed, so that the direction of polarization of the emerging beam is changed by 90°. In this case the two beams interfere destructively at the chosen edge. A second image is digitized and subtracted from the first. The resulting image intensity has the form

$$I_{\text{det}} = 2R_1 R_2 \sin \Delta\phi + \text{offset}. \tag{4.33}$$

4.4 Differential Interference Contrast Imaging

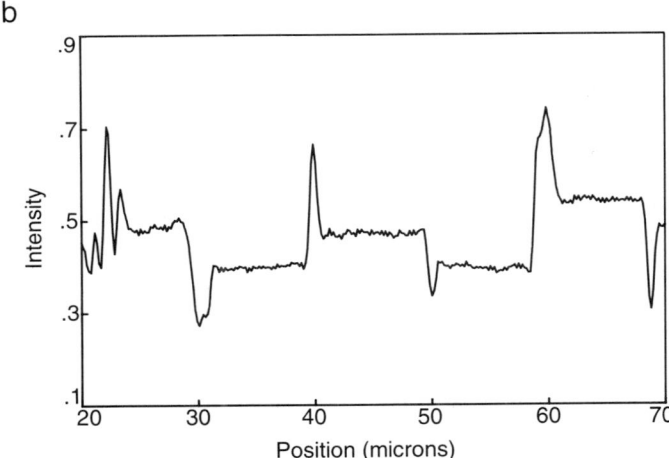

Figure 4.16 Photograph of the same integrated circuit shown in Fig. 4.15 in the RSOM operating in DIC imaging mode with $\phi_{wp} = \pi/2$: (a) top image, (b) line scan through the image. [From T. R. Corle and G. S. Kino, "Differential interference contrast imaging on a real time confocal scanning optical microscope," Appl. Opt. **29,** 3769–3774 (1990), with permission.]

On flat portions of the sample, $\sin \Delta\phi = 0$, so that the subtraction removes the average reflectivity changes from the image, increases the edge contrast, and generates an image that is linearly proportional to the phase.

Figure 4.17(a) is a photograph and Fig. 4.17(b) a line scan through a subtracted DIC image of the overlay structure shown in Figs. 4.15 and 4.16. It will be observed that the reflectivity changes are not present in the DIC image; only the edges are visible. Furthermore, since the intensity of an edge in the image is linearly proportional to the phase change $\Delta\phi$, up and down edges can be distinguished from one another.

The addition of DIC imaging capability to a CSOM or an RSOM allows images to be produced of different layers of a sample which otherwise could not be separated in a standard microscope. An example is a study of the passivation layer deposited over a silicon wafer to protect the circuit elements from deterioration due to environmental affects. The passivation layer is typically made of a material similar to glass and hence has low reflectivity compared with the top layer of the underlying circuit, which is usually metal or polysilicon. In addition, the passivation layer is only ~ 1 μm thick, so that in a standard microscope it cannot be imaged independently of the substrate. Figure 4.18(a) is a photograph of the passivation layer over a *charge-coupled device* (CCD) array taken with the RSOM in DIC ($\phi_{wp} = \pi/2$) mode. A $100\times/0.9N.A.$ objective with a commercial Wollaston prism was used with broadband illumination from a mercury arc lamp. The minute defects and surface roughness in the passivation layer can easily be seen. These defects can influence the intensity of the light which strikes the CCD array.

Figure 4.18(b) is an RSOM photograph of the underlying circuit element taken with the microscope in its normal imaging mode. It would not be possible to obtain this pair of photographs with a conventional optical microscope. The reflection from the underlying circuit elements would obscure the weak image from the passivation layer. Because of the depth sectioning capability of the RSOM, the passivation layer, and the underlying circuit can be imaged independently. These photographs also indicate the ease with which the RSOM can switch between normal and DIC imaging modes. All that is required is to insert or remove the Wollaston prism.

4.4.4 Split Detector DIC Imaging

A Standard Microscope with a Split Detector Alternatives to the Nomarski method of DIC imaging have been extensively developed for the scanning optical microscope.[35,36] One of the earliest, proposed by

4.4 Differential Interference Contrast Imaging

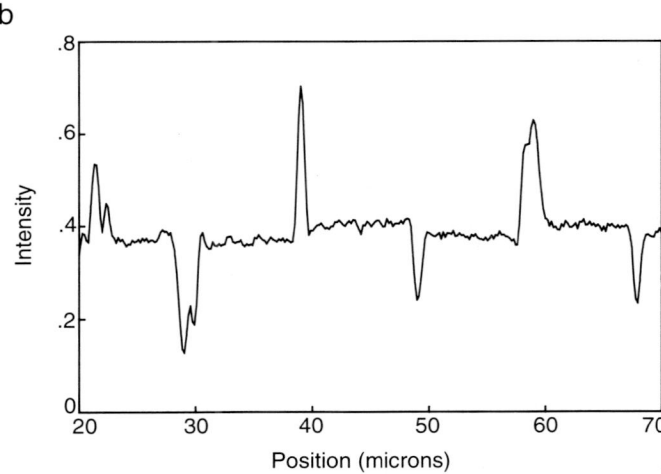

Figure 4.17 A photograph and line scan through a subtracted DIC image of the box-in-box overlay structure made on the RSOM [From T. R. Corle and G. S. Kino, "Differential interference contrast imaging on a real time confocal scanning optical microscope," Appl. Opt. **29**, 3769–3774 (1990), with permission.]

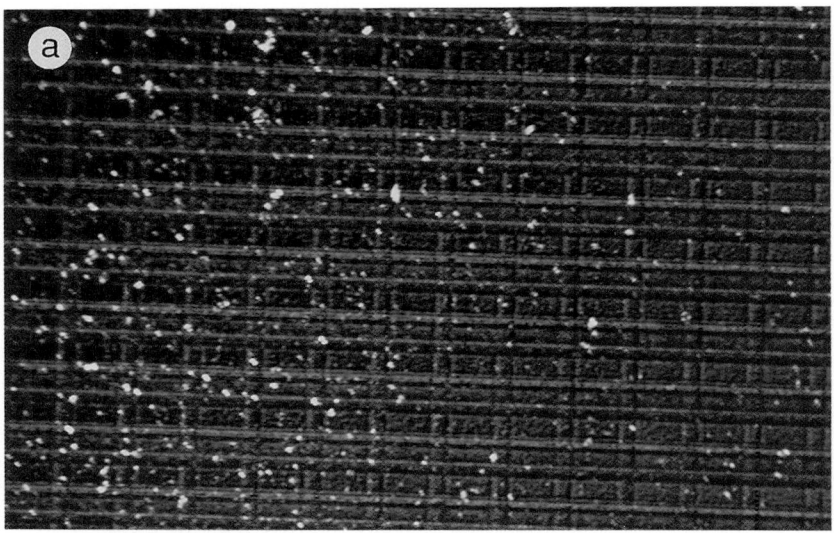

Figure 4.18 A photograph taken with the RSOM in DIC ($\phi_{wp} = \pi/2$) mode of (a) the passivation layer over a CCD array and (b) the underlying circuit elements. [From T. R. Corle and G. S. Kino, "Differential interference contrast imaging on a real time confocal scanning optical microscope," Appl. Opt. **29**, 3769–3774 (1990), with permission.]

4.4 Differential Interference Contrast Imaging

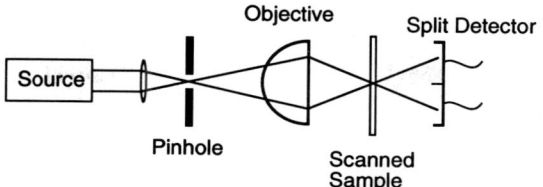

Figure 4.19 Schematic of the DIC imaging technique developed by Dekkers and de Lang applied to a scanning optical microscope.

Dekkers and de Lang,[37] was to replace the collector lens of a scanning microscope with a split detector, as shown in Fig. 4.19. A differential image was generated by subtracting the signals from the two halves of the detector. On featureless portions of the sample, the signals from the two halves of the detector are identical, or differ by a constant if the detector is not well aligned. Scanning over a phase edge, however, bends the light toward one side of the detector or the other, generating a signal in the subtracted image. Like most differential imaging techniques, this one is directional. Phase edges parallel to the split between the detectors will be highlighted more strongly than edges perpendicular to this split. The strength of an edge in the image is thus determined by both the phase change at the edge and its component in the direction of the detector split.

The advantage of Dekkers and de Lang's technique is that a conventional image can be simultaneously generated by adding the signal from the two halves of the detector, allowing the instrument to work simultaneously in both conventional and differential interference contrast modes. Since no pinholes, and no other method of depth discrimination is used, the images produced will be much like those from a standard optical microscope, albeit with the added benefits and limitations of a scanned image. Their microscope also requires only one beam incident on the sample, which improves the resolution compared with most two-beam DIC microscopes.

A Split Detector Microscope with Shallow and Extended Depth of Focus Benchop extended Dekkers and de Lang's idea to build a microscope capable of differential imaging with both a shallow and extended depth of focus.[38,39] A schematic of the instrument is shown in Fig. 4.20. A large depth of focus differential image was made by using a split detector in the pupil plane of the objective similar to that of Dekkers and de Lang. A depth-sensitive differential image is made by using a double wedge-shaped lens to focus the light from the two halves of the objective pupil

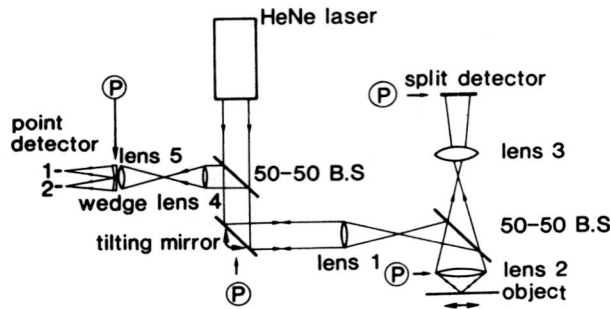

Figure 4.20 Schematic showing Benchop's DIC microscope. [From J. P. H. Benchop, "Confocal differential phase contrast in scanning optical microscopy," *Scanning Imaging Technology*, T. Wilson and L. Balk, editors, SPIE **809**, 90–96 (1987), with permission.]

onto two point detectors. Like Dekkers and de Lang's microscope, a conventional image is obtained by adding the signals from the two detectors, or the two halves of the split detector, and a differential image is obtained by subtracting the signals.

In Fig. 4.21, Benchop compares the four imaging modes of his microscope using a chrome-on-glass substrate at different focus positions. Figure 4.21(a) is a conventional image obtained by adding signals from two halves of the pupil plane detector; Fig. 4.21(b) is a CSOM image obtained by adding the signals from the two point detectors; Fig. 4.21(c) is an extended focus differential image obtained by subtracting the signals from the pupil plane detector; and Fig. 4.21(d) is the difference signal of the two point detectors, a shallow depth-of-focus differential image. As the microscope is defocused, the confocal signal goes to zero while the image obtained by pupil detection blurs. The same is true for the differential images. It should be noted in Fig. 4.21(c) that the out-of-focus signal from the pupil plane detectors can be larger than the signals from an in-focus image.

By simultaneously generating differential images with and without a shallow depth response, Benchop was able to determine the strengths of each technique. He concluded that the main advantage of the large-area detectors, when compared to point detectors, is the improved throughput and the improved signal-to-noise ratio. This advantage, however, is usually outweighed by the fact that out-of-focus phase or amplitude structures can create large modulations of the differential signal.

In Benchop's microscope, only a single spot is placed on the sample; thus it has better transverse resolution than a system using a Wollaston prism. The disadvantage of this configuration is, however, that to obtain

4.4 Differential Interference Contrast Imaging

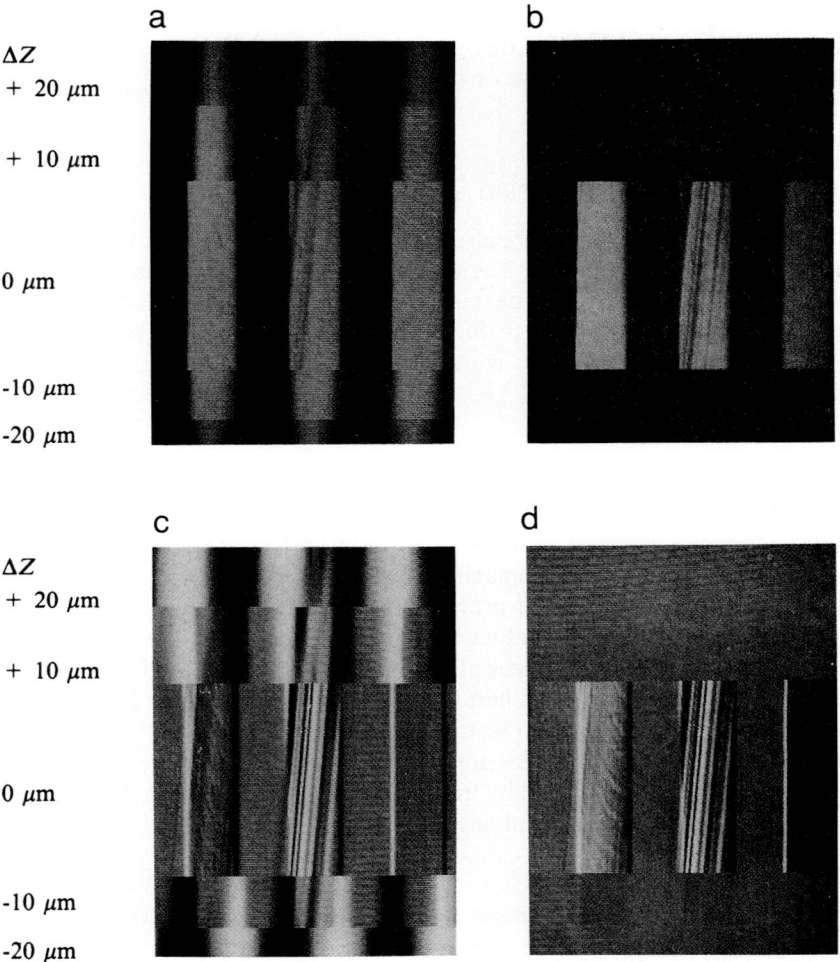

Figure 4.21 Comparison of the four imaging modes available on Benchop's microscope for a chrome-on-glass substrate at different focuses: (a) a conventional image, (b) a CSOM image, (c) an extended focus differential image, and (d) a shallow focus differential image. [From J. P. H. Benchop, "Confocal differential phase contrast in scanning optical microscopy," *Scanning Imaging Technology*, T. Wilson and L. Balk, editors, SPIE **809**, 90–96 (1987), with permission.]

a shallow depth response, two point detectors are used, each of which must be aligned. In addition, this technique cannot be readily adapted to a standard optical microscope or the RSOM.

4.4.5 Differential Probe Beam DIC Imaging

Wilson and Sheppard have also developed several unique methods of DIC imaging for the CSOM. Their work groups DIC imaging into three areas: differential contrast, differential amplitude contrast, and differential phase contrast.[40] These three techniques all use a single probe beam which is scanned across the sample, which differentiates this from the classical Nomarski approach.

Differential Contrast Differential contrast imaging, by Wilson and Sheppard's definition, is the simplest form of DIC imaging. Differential contrast images respond to both amplitude and phase changes on the sample. They are formed by placing any type of differential probe beam on the sample. A differential probe beam can be generated by splitting the objective pupil into two halves and adding a π phase shift to one half of the beam. Alternatively, a linear phase filter can be added to the objective pupil.[41] A third method of generating a differential contrast image is to subtract a CSOM image from a standard microscope image.[42] The disadvantage of differential contrast imaging is that the amplitude and phase cannot be separately detected by the microscope, thus limiting the amount of information which is available to the user.

Differential Amplitude Imaging Differential amplitude contrast images highlight edges at which there is an amplitude or intensity change in the reflected light. A differential amplitude contrast image can be generated by removing the point detector in a CSOM and replacing it with a large-area split detector, as shown in Fig. 4.22.[40] This technique differs from Dekkers and de Lang's because the detector is at the focal point of the lens and not in the pupil plane.[41,43,44] A differential image is formed by subtracting the signals from the two halves of the detector, while a conventional image is obtained when the two signals are added. Differential amplitude images can also be formed using a two-mode optical fiber[45] or an AO modulator to place two spots on the sample.[34] The advantage of using an AO modulator is that ac detection techniques are directly employed, thus increasing the sensitivity of the system.

4.4 Differential Interference Contrast Imaging

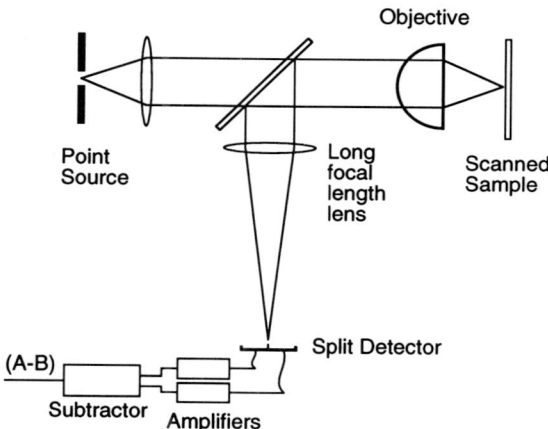

Figure 4.22 Schematic of the differential amplitude contrast scanning microscope developed by Wilson and Sheppard. [From T. Wilson, "Confocal microscopy" *Confocal Microscopy*, T. Wilson, editor (Academic Press, 1990), with permission.]

Differential Phase Images Differential phase-contrast images highlight edges at which there is a phase change on the sample.[46,47] Mathematically, in these systems the image is a multiplication of the square of the reflected intensity and the derivative of the phase change, $I(x) = 2R^2(x) \, d\phi/dx$. This equation describes the form of the image generated in Dekkers and de Lang's microscope. Hamilton, Wilson, and Sheppard have extended this technique by using a quadrant detector to receive the image, allowing the direction of differentiation to be electronically selected.[48,49]

4.4.6 Differential Imaging with an AO Modulator

Laeri and Strand at IBM combined a DIC microscope with a heterodyne interferometer to improve its sensitivity.[50] They modified the illuminator of a Standard Leitz Orthoplan DIC microscope to incorporate a laser illumination system, a pair of polarizing beamsplitting cubes, and two AO modulators, as shown in Fig. 4.23. The frequency of each spot on the sample was shifted a different amount by the AO cells. In their instrument the modulators were driven at frequencies of 30 and 50 MHz. After passing through a standard DIC microscope, the two beams interfered at the detector, producing a signal at twice the beat frequency, 40 MHz. The phase of the 40-MHz signal is the phase difference of the two beams after they traverse the optical system.

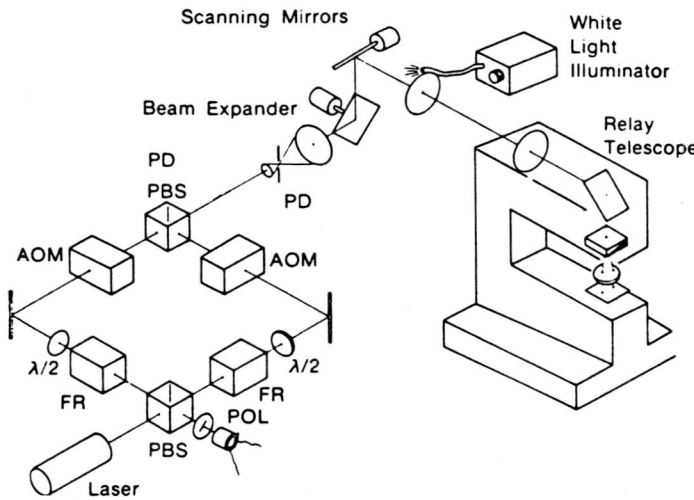

Figure 4.23 Schematic of the combined DIC microscope and heterodyne interferometer built by Laeri and Strand. [From F. Laeri and T. C. Strand, "Angstrom resolution optical profilometry for microscopic objects," Appl. Opt. **26**, 2245–2249 (1987), with permission.]

Although the components were mounted in a compact housing with sealed beam paths, fluctuations in the optical path length introduced unacceptable noise into the measurements. For this reason a reference signal was introduced by measuring both the illuminating and reflected beams at a point internal to the microscope. With this modification they were able to detect the phase edge of a monomolecular layer of cadmium arachidate approximately 2.7-nm-thick on a silicon wafer.

A Single-Pinhole Scanning DIC Microscope Using an AO Modulator
A number of other DIC imaging techniques using AO modulators have been developed by See and Vaez-Iravani.[51] Among these is a very simple single-pinhole scanning DIC microscope. This instrument used a standing wave AO cell in the illumination path to sinusoidally move the beam on the object by a fraction of the spot size, as shown in Fig. 4.24. Structural variations of the sample resulted in a signal at the driving frequency whose strength was proportional to the reflectivity variations of the sample.

The microscope can be modified to detect phase by adding a second AO modulator in the reference arm. In this case the signal detected by the photodiode consists of a carrier at twice the reference frequency ω_r, which is phase modulated by the signal information ω_s.

4.4 Differential Interference Contrast Imaging

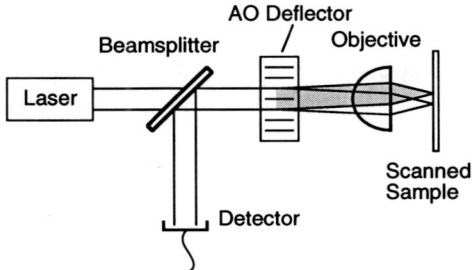

Figure 4.24 Schematic of the AO DIC microscope built by See and Vaez-Iravani.

The instruments described above, as originally implemented, did not use point detectors and so did not have a shallow depth response. A point detector could, in principle, be added to many of the designs without affecting the performance, resulting in microscopes with good depth discrimination.

4.4.7 Differential Imaging with an Optical Fiber CSOM

A technique for obtaining differential images in an optical fiber confocal scanning microscope has been demonstrated by Juskaitis and Wilson.[52] In this instrument a two-mode optical fiber and a split detector were used in place of the point detector of a CSOM, as shown in Fig. 4.25. The small size of the fiber core is equivalent to the use of a single pinhole, so that the microscope discriminates against light reflected from out-of-focus planes in the image. With the addition of scanning, the use of a fiber-optic system becomes a simple way to construct a confocal microscope. Furthermore, the combination of a split detector and two-mode fiber allows coherent signal processing to be performed on the light so that a differential image can be generated. Several different types of images are possible depending on which of the fiber modes are detected.

To understand the origin of the differential signals in this microscope it is necessary to understand the modes of propagation for the light in the fiber. Juskaitis and Wilson make use of the two lowest order modes of the fiber, the LP_{01} and LP_{11} modes. The LP_{01} mode has an amplitude which does not vary with the azimuthal angle φ, while the LP_{11} mode has an amplitude which varies as $\cos\varphi$. Therefore, it reverses in sign on each side of the fiber and has zero amplitude at $\varphi = \pm\pi/2$. We can approximate the mode profile in the LP_{11} mode, $\mathbf{e}_2(x,y)$, as the differential with respect to x of the profile $\mathbf{e}_1(x,y)$ of the LP_{01} mode, i.e., $\mathbf{e}_2(x,y) \sim d\mathbf{e}_1(x,y)/dx$.

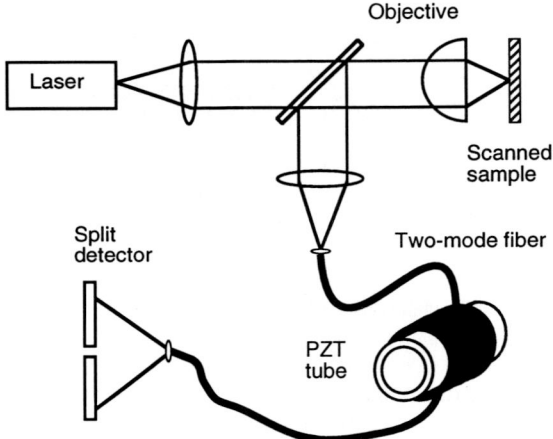

Figure 4.25 Schematic of the fiber-optic differential microscope developed by Juskaitis and Wilson.

Following the explanation of Juskaitis and Wilson,[52,53] if the first two modes of the fiber are excited with amplitudes a_1 and a_2, respectively, then the field at the end of a fiber of length l is given by

$$U(l) = a_1 e^{-j\beta_1 l} \mathbf{e}_1(x,y) + a_2 e^{-j\beta_2 l} \mathbf{e}_2(x,y). \quad (4.34)$$

In Eq. (4.34) β_1 and β_2 are the propagation constants of the two fiber modes. When the detector is placed in the far field of the output beam from the fiber, the detected intensity of the light exiting the fiber is given by integrating the Fourier transform of the field pattern at the end of the fiber over the surface area of the detector,

$$I_{\text{det}} = |a_1|^2 \int |E_1^2| D ds + |a_2|^2 \int |E_2^2| D ds + 2 \operatorname{Re}\left[a_1 a_2^* e^{-j\Delta\beta l} \int E_1 E_2^* D ds \right]. \quad (4.35)$$

In Eq. (4.35) $\Delta\beta = \beta_2 - \beta_1$ is the difference in propagation constant of the two modes, E_1 and E_2 are the Fourier transforms of $\mathbf{e}_1$ and $\mathbf{e}_2$, respectively, D is the intensity sensitivity of the detector, and Re[.] denotes the real part. The integral ds covers the entire surface of the detector.

The odd and even nature of the mode structure imply that E_1 is real and even and has an amplitude that is azimuthally invariant, whereas E_2 is imaginary and odd and has an amplitude that varies as $\cos\varphi$. Thus a detector with a uniform response ($D = 1$) will not detect the cross-product

4.4 Differential Interference Contrast Imaging

term. If, however, a split detector is used and the signals from the two halves are subtracted, only the cross-product term will remain,

$$I_{\text{det}} = 2\,\text{Re}[a_1 a_2^* e^{-j(\Delta\beta l - \pi/2)}]. \tag{4.36}$$

Although it is not readily apparent in its present form, Eq. (4.36) represents a mixture of differential amplitude and differential phase images. Typically a high-power lens is used to focus most of the light reflected from the sample and collected by the objective into the fiber core, as shown in Fig. 4.25.[54] If the numerical aperture of this lens is large enough so that the main lobe of the focused beam at the fiber entrance face is smaller than the fundamental mode of the fiber, then to a good approximation, the amplitude of the field at the end of the fiber is given by a convolution of the objective response, h^2, with the reflectivity of the sample, r, at that point,

$$a_1 = h^2 * r \tag{4.37}$$

and

$$a_2 = h\frac{\partial h}{\partial x} * r. \tag{4.38}$$

In the above equations we have neglected the constant denoting the relative amplitudes of a_1 and a_2.

If we write the field amplitude a_1 in terms of its modulus b and phase ϕ in the form $a_1 = be^{j\phi}$, it follows from Eqs. (4.36)–(4.38) that

$$I_{\text{det}} \sim \frac{\partial b}{\partial x}\sin\Delta\beta l + 2b^2\frac{\partial \phi}{\partial x}\cos\Delta\beta l. \tag{4.39}$$

The modulus b can be thought of as the reflected amplitude from the sample, while ϕ is the phase. In general the image will be a mixture of these two modes. If, however, the length of the fiber is chosen such that only the first term is detected, then a differential amplitude image will be produced. By changing the length of the fiber, the second term can be selected and a differential phase image will be generated. In Juskaitis and Wilson's microscope the beat length of the two modes was approximately 100 μm. To select between differential phase and differential amplitude images, they used a phase modulator consisting of a 3-m-long section of polarization-preserving fiber wrapped around a piezoelectric (PZT) cylinder. By applying a voltage to the cylinder, the fiber could be stretched, thus changing its length and selecting one of the mode patterns.

In summary, by detecting only the lowest-order mode LP_{01} fiber mode, a standard confocal image of the sample is obtained. A differential contrast image is generated by detecting the second-order or LP_{11} mode. If the two modes are present, either a differential amplitude or differential phase image can be made by stretching the fiber.

Juskaitis and Wilson have also noted that, by slightly misaligning the fiber and detecting the interference term between the two modes, a differential depth response is obtained similar to that for the EO cell of Section 4.3.3.[55]

The fiber-optic differential fiber microscope has much in common with Benchop's differential CSOM while providing the additional advantage of simplicity and confocal imaging. When the signals from two halves of the detector are added, a true confocal image is obtained. The microscope is, therefore, capable of optically excluding light from out-of-focus planes in the image. In addition to generating simultaneous confocal and differential images, the imaging properties can be easily shifted between differential amplitude and differential phase by stretching the fiber.

In this chapter we have reviewed a variety of phase imaging microscopes. Most of these instruments produce images whose intensity variations are proportional to the phase of the reflected or transmitted light from the sample. In that sense a phase or differential image can be directly observed in the microscope eyepieces or on a television monitor that is connected to the system detectors. In the next section, interference microscopes will be discussed. In these instruments the image consists of an interferogram of the sample in which the intensity changes are proportional to both the amplitude and phase of the light. In interference microscopes, electronic processing is used to extract the amplitude and phase information from the interferogram.

4.5 Phase Imaging with an Interference Microscope

There are many varieties of interference microscopes available today. The different hardware configurations have been discussed in Chapters 1 and 2. The trait they share is that the image is an interferogram of the sample from which the phase and amplitude information is extracted during postprocessing of the image. In this section we will briefly review two methods for extracting the phase information from the interferogram, the integrating bucket and the Fourier transform technique.

4.5.1 The Integrating Bucket Technique

One of the most commonly used techniques for separating the phase and amplitude in an interferogram is called the integrating bucket technique. It was developed by Bhushan at IBM and Wyant at the University

4.5 Phase Imaging with an Interference Microscope

of Arizona.[56,57,58] Instruments that use this method of measurement for analyzing data from a Mirau interferometric microscope are commercially available from Wyco Corporation and from Zygo Corporation.

In the integrating bucket technique, as first proposed, the phase of the reference beam is changed at a constant rate, $\phi(t)$. Once the phase change has gone through $\pi/2$, the detector is read out. This procedure is repeated three times, producing three different images. The phase of the light reflected from the sample is given by the arctan of the normalized difference of the images.

The image intensity of a single pixel in an interferometric microscope with a time-varying phase component can be expressed as

$$I = I'_1 + I'_2 \frac{\sin[kz(1 - \cos\theta_0)]}{kz(1 - \cos\theta_0)} \cos[kz(1 + \cos\theta_0) + \phi(t)]. \quad (4.40)$$

If we assume that the sin X/X term in Eq. (4.40) remains essentially constant for small changes in z corresponding to a change in phase in the second term of $3\pi/2$, we can write Eq. (4.40) in the form

$$I = I_1 + I_2 \cos[\phi(x,y) + \phi(t)], \quad (4.41)$$

where I_1 and I_2 are constants, and

$$\phi(x,y) = k(1 + \cos\theta_0) z(x,y). \quad (4.42)$$

The intensity at each point in the image is integrated while $\phi(t)$ varies linearly with time from 0 to $\pi/2$, $\pi/2$ to π, and π to $3\pi/2$. The resulting images will be relatively noise free and will take the form

$$I_A = I_1 + I_2[\cos\phi(x,y) - \sin\phi(x,y)], \quad (4.43)$$
$$I_B = I_1 + I_2[-\cos\phi(x,y) - \sin\phi(x,y)], \quad (4.44)$$
$$I_C = I_1 + I_2[-\cos\phi(x,y) + \sin\phi(x,y)]. \quad (4.45)$$

The phase at any point of the images $\phi(x,y)$ is then given by the relation

$$\phi(x,y) = \tan^{-1}\left[\frac{I_C - I_B}{I_A - I_B}\right]. \quad (4.46)$$

The subtraction and division in Eq. (4.46) cancels out the effects of fixed pattern noise and gain variations across the detector.

The integrating bucket technique has proven to be a practical and versatile method for obtaining the phase information from an interference microscope image. More recently the technique has been modified slightly to digitize an image after changing the phase of the interference signal by $\pi/2$. In this modification five frames are digitized, I_1 through I_5, corresponding to $\phi(t) = -\pi, -\pi/2, 0, \pi/2$, and π. The phase of a point (x,y) on the object is then given by[59]

$$\phi(x,y) = \tan^{-1}\left[\frac{2(I_2 - I_4)}{2I_3 - I_5 - I_1}\right]. \quad (4.47)$$

After measuring the phase, the height of a surface feature can be calculated from the relation

$$z = \frac{f}{2}\frac{\phi(x,y)}{k} = f\frac{\lambda}{2}\frac{\phi(x,y)}{2\pi}, \quad (4.48)$$

where λ is the wavelength of the source illumination. The constant f is a numerical correction factor. The term $f\lambda/2$ gives the approximate spacing of the fringes in a focused beam interference microscope. In an interference microscope, increasing the numerical aperture of the objective lens increases the fringe spacing and thus decreases the height-resolving ability of the microscope.

One form of the correction factor can be seen from Eq. (4.42) to be

$$f = \frac{2}{1 + \cos\theta_0}. \quad (4.49)$$

In practice, when the method of phase measurement described above is used, this simple correction factor may not be adequate. One reason is that the pupil function of the objective may have some apodization. Another is that the amplitude of the signal varies with distance as $\sin X / X$, where $X = z(1 - \cos\theta_0)$, and so it may not be adequate to assume that this factor is a constant over the measurement range. Consequently, with the integrating bucket technique, both the amplitude of the envelope and the phase change itself enter into the measurement.

Various empirical corrections have been employed by different researchers.[59,60,61] Examples are

$$f = \frac{\ln(\cos\theta_0)}{\cos\theta_0 - 1} \quad (4.50)$$

and

$$f = 1 + \frac{\sin^2\theta_0}{4}. \quad (4.51)$$

The fringe spacing given by all of these correction factors reduces to the spacing for an interferometer in the parallel beam limit, $\lambda/2$. The relative merits of these correction factors have been summarized and compared to experimental data in a paper by Creath.[59]

4.5 Phase Imaging with an Interference Microscope

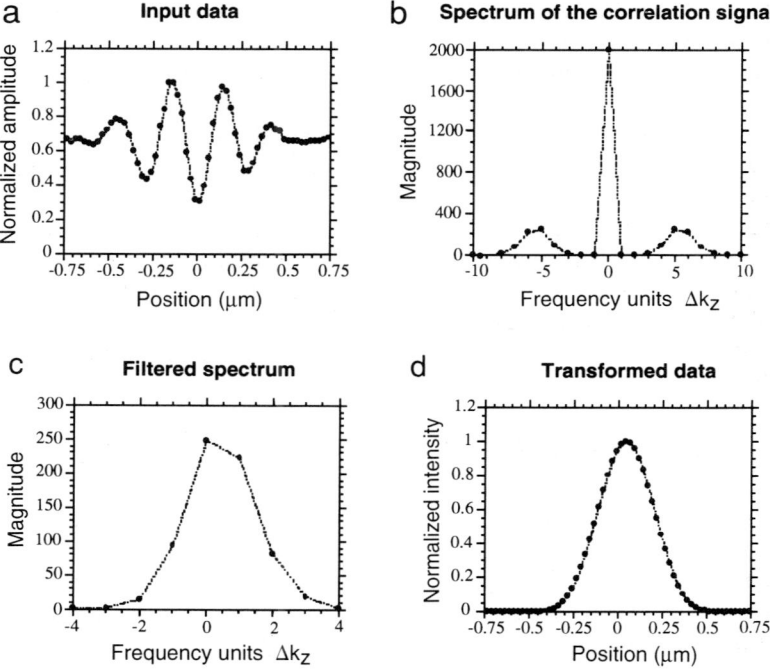

Figure 4.26 A flowchart summarizing the Fourier transform data analysis technique developed by Chim and Kino: (a) input data from the microscope, (b) spectrum of the input data, (c) filtered spectrum, and (d) amplitude of the filtered data in the space domain. [Courtesy: S. S. C. Chim, "The Mirau correlation microscope—a new tool for optical metrology," Ph.D. Dissertation, Department of Electrical Engineering, Stanford University, Stanford, California, USA (June 1991).]

4.5.2 The Fourier Transform Technique

An alternative to the integrating bucket technique has been developed by Chim and Kino, who demonstrated it on the MCM.[61] Their method is to extract the phase at each point (x,y) of the image directly from the depth response of the microscope by using a Fourier transform technique. To collect the data, intensity measurements are made at each pixel in the image at 64 different focus positions, as shown in Fig. 4.26(a). The resulting signal obtained for each pixel consists of the fringes whose amplitude falls off with distance $|z|$. The waveform obtained is similar to that of a modulated carrier, in which a constant-frequency sinusoidal carrier wave is modulated by a relatively low-frequency waveform. Since the carrier can be represented as the sum of positive and negative frequency components, the purpose of this technique is to eliminate one of these components and measure the phase of the other one directly.

To carry out this process, the data are Fourier transformed to yield the frequency spectrum of the z-scan, shown in Fig. 4.26(b). One of the frequency packets in the transform is eliminated and the remaining spatial frequencies are centered at the origin, as shown in Fig. 4.26(c). The phase of the filtered Fourier transform is a direct measurement of the phase change of the light reflected from the sample. The filtered transform can also be inverse transformed (Fig 4.26(d)) to obtain the envelope of the signal corresponding to the amplitude of the reflected light.

Put mathematically, in Chapter 3 it was shown that the expression for the correlation signal produced by an interference microscope is of the form

$$I(z) = I_2 \frac{\sin[k_0 z(1 - \cos\theta_0)]}{k_0 z(1 - \cos\theta_0)} \cos[k_0 z(1 + \cos\theta_0) - \phi_S]. \quad (4.52)$$

In Eq. (4.52), z is the defocus distance between the sample and the reference mirror. If the sample has a surface profile, $z_s(x,y)$, then the axial distance between the focal plane and any point (x,y) on the surface of the object is $(z + z_s)$, where z is the average distance of the sample from the focal plane. With this substitution, Eq. (4.52) becomes

$$I(z) = I_2 \frac{\sin[k(z + z_s)(1 - \cos\theta_0)]}{k(z + z_s)(1 - \cos\theta_0)} \cos[k(z + z_s)(1 + \cos\theta_0) - \phi_S]. \quad (4.53)$$

The signal has the form of an amplitude-modulated carrier, modulated in space rather than time.

We can take the Fourier transform of this signal by writing

$$F(I) = \int_{-\infty}^{\infty} I(z) e^{2j\pi s z}\, dz, \quad (4.54)$$

where s is the spatial frequency.

By using the shift and similarity theorems, the following result is obtained:

$$F(I) = K\delta(0) * \frac{K}{(1 - \cos\theta_0)} \Pi\left(\frac{s}{k(1 - \cos\theta_0)}\right) * \frac{e^{j2\pi s\left[(z+z_s) - \frac{\Phi_s \lambda}{2\pi(1+\cos\theta_0)}\right]}}{k(1 - \cos\theta_0)}$$

$$\left[\delta\left(\frac{s\lambda}{2(1 + \cos\theta_0)} + \frac{1}{2}\right) + \delta\left(\frac{s\lambda}{2(1 + \cos\theta_0)} - \frac{1}{2}\right)\right], \quad (4.55)$$

4.5 Phase Imaging with an Interference Microscope

where $\delta(s)$ is the delta function, $\Pi(s)$ is the rectangle function, and the symbol * denotes the convolution operation. It is apparent from Eq. (4.55) that the Fourier transform of the interferogram contains two packets of spatial frequencies centered around

$$s = \pm(1 + \cos\theta_0)/\lambda, \qquad (4.56)$$

as shown from experimental results with a finite bandwidth signal in Fig. 4.26(b).

Chim and Kino separate the phase and amplitude terms by eliminating one of the packets and centering the remaining frequencies about the origin,

$$F_0(I) = K \prod \left(\frac{s}{k(1-\cos\theta_0)}\right) e^{js\left[(z+z_s) - \frac{\phi_s \lambda}{2\pi(1+\cos\theta_0)}\right]}, \qquad (4.57)$$

where K is a constant.

The intensity of the reflected light is given by squaring the amplitude term in Eq. (4.57), $I = F_0 F_0^*$. The phase of the filtered Fourier transform is a direct measurement of the phase change of the light reflected from the sample. It is given by the argument of the exponential, which can be extracted by taking the arctan of the quotient of the difference and sum of the Fourier transform with its complex conjugate

$$\phi(x,y) = \tan^{-1}\left(\frac{F_0 - F_0^*}{F_0 + F_0^*}\right). \qquad (4.58)$$

This operation yields the result

$$\phi(x,y) = 2\pi s \left[(z + z_s) - \frac{\phi_s \lambda}{2\pi(1+\cos\theta_0)}\right]. \qquad (4.59)$$

Substituting Eq. (4.55) for s into Eq. (4.58) gives the formula

$$\phi(x,y) = \frac{2\pi}{\lambda}(1 + \cos\theta_0)(z + z_s) - \phi_s. \qquad (4.60)$$

Solving for the height change on the sample we obtain the by now familiar equation

$$z + z_s = \frac{\phi(x,y) + \phi_s}{k(1 + \cos\theta_0)}. \qquad (4.61)$$

It will be noted that this result is independent of the amplitude of the envelope of the interference signal. The term ϕ_s in Eq. (4.61) is a phase change that depends on the material properties of the sample. A rigorous examination of the measurement conditions required would lead to the conclusion that the only way to accurately measure height on an interference microscope is to be sure that the material is identical on both sides of the phase step so that ϕ_s is a constant. However, in practice, the phase shifts of most metals are approximately equal in the visible range, so this requirement can often be relaxed. In this case z_s will be directly related to $\phi(x,y)$ through Eq. (4.61). If both the height and material properties of the sample change, the height difference can be determined from the displacement of the envelope of the correlation function, the $\sin(x)/(x)$ term in Eq. (4.53), or alternatively ϕ_s can be calibrated using a sample with a known step height made of the same materials.

An example of how phase imaging can be used for measurement applications is discussed in Chapter 5.

4.6 Conclusion

Numerous phase measurement techniques for the optical microscope have been and will continue to be devised. We have highlighted a few of those which are used on a scanning optical microscope and interference microscope. The ability to perform accurate phase measurements is extending the capabilities of optical microscopy to a range of new applications, such as the measurement of surface roughness and film thickness, which are of great importance.

References

1. F. Zernike, "Das phasen kontrastverfahren bei der mikroskopischen Beobachtung," Z. Tech. Phys. **16**, 454–457 (1935).
2. H. A. Ferwerda, "Fritz Zernike–life and achievements," Opt. Eng. **32**, 3176–3181 (1993).
3. M. Pluta, "Addendum to Fritz Zernike–life and achievements," Opt. Eng. **32**, 3182–3183 (1993).
4. H. Gundlach, "Phase contrast and differential interference contrast instrumentation and applications in cell, developmental, and marine biology," Opt. Eng. **32**, 3223–3228 (1993).
5. M. Spencer, *Fundamentals of Light Microscopy* (Cambridge University Press, 1982).
6. J. R. Benford and R. L. Seidenberg, "Phase contrast microscopy for opaque specimens" JOSA **40**, 314–316 (1950).
7. D. Goodman, "Inadequate explanations of the Zernike phase-contrast method," presented at the OSA Annual Meeting (October 22, 1987).
8. C. P. Saylor, A. T. Brice, and F. Zernike, "Color phase-contrast microscopy: Requirements and applications," JOSA **40**, 329–334 (1950).

9. E. Inglestam, "Some quantitative measurements of path differences and gradients by means of phase contrast and new interferometric devices," JOSA **47**, 536–544 (1957).
10. P. K. Mondal, "Phase contrast microscopy with partially coherent light," Optica Acta **15**, 65–82 (1968).
11. D. C. Leiner and D. T. Moore, "Real-time phase microscopy using a phase locked interferometer," Rev. Sci. Inst. **49**, 1702–1705 (1978).
12. G. J. Brakenhoff, "Imaging modes in confocal scanning light microscopy (CSLM)," Journal of Microscopy **117** Pt. 2, 233–242 (1979).
13. D. K. Hamilton and C. J. R. Sheppard, "A confocal interference microscope," Optica Acta, **29**, 1573–1577 (1982).
14. T. Wilson, "Images of phase edges in conventional and scanning optical microscopes," Appl. Opt. **20**, 3238–3244 (1981).
15. C. J. R. Sheppard, A. R. Carlini, and H. J. Matthews, "Three-dimensional imaging of phase steps," Optik **80**, 91–94 (1988).
16. H. J. Matthews, D. K. Hamilton, and C. J. R. Sheppard, "Surface profiling by phase-locked interferometry," Appl. Opt. **25**, 2372–2374 (1986).
17. G. S. Kino, T. R. Corle, and G. Q. Xiao, "New types of scanning optical microscopes," *Integrated Circuit Metrology, Inspection and Process Control II*, Kevin M. Monahan, editor, SPIE **921**, 116–122 (1988).
18. G. S. Kino, C.-H. Chou, T. R. Corle, and P. C. D. Hobbs, "Scanning differential contrast microscopy," *Review of Progress in Quantitative Nondestructive Evaluations Vol. 6B*, D. O. Thompson and D. E. Chimenti, editors, 1315–1326 (Plenum Press, 1987).
19. T. R. Corle, J. T. Fanton, and G. S. Kino, "Distance measurements by differential confocal optical ranging," Appl. Opt. **26**, 2416–2420 (1987).
20. T. R. Corle and G. S. Kino,. "Phase imaging in scanning optical microscopes, *Scanning Imaging,* T. Wilson, editor, Proc. SPIE **1028**, 114–121 (1988).
21. K. Osato, G. S. Kino, and S. Kubota, "An electro-optic focusing system for a high-density optical disk," Opt. Lett. **18**, 1244–1246 (1993).
22. M. Born and E. Wolf, *Principles of Optics*, 6th ed. (Pergamon, 1983).
23. R. L. Jungerman, P. C. D Hobbs, and G. S. Kino, "Phase sensitive scanning optical microscope," Appl. Phys. Lett. **45**, 846–848 (1984).
24. G. E. Sommargren, "Optical heterodyne profilometry," Appl. Opt. **20**, 610–618 (1981).
25. C. C. Huang, "Optical heterodyne profilometer," Opt. Eng. **23**, 365–370 (1984).
26. P. C. D. Hobbs, "Heterodyne interferometry with a scanning optical microscope," Ph.D. Dissertation, Department of Applied Physics, Stanford University, Stanford, California, USA (August 1987).
27. P. C. D. Hobbs and G. S. Kino, "Generalizing the Confocal Microscope Via Heterodyne Interferometry and Digital Filtering," Journal of Microscopy **160** 245–264 (1990).
28. S. Hell, and E. H. K. Stelzer, "Properties of a 4Pi-confocal fluorescence microscope," JOSA, **A9**, 2159–2166 (1992).
29. M. G. Nomarski, "Microinterférometre differentiel à ondes polarisées," J. Phys. Radium, **16**, 9S–13S (1955).

30. T. R. Corle and G. S. Kino, "Differential interference contrast imaging on a real time confocal scanning optical microscope," Appl. Opt. **29**, 3769–3774 (1990).
31. C. J. Cogswell and C. J. R. Sheppard, "Confocal brightfield imaging techniques using an on-axis scanning optical microscope," *Confocal Microscopy* T. Wilson, editor (Academic Press, 1990).
32. M. B. Dowells, C. A. Hultman, and G. M. Rosenblatt, "Determination of slopes of microscopic surface features by Nomarski polarization interferometry," Rev. Sci. Inst. **48**, 1491–1497 (1977).
33. M. Vaez-Iravani, and C. W. See, "Linear and differential techniques in the scanning optical microscope," *Scanning Microscopy Techniques and Applications*, E. Clayton Teague, editor, SPIE **897**, 43–54 (1988).
34. C. W. See and M. Vaez-Iravani, "Differential amplitude scanning optical microscope: theory and applications," Appl. Opt. **27**, 2786–2792 (1988).
35. P. Chrusch, Jr. and M. Vaez-Iravani, "Differential optical microscopy based on higher-order Gaussian-Hermite beam patterns," Appl. Opt. **31**, 7344–7347 (1992).
36. C. J. Cogswell and C. J. R. Sheppard, "Confocal differential interference contrast (DIC) microscopy: including a theoretical analysis of conventional and confocal DIC imaging," Journal of Microscopy **165** Pt. 1, 81–101 (1992).
37. N. H. Dekkers, and H. de Lang, "Differential phase contrast in a STEM," Optik, **41**, 452–456 (1974).
38. J. P. H. Benchop, "Confocal differential phase contrast in scanning optical microscopy," *Scanning Imaging Technology* T. Wilson and L. Balk, editors, SPIE **809**, 90–96 (1987).
39. J. P. H. Benchop, "Signal detection and interpretation in scanning optical microscopy," Ph.D. Dissertation, Universiteit Twente, The Netherlands (1989).
40. T. Wilson, "Confocal microscopy" *Confocal Microscopy*, T. Wilson, editor (Academic Press, 1990).
41. T. Wilson and C. J. R. Sheppard, "Coded apertures and detectors for optical differentiation," *International Optical Computing Conference II*, W.T. Rhodes, editor, SPIE **232**, 203–209 (1980).
42. T. Wilson and D. K. Hamilton, "Difference confocal scanning microscopy," Opt. Acta, **31**, 452–465 (1984).
43. T. Wilson and D. K. Hamilton, "Differential amplitude contrast imaging in the scanning optical microscope," J. Appl. Phys., **B32**, 187–191 (1983).
44. D. K. Hamilton and T. Wilson "Edge enhancement in scanning optical microscopy by differential detection." JOSA, **A1**, 322–323 (1984).
45. R. Juskaitis and T. Wilson, "Differential confocal scanning microscope with a two-mode optical fiber," Appl. Opt. **31**, 898–902 (1992).
46. M. R. Atkinson, A. E. Dixon, and S. Damaskinos, "Surface-profile reconstruction using reflection differential phase-contrast microscopy," Appl. Opt. **31**, 6765–6771 (1992).
47. D.K. Hamilton and C.J.R. Sheppard, "Differential phase contrast in scanning optical microscopy," Journal of Microscopy **133** Pt. 1, 27–39 (1984).
48. D.K. Hamilton and C.J.R. Sheppard, "Two-dimensional phase imaging in the scanning optical microscope," Appl. Opt **23**, 348–352 (1984).

49. T. Wilson, "Enhanced differential phase contrast imaging in scanning microscopy using a quadrant detector," Optik, **72**, 109–114 (1990).
50. F. Laeri and T. C. Strand, "Angstrom resolution optical profilometry for microscopic objects," Appl. Opt. **26**, 2245–2249 (1987).
51. C. W. See and M. Vaez-Iravani, "Linear imaging in scanning polarization/interference contrast microscopy," Elect. Lett. **22**, 1079–1080 (1986).
52. R. Juskaitis and T. Wilson, "Differential confocal scanning microscope with a two-mode optical fiber," Appl. Opt. **31**, 898–902 (1992).
53. T. Wilson, "Image formation in two-mode fiber-based confocal microscopes," JOSA A **10**, 1535–1543 (1993).
54. S. Kimura and T. Wilson, "Confocal scanning optical microscopes using single mode fiber for signal detection," Appl. Opt. **30**, 2143–2150 (1991).
55. R. Juskaitis and T. Wilson, "Surface profiling with scanning optical microscopes using two-mode optical fibers," Appl. Opt. **31**, 4569–4574 (1992).
56. J. C. Wyant, C. L. Koliopoulous, B. Bhushan, and D. Basida, "Development of a three-dimensional non contact digital optical profiler," J. Tribology **108**, 1–8 (1986).
57. B. Bhushan, J. C. Wyant, and C. L. Koliopoulous, "Measurement of surface topography of magnetic tapes by Mirau interferometry," Appl. Opt. **24**, 1489–1497 (1985).
58. J. F. Biegen, "Step height measurement range extended for an interference microscope utilizing the obliquity effect," *Surface Characterization and Testing II,* J.E. Greivenkamp and M. Young, editors, SPIE **1164**, 85-90 (1989).
59. K. Creath, "Calibration of numerical aperture effects in interferometric microscope objectives," App. Opt. **28**, 3333–3338 (1989).
60. J. F. Biegen "Calibration requirements for Mirau and Linnik microscope interferometers," Appl. Opt. **28**, 1972–1974 (1989).
61. S. S. C. Chim, "The Mirau correlation microscope–a new tool for optical metrology," Ph.D. Dissertation, Department of Electrical Engineering, Stanford University, Stanford, California, USA (1991).

CHAPTER 5

Applications

5.1 Introduction

Confocal scanning optical microscopes (CSOMs) have become accepted as important imaging tools in such diverse fields as biology, geology, and materials science.[1] Commercial instruments are now available targeting specialized applications in semiconductor inspection and metrology and a wide range of uses in biology. In this chapter we will discuss a few of these applications.

Semiconductor metrology and film thickness measurements are covered at length in this chapter in order to give an adequate treatment of a subject which is of great importance to fields using devices with submicron-sized features. The chapter is divided into three sections: 5.2, Semiconductor Metrology; 5.3, Film Thickness Measurements; and 5.4, Biological Imaging.

In Section 5.2, both critical dimension and overlay misregistration measurements are discussed. The need for precision, accuracy, and the influence of polarization on dimensional measurements are examined.

In Section 5.3, the form of the interference signal from a focused beam *constant angle of reflection interference spectroscope* (CARIS) and *variable angle monochromatic fringe observation* (VAMFO) spectroreflectometers are derived. The expressions for the interference between the reflections from the top and bottom surfaces of a film are shown to be similar to the interference term from an interference optical microscope when the spacing between the sample and the reference is varied. This section concludes with an illustration of how, in an interference microscope, the phase can be used for film thickness measurements of opaque films.

Section 5.4 discusses biological imaging. Biological imaging applications include reflection, transmission, phase, and fluorescent imaging.

Since these topics have received excellent coverage in other texts, this section will be relatively short. In addition to practical examples of everyday use, the imaging properties of the various types of CSOMs will be examined in the context of these applications.

5.2 Semiconductor Metrology

5.2.1 Microlithography Measurements

One of the fastest growing solid-state applications of the CSOM and interference microscopes has been integrated circuit inspection and metrology.[2,3,4,5,6] Several companies build CSOMs specifically designed for integrated circuit inspection. These instruments range from desktop microscopes for visual inspection to complete metrology systems incorporating automated wafer handling, pattern recognition, and data analysis for automated in-line production measurements. The microscopes are used to measure the patterned photoresist on semiconductor wafers during manufacturing.

The basic process involved is to spin a thin layer of photoresist, a photosensitive materials, on to the semiconductor substrate. When this layer is exposed to light through a mask, a pattern corresponding to the features required is laid down in the photoresist. In one type of photoresist, the regions that have been exposed can then be chemically dissolved and metal films or doping material laid down in the holes so formed. Alternatively, holes or trenches can be etched into the substrate through the exposed regions of the photoresist. The photoresist is then removed and the process repeated, often many times, with different patterns to form a complete integrated circuit.

Measurement of the patterned photoresist is important because it allows the process engineer to simultaneously monitor for defects, misalignment, or other artifacts that may affect the manufacturing line. Such metrology is also cost effective because the process is reversible at this stage. If a mistake has been made the patterned photoresist can be removed, new photoresist applied, and the wafer placed back on the manufacturing line with little impact on the cost or quality of the finished circuit.

Confocal scanning microscopes and interference microscopes are used for several different measurements on integrated circuits. The measurement issues of importance are *critical dimension* (CD) and *overlay measurements* (OL). CD measurements determine the dimensions of structures or linewidths printed by a microlithography tool. Overlay measurements determine the overlay or registration of the photoresist structures to underlying layers of the circuit. In addition, work has been

5.2 Semiconductor Metrology

done to characterize the electrical properties of a circuit using *optical beam induced currents* (OBIC).[7] Interference microscopes add the ability to measure the thickness of opaque and thick transparent films to these capabilities. This added capability is especially useful when measuring phase-shift masks, a type of mask in which neighboring features on the mask transmit the light with different phase delays so that the structures being printed have sharper edges than if they were printed from a simple light-absorbing mask.

5.2.2 Precision, Linearity, and Accuracy in Semiconductor Metrology

Precision The precision (or repeatability) of a measurement is defined as the variation of the measured values in a particular instrument, when the measurements are made repeatedly under the same conditions. The measurement precision is specified by either the 3 standard deviation (3σ) variation of the measurements or the total range of the measured data with an allowance for a small number of fliers. These definitions assume that the number of measurements is large, the sample is stable over time, and the measurement results are normally distributed.[8]

In a semiconductor manufacturing process, the precision requirement for the metrology tool is determined by the size of the features of the circuit. The semiconductor industry is currently manufacturing devices with 0.5-μm-wide lines.[9] One example is the 16-megabit DRAM devices. The 0.5-μm lines need a linewidth control of 10% or better, i.e., 0.050 μm. In order to measure these dimensions, the metrology tool should have a 3σ precision of better than one third of the linewidth control or 0.016 μm. Similar specifications have been adopted for overlay measurements.

Linearity In addition to the measurements being repeatable, it is also desirable for the measured width of a sample to vary linearly with the true width or with the width obtained from a reference instrument, such as a *scanning electron microscope* (SEM) over a defined range of values. The agreement is quantified by plotting the data from the two instruments and calculating either the correlation coefficient (R^2), the maximum deviation of any point from the best-fit line, or the standard error from the least squares best-fit line. The ability to make linear measurements assures the device manufacturer of an instrument's ability to track changes in the linewidth or overlay over the range of normal process variations.

Accuracy Specifying the accuracy of a measurement is a much more difficult problem. Accuracy is usually thought of as the agreement of an experimental measurement with the "true value" or in practice an accepted reference value.[10] The source of the reference value in the United States is usually an international standard maintained by the *National Institute of Standards and Technology* (NIST). When these are not available, manufacturers rely on a set of in-house standards to monitor their metrology tools.

To make accurate measurements requires both a set of standards for calibration and a comprehensive theoretical understanding of the instrument so that exact dimensions can be calculated from the detected signals. Neither a linewidth calibration standard for thick samples nor a complete theoretical understanding of any linewidth metrology tool currently exists for use in submicron metrology, although advances have recently been made in both areas.[11,12] To reduce the accuracy problem many manufacturers maintain a set of "golden wafers" for each level of their process. These samples are used to maintain internal measurement consistency by checking that each machine's performance remains stable over time and that different machines make similar measurements. Fortunately, the ability to make absolutely accurate measurements is not critical to integrated circuit manufacturing. Relative changes in linewidth and overlay during the manufacturing process, rather than the actual dimensions, are generally of greater interest. Thus the repeatability and linearity of the measurements are the most important parameters when comparing measurement tools.

5.2.3 Critical Dimension Measurements

The electrical performance of an integrated circuit is strongly influenced by the size of the smallest features on the chip. These critical dimension features are usually the transistor gates that control the switching of the transistors from on to off. If the gates are too small, breakdowns can occur and the chip will be unreliable. If they are too large, the circuit will operate at a slower speed than its design value. For many years standard optical microscopes were used to measure and control critical dimension structures during integrated circuit manufacturing.[13,14,15] Their success in this task was based on the pioneering work of Nyyssonen, Kirk, and others in the field.[16,17] They developed the basic theoretical models of how common semiconductor linewidth structures are imaged by an optical microscope. Many of their ideas formed the analytical base

5.2 Semiconductor Metrology

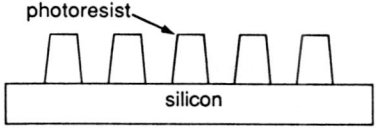 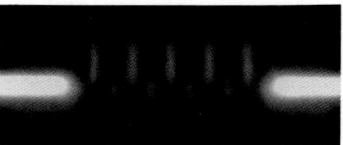

Figure 5.1 Cloud plots of an array of 0.6-μm-wide equal lines and spaces in 1.0-μm-tall photoresist on silicon. [Reprinted from T. R. Corle, "Submicron metrology in the semiconductor industry," Solid State Electronics **35**, 391–402. Copyright (1992), with kind permission from Elsevier Science Ltd., The Boulevard, Langford Lane, Kidlington OX5 1GB, UK.]

of the Questar optical metrology tool. This instrument used a standard optical microscope to inspect semiconductor wafers in high-volume production applications.

In recent years the SEM has become the accepted tool for linewidth metrology. The transition from the optical microscope to the SEM was driven by the superior resolution of the SEM, which was needed to measure the smaller linewidths coming into production. Although the SEM has excellent resolution, in operation the sample must be placed in a vacuum, increasing the cost and complexity of the measurement. In addition, because the electron beam does not penetrate the top surface of the sample, the SEM cannot measure overlay on buried structures. These factors have created a niche market for other linewidth metrology tools; among them are the confocal scanning and interference optical microscopes.[18,19,20,21,22]

Whether the microscope is a confocal scanning or an interferometric optical microscope, the process of linewidth measurement is essentially the same. The first step is to place the structure in the field of view. This step can be as simple as manually placing the wafer on the stage and then visually moving it to the desired location. Alternatively, as is more commonly done in a manufacturing environment, the wafer can be placed using robotic wafer handling, automatic wafer alignment, and pattern recognition to move to the measurement site.

Once the structure is in the field of view, a cross-sectional image or "cloud plot" is generated by scanning the sample in the focus direction and recording how the intensity varies at each pixel of a selected line scan in the image. This cross-sectional image is used to help determine the exact focus position for the measurement. Figure 5.1 is a cross-sectional image or cloud plot of equal lines and spaces 0.6-μm-wide in 1.0-μm-tall photoresist on silicon. These images were made on an RSOM developed

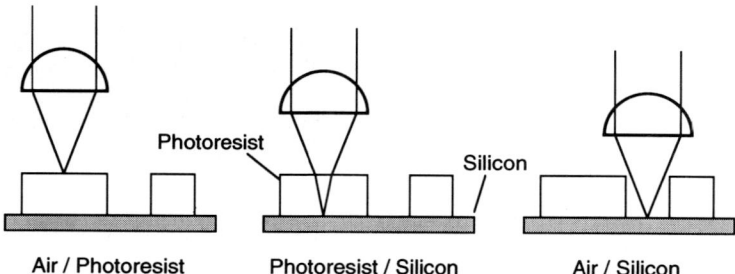

Figure 5.2 Schematic illustrating the three material interface focus locations for a photoresist on silicon sample.

at the Prometrix Corporation. The instrument uses a 50 ×/0.95 N.A. objective lens and the 400–500-nm wavelength band of a mercury arc lamp for illumination.

Focus Position The exact focus position at which a linewidth measurement is made is an important factor for determining the ability of the instrument to track a manufacturing process. Photoresist lines are typically narrower at the top than they are at the base, as shown in Fig. 5.1. Since the width at the base determines the area of the circuit that is masked, and thus protected from further processing, it is desirable to measure the bottom width. Determining the exact position of the bottom of the photoresist lines in a cross-sectional image can be a challenging task. For example, in the photoresist on silicon sample shown in Fig. 5.1, three focus positions are of special interest, as shown in Fig. 5.2. They correspond to the three interfaces between different materials in the sample: air-photoresist (air/PR), photoresist-silicon (PR/Si), and air-silicon (air/Si).[23] The air/PR interface is located at the top of the photoresist lines and hence is often called top focus. The PR/Si and air/Si interfaces are located at the bottom of the lines. Measurements made at these positions define the bottom linewidth. Although the PR/Si and air/Si interfaces are physically in the same plane on the sample, they appear at different focus locations in the cross-sectional image shown in Fig. 5.3. The reason for this discrepancy is that refraction of the light at the top surface of the photoresist causes the light to focus at a different position within photoresist than the nominal focus position in air. This affect causes the PR/Si interface to appear higher than the air/Si interface in the cross-sectional image.

5.2 Semiconductor Metrology

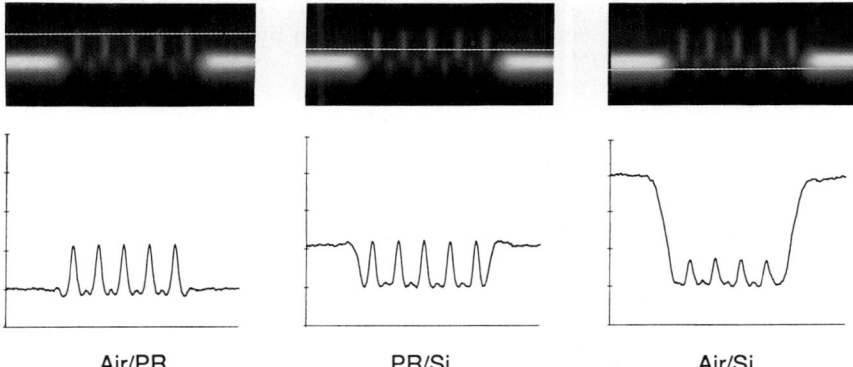

Figure 5.3 Cloud plots with line scans at the three material interfaces of a sample containing 0.6-μm-wide lines in 1.0-μm-thick photoresist on silicon. [Reprinted from T. R. Corle, "Submicron metrology in the semiconductor industry," Solid State Electronics **35**, 391–402. Copyright (1992), with kind permission from Elsevier Science Ltd., The Boulevard, Langford Lane, Kidlington OX5 1GB, UK.]

The position of the PR/Si interface in the cross-sectional image can be calculated using ray optics. If a ray of light intersects the photoresist at an angle θ it will be refracted and cross the optical axis of the system a distance d from the surface of the film, as shown in Fig. 5.4. The ratio

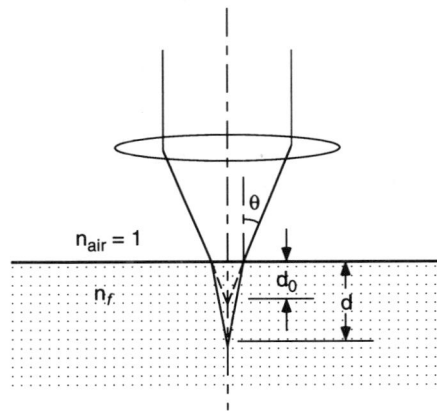

Figure 5.4 Schematic illustrating refraction of a focused beam by photoresist.

between the distance d, and the point where the ray would intersect the optical axis if no film were present, d_0, is given by

$$\frac{d}{d_0} = \frac{\sqrt{n_f^2 - \sin^2\theta}}{\cos\theta}. \tag{5.1}$$

In Eq. (5.1), n_f is the index of refraction of the film material and the material outside the film is assumed to be air. The location of the PR/Si interface in the image can be determined by calculating the ratio between the actual photoresist thickness, t_{resist}, and the distance between the PR/Si and air/Si focus positions, t_{image}. The quantity t_{image} can be thought of as the average shift in the focal position of all the rays.[24] The ratio is given, to a good approximation, by the normalized integral of Eq. (5.1) over all angles in the focused beam,

$$\frac{t_{resist}}{t_{image}} = \frac{\int_0^{\theta_0} \frac{\sqrt{n_f^2 - \sin^2\theta}}{\cos\theta} \sin\theta\, d\theta}{\int_0^{\theta_0} \sin\theta\, d\theta}. \tag{5.2}$$

In Eq. (5.2), θ_0 is the angle aperture of the objective ($\sin\theta_0$ = numerical aperture). For photoresist with n_f = 1.63 and a lens with numerical aperture $\sin\theta_0 = 0.9$, the ratio of the photoresist thickness to the cross-sectional image distance is t_{resist}/t_{image} = 2.12. Thus the apparent location of the photoresist silicon interface is approximately halfway between the physical interface and the top of the photoresist. The correction factor is, to first order, independent of the film thickness. Thus the ratio need only be calculated once for each material index.

The method of calculating the correction factor outlined above is not strictly correct since it does not account for diffraction within the film or possible waveguiding effects. It does, however, offer a quick way of calculating the distance to within a few percent. More exact calculations involve solving the full diffraction integrals. These calculations for a photoresist film covering an infinite half-plane give a thickness ratio t_{resist}/t_{image} = 2.09.[25] Once the waveforms are obtained, a measurement algorithm is applied to extract the linewidth.

5.2.4 Experimental Results

The CSOM has demonstrated its ability to make demanding critical dimension measurements on both wafers and photomasks.[26] Figure 5.5 shows images with the microscope focused on top and bottom of the trench, a cloud plot, and line scans from the top and bottom focus posi-

5.2 Semiconductor Metrology

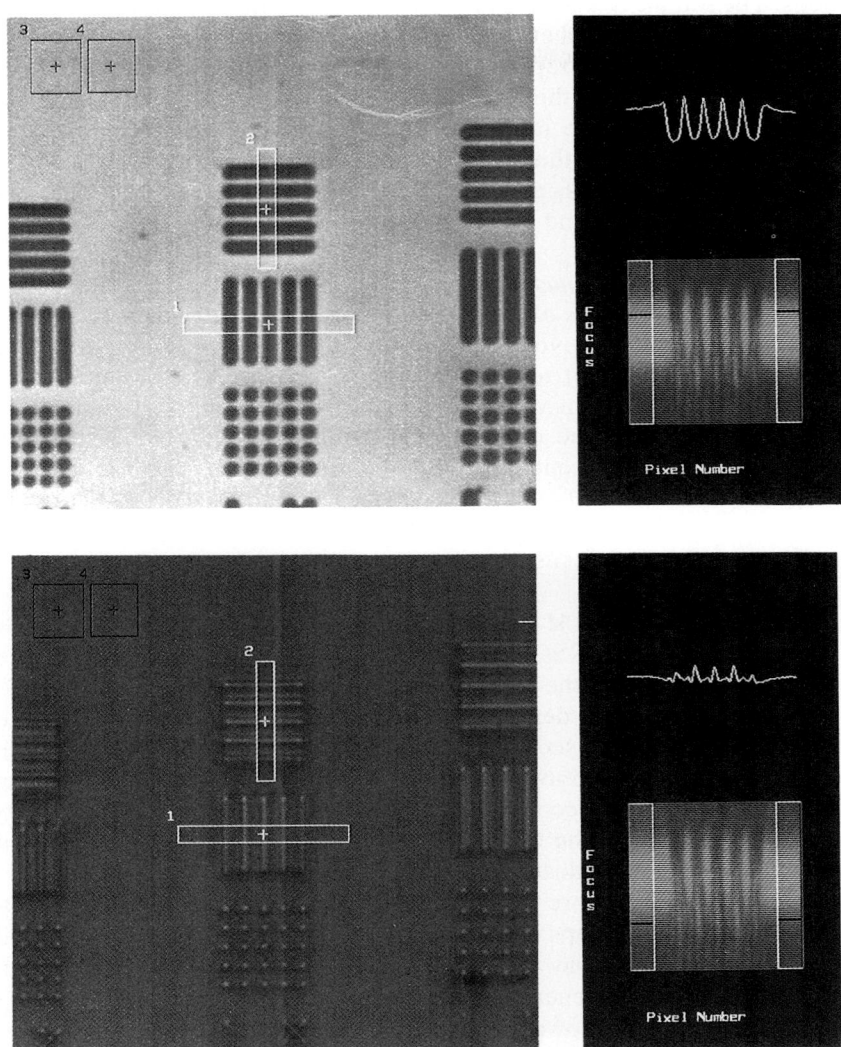

Figure 5.5 Top and bottom focus images with accompanying optical cloud plots of 0.5-μm-wide trenches in 1.0-μm-thick photoresist on silicon.

tions. The sample is a group of 0.5-μm-wide trenches in 1-μm-thick photoresist on silicon. The photoresist uniformly covers the visible area in the field of view. Both the top and bottom of the trenches are clearly imaged in the microscope. In the lower image, which represents bottom focus, it can be seen that the outermost trenches are not completely clear of photoresist. This hypothesis is confirmed in the cloud plots. The base of the rightmost trench has a weaker reflected signal intensity than the base of the center trench, and it is displaced upward.

Linearity Measurements Figure 5.6 compares the measurements of width made on a series of isolated trenches in 1.25-μm-thick photoresist on silicon using a low-voltage SEM and the *real-time scanning optical microscope* (RSOM) at top and bottom focus. The RSOM used a 50×/0.95 *N.A.* objective lens with illumination from a mercury arc lamp in the 400–500-nm range. The measurement results from the two instruments are in excellent agreement, with a slight offset due to the fact that different edge thresholds were chosen in the two measurements. The standard error of the least squares best-fit line to the data is 0.018 μm at top focus and 0.020 μm at bottom focus.

Comparison of SEM and RSOM The advantage of measuring isolated trenches in the RSOM is that the measurements can be performed more quickly than in the SEM. Furthermore, in certain cases, the SEM measurements may be degraded by sample charging, which is aggravated by the trapping of the secondary electrons in the bottom of the trenches. For this reason and to avoid damage to the semiconductor, low-voltage SEMs, for which the secondary emission coefficient is greater than unity, are normally used. The SEM, where it can be used, is the commonly accepted standard against which other measurements are calibrated.

Isolated trenches are not the only measurement structures of interest. Isolated lines and groups of lines, called dense arrays, are also commonly measured. A dense array typically consists of a group of five or nine lines. The measurement is generally made on the center line of the group. These measurements are known to track the microlithographic process more accurately than isolated linewidth measurements and consequently are commonly measured in production.

5.2.5 Polarization-Enhanced Imaging of Dense Arrays

Dense arrays are difficult to measure in an optical microscope. This difficulty arises from several causes. First, dense arrays of lines act like diffraction gratings by scattering a wide beam of light into a series of

5.2 Semiconductor Metrology

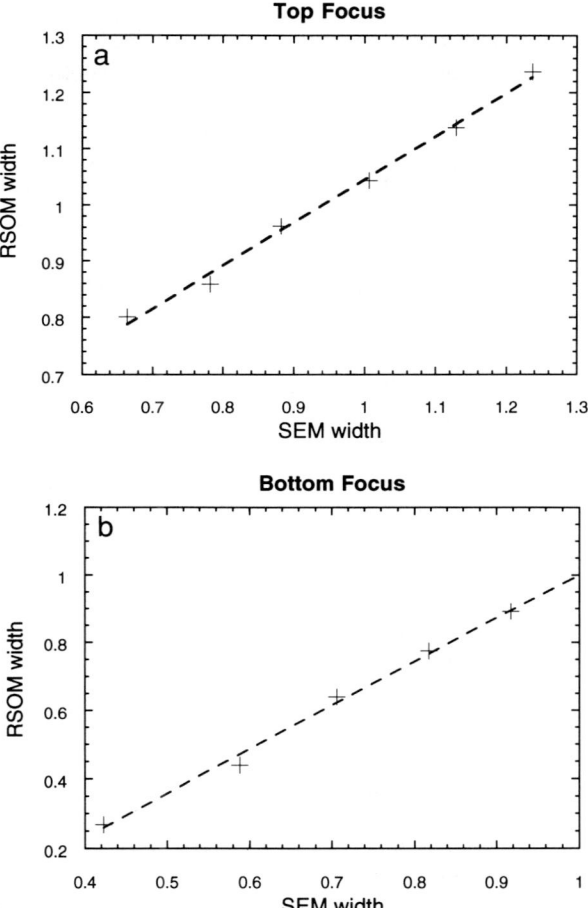

Figure 5.6 Comparison between critical dimension measurements of a series of isolated trenches of different nominal sizes made using an SEM and the RSOM. [Reprinted from T. R. Corle, "Submicron metrology in the semiconductor industry," Solid State Electronics **35**, 391–402. Copyright (1992), with kind permission from Elsevier Science Ltd., The Boulevard, Langford Lane, Kidlington OX5 1GB, UK.]

diffraction lobes.[27] The scattering angle and intensity in the diffraction lobes depend on the pitch of the array as well as the width of the individual lines. A further problem is that measurements on isolated and dense lines can exhibit a resonance phenomenon.[28] The resonance occurs because the index of refraction of the photoresist is greater than that of the surrounding air. The lines act as waveguides for the light, resonating at specific illumination wavelengths depending on their height and width. Isolated trenches do not exhibit this resonance phenomenon because the index of refraction of the photoresist is greater than the index of refraction of the air in the trench, and thus the resonance modes are coupled away into the photoresist. Finally, as linewidths shrink, the light signal received from the substrate region, or space between the lines, is reduced in intensity. When the linewidth is less than 0.5 μm, little or no reflected light is received from the substrate region of the array. Making accurate linewidth measurements of the bottom of a trench once this signal has disappeared is obviously not possible.

The image quality when measuring a group of lines can be improved in a CSOM by using polarization to increase the signal intensity from the space between the lines. The improvement in the image is greatest for linewidths narrower than 1.0 μm and leads to more accurate linewidth measurements in the CSOM.[29] The technique, called *polarization-enhanced* (PE) imaging, takes advantage of the fact that polarized light undergoes a change in polarization when reflected from a grating.[30,31,32,33] The change in the plane of polarization of the reflected light is due to the fact that polarized light guided by the trenches will have different phase velocities depending on whether the polarization is parallel or perpendicular to the dense array. However, light reflected from the top of the resist shows very little rotation of the plane of polarization. Consequently by illuminating the sample with elliptically polarized light of the correct orientation, the detected intensity from the bottom of a trench can be increased relative to that from the top surface of the resist.

In the normal operation of an RSOM containing an optical isolator, the optical axis of the *quarter-wave plate* (QWP) forms an angle of 45° to the direction of polarization of the incident light. In this position the QWP converts the linearly polarized incident light to circularly polarized light at the sample. Reflection from a mirror sample changes the polarization state to circular polarization in the opposite sense (i.e., right-handed circular polarization becomes left-handed circular polarization). The reflected light is converted back to linear polarization when it passes through the QWP a second time. The polarization direction is now, however, rotated by 90° with respect to the illumination polarization. This light then travels back through the Nipkow disk and the analyzer to form an image on the *charged coupled device* (CCD) camera.

If the sample is not a perfect mirror the polarization of the reflected light is elliptical rather than circular at the QWP. In this case the light incident on the analyzer will not be linearly polarized. The analyzer will absorb a portion of this light rather than transmitting it to the CCD camera.

5.2 Semiconductor Metrology

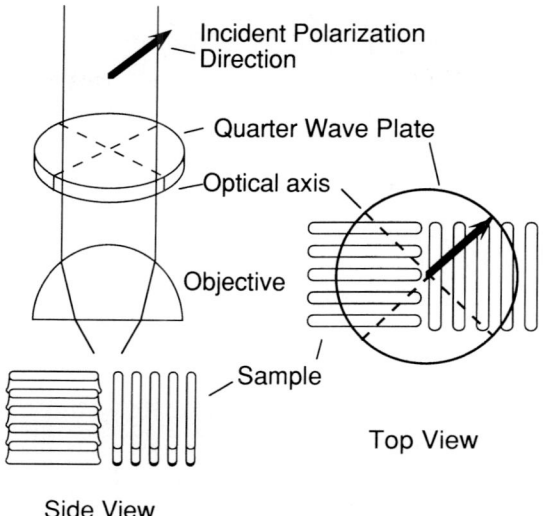

Figure 5.7 Illustration showing the relationship of the lines on the sample to the polarization axis for PE imaging. [From T. R. Corle, L. C. Mantalas, T. R. Kaack, and L. J. LaComb, Jr., "Polarization enhanced imaging of photoresist gratings in the real-time scanning optical microscope," Appl. Opt. **33**, 670–677 (1994), with permission.]

A potentially useful signal generated by the nonideal behavior of the sample has been lost. In the simplest sense, illuminating the sample with elliptically polarized light enables the light that has had its polarization state changed by the sample to be nearly circular when it returns to the QWP and thus be imaged by the CCD camera.

An RSOM can be modified for PE imaging by mounting the QWP so that it can be rotated in a plane parallel to its optical axis. With this modification, the sample can be illuminated and imaged with light of any desired elliptical polarization state by rotating the QWP. In addition, the sample must be oriented so that when linearly polarized light from the QWP strikes the sample, the major linear features of the sample form an angle of 45° to the polarization direction, as shown in Fig. 5.7. If the linewidth features are oriented either perpendicular or parallel to the illumination polarization direction, little or no image enhancement is observed.

The rotational position of the QWP in which it passes linearly polarized light through the system unchanged defines the null position, or 0° position, of the imaging system. At this angle little or no light is received at the camera from a bare substrate.

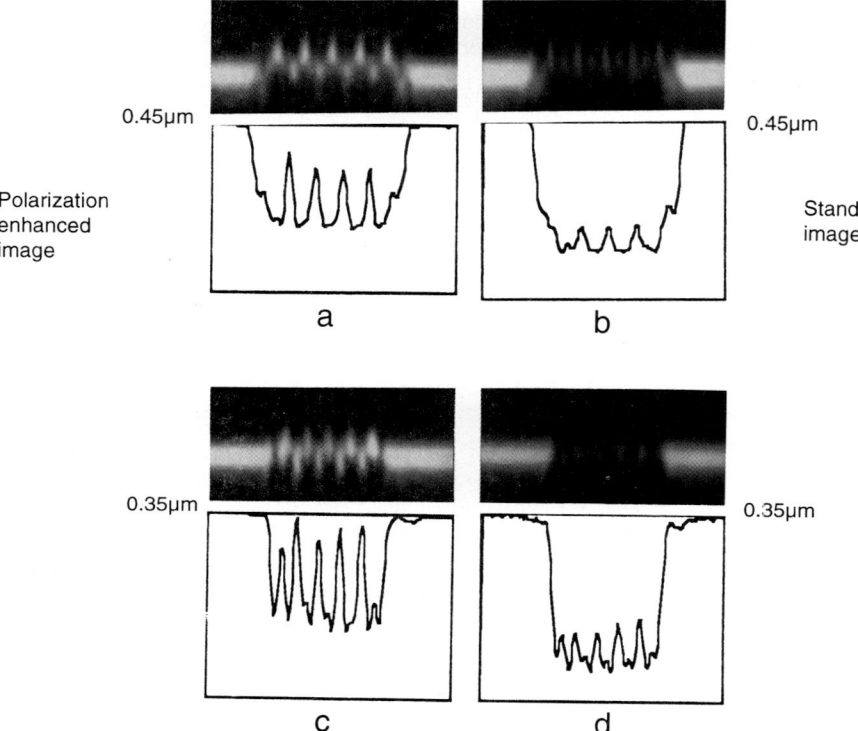

Figure 5.8 Cloud plots and line scans through a dense array of lines in 1.0-μm-thick photoresist on silicon; (a) a PE image and line scan of a 0.45-μm line-space array, (b) a standard RSOM image and line scan of the same 0.45-μm structure, (c) a PE image and line scan of a 0.35-μm line-space array, (d) a standard RSOM image and line scan of the same 0.35-μm structure. [From T. R. Corle, L. C. Mantalas, T. R. Kaack, and L. J. LaComb, Jr., "Polarization enhanced imaging of photoresist gratings in the real-time scanning optical microscope," Appl. Opt. **33**, 670–677 (1994), with permission.]

Experimental Results Figure 5.8 shows four cross-sectional images of a dense array of lines in 1.0-μm–thick photoresist on silicon, imaged both with and without polarization enhancement on an RSOM developed at the Prometrix Corporation.[34,35] The test structure is usually printed in doped polysilicon or metal on silicon dioxide on silicon, although other conducting films and insulators may be used.

Figure 5.8(a) is a PE image and line scan of a 0.45-μm line-space array made by rotating the QWP +17° from the null position. Figure 5.8(b) is a standard RSOM image and line scan of the same structure. Figure 5.8(c)

5.2 Semiconductor Metrology

is a PE image and line scan of a 0.35-μm line-space array made at the same QWP position. Figure 5.8(d) is a standard RSOM image and line scan. The line scans are at focus positions that correspond to the substrate. Comparing the cross-sectional images and the line scans, it will be observed that there is a significant improvement in the PE images. Both the photoresist lines and spaces can be easily distinguished in the PE image, whereas the substrate signal is not visible in the standard RSOM image.

In both the standard RSOM and the PE-RSOM, it is possible to image the tops of photoresist lines with widths as small as 0.15 μm. In this sense, either instrument is able to resolve the structures. The PE-RSOM, however, also provides a signal from the substrate region of the sample, improving its ability to make accurate linewidth measurements.

Polarization Enhanced Measurements of Linewidth Figure 5.9 shows the improvement in the measurement results when PE imaging is employed. The measured data were taken on a range of linewidths in a focus exposure wafer printed in 1.0-μm-thick photoresist on titanium on silicon dioxide on silicon. In a focus-exposure wafer the focus position and exposure energy of the microlithography tool are varied at each site on the wafer in order to determine the optimum focus position and exposure energy needed to print subsequent wafers. A titanium substrate was chosen because it is easy to print narrow lines on, and it is conductive so that the titanium can be etched from between the photoresist lines and the linewidths measured using an electrical tester for calibration purposes. For this test, a batch of wafers was printed and then divided. Half the wafers were etched, stripped of photoresist, and measured on an electrical tester, the Prometrix EM1 Production Monitoring and Yield Enhancement System. The remaining wafers were measured on a polarization-enhanced RSOM, the Prometrix ConQuest 2000 Confocal CD/Overlay Metrology System.

Electrical testers are essentially four-point probes optimized for CD measurements. One pair of leads acts as a current source; the second set measures the voltage drop across a test structure.[36,37] From the measured voltage and current and Ohm's law, $V = IR = IR_s \ell/w$, where R_s is the resistance per square of the film, ℓ is its length, and w its width, the resistance of the test structure is calculated. Since the width is determined by etching through the photoresist, and the resistance per square R_s can be determined with a four-point probe, electrical linewidth measurements are highly accurate and are therefore an excellent check on the optical measurements.

The correlation between the electrical and optical measurements, shown in Fig. 5.9(a), is excellent. Figure 5.9(b) shows the optical measure-

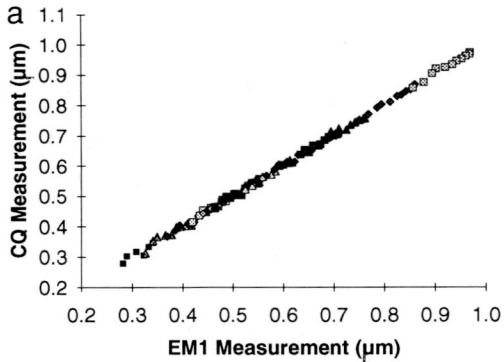

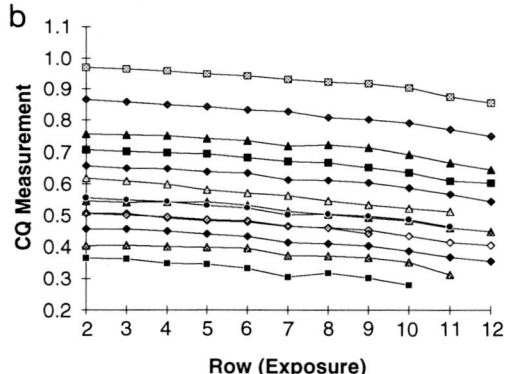

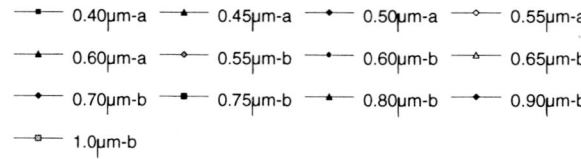

Figure 5.9 Linewidth measurements made over a range of different size lines on a focus exposure wafer in 1.0-μm-thick photoresist on titanium on silicon dioxide on silicon. [From T. R. Corle, L. C. Mantalas, T. R. Kaack, and L. J. LaComb, Jr., "Polarization enhanced imaging of photoresist gratings in the real-time scanning optical microscope," Appl. Opt. **33**, 670–677 (1994), with permission.]

5.2 Semiconductor Metrology

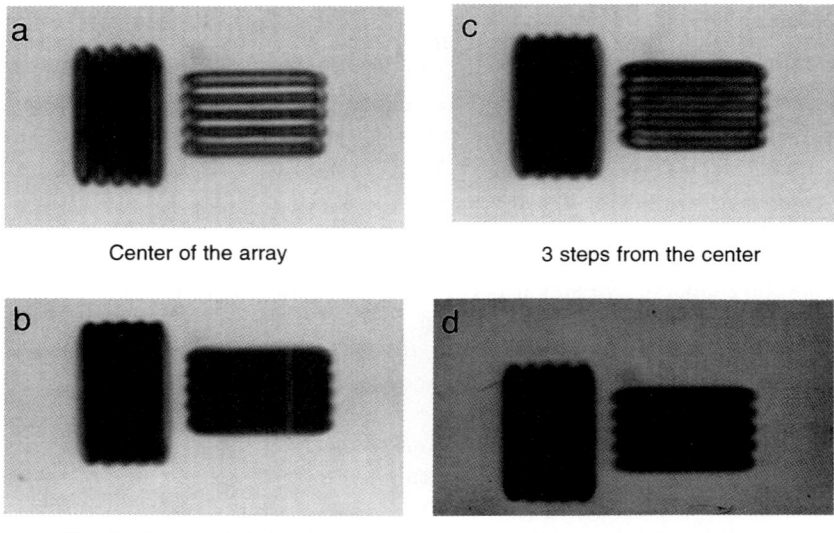

Figure 5.10 Four images of a 0.35-μm dense array of lines in 1.0-μm-tall photoresist on silicon: (a) a site near the center of a focus exposure matrix on a silicon wafer, (b) a site three exposure steps and three focus steps away from the center, (c) a site three exposure steps and four focus steps away from the center, (d) standard RSOM image of the site shown in (a). [From T. R. Corle, L. C. Mantalas, T. R. Kaack, and L. J. LaComb, Jr., "Polarization enhanced imaging of photoresist gratings in the real-time scanning optical microscope," Appl. Opt. **33**, 670–677 (1994), with permission.]

ments as a function of exposure. Two different test setups were employed. In the 0.40 μm–0.60 μm region (segment a) a QWP angle of −12° was used with an 80-nm illumination bandwidth centered at 440 nm and a 50×/0.95 N.A. lens. The region from 0.55 μm to 1.0 μm (segment b) was measured with a QWP angle of +15° using the same illumination and a 50×/0.55 N.A. lens. The QWP angle that produces the best measurements is selected based on both the structure size and the algorithms used to make the measurement. The standard error of the best-fit line between the optical and electrical measurements is only 9 nm. There is also good overlap of the measurements in the transition region.

Other Inspection Applications PE imaging has also been used to determine when the spaces between photoresist lines are clear of photoresist all the way to the substrate. Figure 5.10 shows four images of a 0.35-μm line-space array in 1.0-μm-thick photoresist on silicon. The image shown in Fig. 5.10(a) is from a location near the center of a focus exposure matrix on a silicon wafer. In this figure, PE imaging was used to accentuate the information from the horizontal lines. When the horizontal lines are imaged, information is lost from the vertical lines. Some edge roughness can be observed in the image; however, the lines are clear of photoresist all the way to the substrate. Figure 5.10(b) is a site located three exposure steps and three focus steps away from the site of Fig. 5.10(a). The spaces between the lines are now very narrow and the lines show increased edge roughness. One more step away in focus, Fig. 5.10(c), and the spaces are completely closed. No signal is returned from the substrate between the photoresist lines. For comparison Fig. 5.10(d) is a standard RSOM image of the structure shown in Fig. 5.10(a). It is difficult to tell from the standard RSOM image to what extent the spaces are clear of photoresist.

Other Methods of Polarization-Enhanced Imaging Several other approaches for using polarization to enhance the imaging capabilities of scanning optical microscopes have been tried. Kimura and Wilson have built a dark field polarization microscope for examining large structures in photoresist.[38] Wijnaendts-van-Resandt has used polarization to increase the signal from the top layer of a 0.8-μm-wide photoresist line on polysilicon in a laser scanning optical microscope.[39] Betzig et al. have used different polarization states in a near-field scanning optical microscope to observe gaps in 1-μm-diameter aluminum rings.[40] In their experiments they used both linear polarizers and waveplates to modify the polarization. The aluminum rings exhibited significant polarization-dependent effects. These effects were attributed to the sharp metallic features of the aluminum rings that enforced a well-defined polarization state because of the boundary condition at the surface of the conductor. They also examined annular slots in a dielectric. A much weaker effect was observed using these samples. The weakness of the effect was probably caused by the geometry of the samples tested.

5.2.6 Calibration

To accurately correlate the optical linewidth measurements of dense arrays made with the RSOM and those made using other techniques, different calibration curves are needed for each nominal linewidth. The reason is that the field scattered from a dense linewidth structure varies with the linewidth but also depends on the pitch of the grating. This difference arises because a beam focused to a finite diameter spot on the sample interacts with neighboring lines. Consequently, if the spot size is comparable to the period of the grating, the image of a single line will be different from that which would be obtained if the period of the grating were much larger than the spot size or the linewidth.

In semiconductor metrology, linewidth measurements must be calibrated so that the processes on different manufacturing lines can be compared. Calibration is also necessary to ensure that the dimensions of the manufactured integrated circuit are consistent with the designed values for the circuit. To perform and maintain these calibrations two types of calibration wafers are commonly available: constant pitch test wafers and focus-exposure wafers.

Constant pitch test wafers can be made at the best focus-exposure conditions for the lithography equipment. In such wafers, the sidewall slopes will be vertical, and the spaces between the lines clear of photoresist all the way to the substrate. The different linewidths used for calibration of the instrument are programmed into the mask. Focus-exposure wafers, on the other hand, used variation in the exposure energy and focus position of the microlithography tool to change the linewidth of the test structures while keeping the pitch constant. These wafers most accurately model potential problems with the exposure tool.

5.2.7 Overlay Misregistration Measurements

Today's complex integrated circuits are composed of many patterned layers of different materials. For a common circuit up to 30 different mask layers containing different patterns must be successfully deposited on the silicon. In addition to controlling the horizontal and vertical dimensions of the structures, it is important that the patterns precisely overlie one another for the circuit to work. Overlay misregistration measurements quantify the pattern offsets between the different layers of the circuit. These measurements track the ability of the stage on the microlithographic exposure tool to move from one part of the wafer to another and the ability of the tool to align the image of the photomask with existing circuit layers.

The importance of OL measurements to semiconductor metrology is increasing due to technological advances.[41] For example, the reduction of exposure wavelength from the mercury G-line (436 nm) to I-line (365 nm) and now the use of deep UV lasers (248 nm) has made it easier to print dimensionally stable lines. Overlay, on the other hand, is controlled by the mechanical stages in the microlithographic tools. These stages have not undergone a similar improvement in resolution. Overlay misregistration measurements cannot currently be made with an SEM because of its inability to penetrate the top surface of an integrated circuit to image buried structures. Scanning optical and interference microscopes thus remain the best choices for overlay measurements.

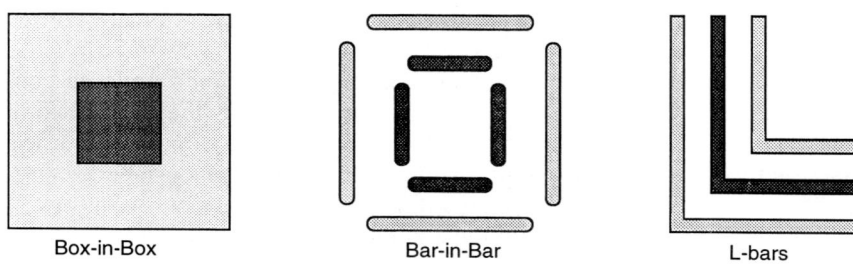

Figure 5.11 Comparison of three different types of overlay structures used in optical metrology: (a) box-in-box, (b) bar-in-bar, and (c) L-bars. [Reprinted from T. R. Corle, "Submicron metrology in the semiconductor industry," Solid State Electronics **35**, 391–402. Copyright (1992), with kind permission from Elsevier Science Ltd, The Boulevard, Langford Lane, Kidlington OX5 1GB, UK.]

To measure the OL error using an optical microscope, a variety of structures are used. Some of the most common are box-in-box, bar-in-bar, and L-bars, Fig. 5.11. These structures range in size from 1 μm for the inner box and 3 μm for the outer box of the box-in-box structure, to 20 μm and 60 μm for the inner and outer boxes. Typically one portion of the structure, such as the inner box, is deposited as part of an early mask step in the manufacturing process. The outer portion of the box is then deposited as part of a later mask step. To measure the overlay misregistration between these two levels of the integrated circuit the position of the center of the inner structure relative to the center of the outer structure is measured. The distance between the center locations is called the overlay error.

An image of a box-in-box overlay structure taken with an RSOM is shown in Fig. 5.12. This structure measures the offset between a photoresist layer covering polysilicon and an underlying oxide layer prior to etching the polysilicon. The two boxes in the top-down image show the measurement region. The line scan in the cross-sectional image is representative of the edge contrast at the selected focus location. To avoid grains on the substrate or illumination nonuniformity affecting the overlay measurements, only the edges of the structure are used to calculate the location of the pattern centers. The use of a CSOM allows the focus position to be precisely and repeatable selected, thus improving the accuracy and repeatability of the measurements.

Overlay measurements on most layers of an integrated circuit are made using broadband light and low numerical aperture lenses. Broadband light helps to minimize coherent interference effects, which can lead to measurement inaccuracies. In addition, shadowing of a focused beam by

5.2 Semiconductor Metrology

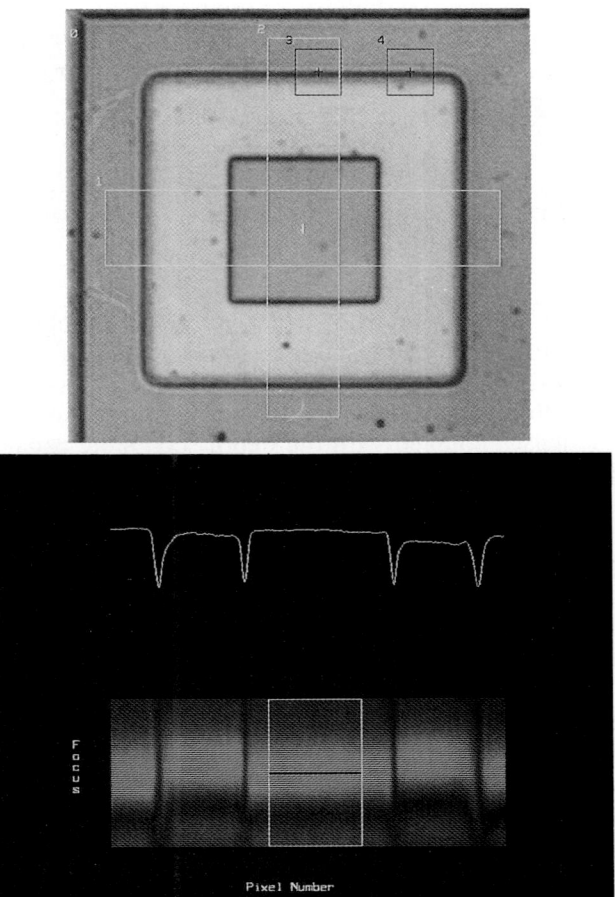

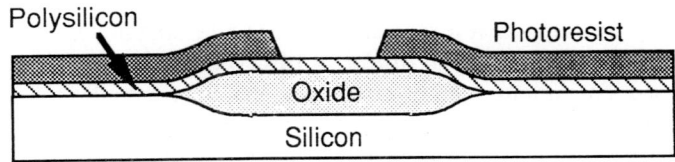

Figure 5.12 Top-down and cross-sectional images of a box-in-box overlay structure.

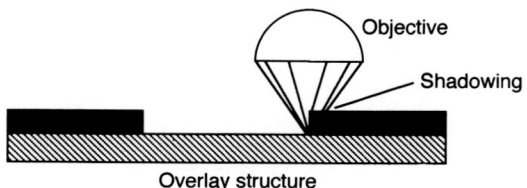

Figure 5.13 Shadowing of a focused beam by the overlay structure.

the top corner of the measurement structure can be a problem, as shown in Fig. 5.13. Consequently, measurements made with low-$N.A.$ lenses are usually less sensitive to focus position and edge roughness than measurements made with higher $N.A.$ lenses. On some layers, however, the use of high-$N.A.$ lenses coupled with the depth sectioning capability of the CSOM is an asset. These OL structures usually consist of a grainy metal beneath a thick layer of patterned photoresist. Grain boundaries within the highly reflective metal layer can cause greater contrast changes than the weakly reflecting photoresist, making these samples difficult to measure in a standard optical microscope. A confocal scanning or interference optical microscope can focus on the photoresist layer while reducing contributions from the defocused metal layer.

Experimental Results When measuring the overlay error of a silicon wafer in production, measurements are typically carried out at various sites around the wafer. Either a five-site or nine-site pattern is common, although as wafer sizes get larger a greater number of sites are being used. To quantify the performance of a new overlay tool its measurements are usually compared to a standard such as another qualified optical tool or, for more accuracy, an electrical tester. Electrical overlay measurements, as described in Section 5.2.5, are essentially four linewidth measurements made on the four sides of an overlay test structure.

Overlay misregistration measurements made on both an RSOM and an electrical tester are shown in Fig. 5.14. To compare the performance of these two instruments a photomask was designed which included both optical and electrical test structures.[42] A set of wafers was printed using this mask and the overlay error measured on an RSOM with the results shown in Fig. 5.14(a). After the optical measurements, the wafers were etched and then measured with an electrical tester with the results shown in Fig. 5.14(b). There is excellent agreement between the two techniques. Measurements were made on a variety of different overlay structures

5.2 Semiconductor Metrology

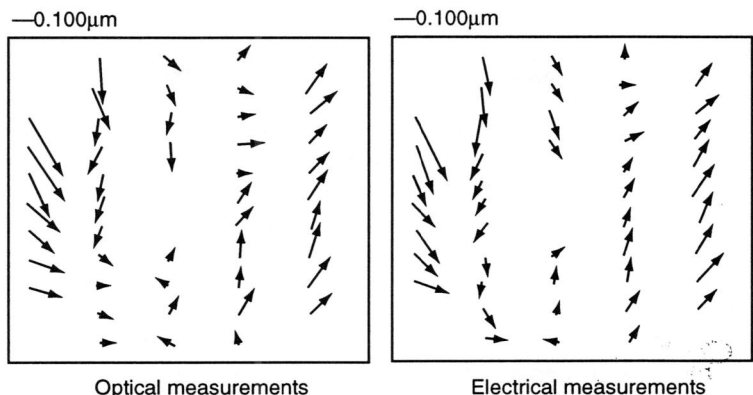

Figure 5.14 A comparison between measurements of overlay misregistration made using (a) an RSOM and (b) an electrical tester. [Reprinted from T. R. Corle, "Submicron metrology in the semiconductor industry," Solid State Electronics **35**, 391–402. Copyright (1992), with kind permission from Elsevier Science Ltd., The Boulevard, Langford Lane, Kidlington OX5 1GB, UK.]

including box-in-box, bar-in-bar, and L-bars. There was no evidence reported in this study that one type of optical overlay structure had a stronger correlation to the electrical measurements than another.[42]

Although these examples are taken from samples measured on an RSOM, the same basic principles apply when measuring overlay with an interference optical microscope. The KLA 5000 series of interference optical microscopes is a widely used overlay measurement tool for production semiconductor manufacturing. In this role its main asset is its shallow depth of focus which allows the exact focal plane for the measurement to be repeatably and accurately determined.

Sources of Error One difficulty encountered when making overlay measurements with an optical microscope is that the results obtained can change when the wafer is rotated by 180°. This phenomenon, known as *tool-induced shift* (TIS), is usually caused by asymmetries or misalignment in the optical system and can be reduced to a few nanometers error by careful alignment.[43] Local variations in photoresist thickness can lead to asymmetric overlay structures, which can cause the 0° and 180° measurements to differ. This effect is called *wafer-induced shift*.[44] With careful design, such errors can be minimized and excellent results obtained.

5.3 Film Thickness Measurements

The three most commonly used thickness measurement techniques for transparent optical films are CARIS, VAMFO, and ellipsometry. The first two methods are best suited to film thicknesses in the range from 200 Å to several micrometers, while ellipsometry is particularly well suited to the measurement of extremely thin films (<200 Å) and for refractive index measurements. The CARIS and VAMFO techniques, which are widely used in semiconductor manufacturing, depend on the interference of waves reflected from the top and bottom surfaces of a film and work on a principle similar to that of the interference optical microscope. The section concludes with a discussion of the techniques used to measure the thickness of an opaque film with the interference microscope, using the phase of the reflected signal as a measurement of step height.

5.3.1 CARIS and VAMFO

In modern semiconductor metrology, as the distances between components on an integrated circuit decrease, the area available for thin-film measurements also decreases. These changes force the metrology tools to measure samples using smaller and smaller measurement spots, which requires a focused optical system. Thin-film measuring microscopes based on the CARIS and VAMFO techniques have thus evolved into interference microscopes in which the interference occurs between the top and bottom of the film surfaces rather than between a separate reference and sample surface.

In the interference microscope the sample is moved up and down and an interference fringe pattern is obtained as a function of sample position from which distance changes may be determined with great accuracy. In film thickness measurements, the distance between the top and bottom surfaces of a transparent thin film cannot be changed. Instead, in the CARIS system, the phase is changed between the two reflections by varying the measurement wavelength and using a spectrometer as a detector as shown in Fig. 5.15(a).[45] The VAMFO system accomplishes this task by illuminating the sample with monochromatic light and varying the ray angle of the beam path through the film as shown in Fig. 5.15(b).[46,47]

The mathematical relationships describing the formation of interference fringes for ideal plane wave illumination have been described elsewhere.[48] Here we will be concerned with the form of the fringes when these techniques are used with a focused beam. The fringe amplitude as a function of film thickness can be calculated using a theory similar to that employed in Chapter 3 for the interferometric microscope. In this

5.3 Film Thickness Measurements

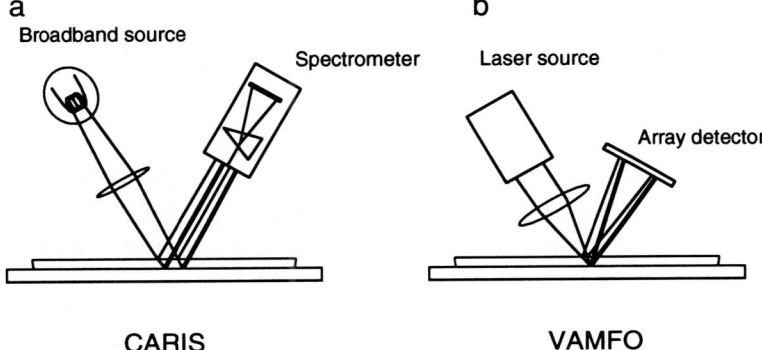

Figure 5.15 Schematic of (a) a film thickness measurement system based on the CARIS technique and (b) a system based on the VAMFO technique.

case, as we are concerned with focused beams, we decompose the incident wave into plane waves incident on the sample at an angle θ and determine the interference between rays reflected from the top and bottom surfaces of the film at the same angle. We then integrate the result over a range of angles corresponding to the acceptance angle of the detector.

We denote the amplitude of the light incident on the sample as $u(x,y)$ and its Fourier spectrum as $U(k_x, k_y)$, respectively, where $k = 2\pi/\lambda$ and k_x and k_y are the x and y components of the wave vector, or transverse spatial frequencies. The reflected amplitude of the plane wave component, with transverse spatial frequencies k_x and k_y, from the top surface of the film is written as

$$U_T(k_x,k_y) = R_T U(-k_x,-k_y), \tag{5.3}$$

where we assume, for simplicity, that R_T is a constant independent of the incident angle. The reflected amplitude from the bottom of the film is similarly given by

$$U_B(k_x,k_y) = R_B U(-k_x,-k_y) e^{j2k_z t}. \tag{5.4}$$

In Eq. (5.4), k_z is the z axis component of the wave vector,

$$k_z = \sqrt{k^2 n_f^2 - k_x^2 - k_y^2}, \tag{5.5}$$

R_B is the reflection coefficient from the film/substrate interface, t is the thickness, and n_f is the refractive index of the film.

These two signals propagate to the detector, where they interfere. The detected intensity is proportional to the square of the sum of the reflected amplitudes, $|U_T + U_B|^2$. The total signal at the detector is given by integrating over all values of k_x and k_y for the beam that passes through the lens pupil to an individual detector element,

$$I_{\text{det}} = \int_{\text{det}} |U_T + U_B|^2 \, dk_x dk_y. \tag{5.6}$$

Equation (5.6) represents the interference signal for the light reflected from the top and bottom surfaces of the film only. The mathematical model does not account for multiple reflection within the film because the contributions of these reflections are small. At low angles of incidence, the reflectivity of silicon is approximately 40%. The reflectivity of the top surface of an SiO_2 film which may cover the silicon surface is approximately 4%. Thus the intensity of the light that has undergone multiple reflections in the film is only about 1.5% of the intensity of the light reflected directly from the substrate.

Since the system is radially symmetric it is easier to work in spherical coordinates, where the radial and axial wave propagation constants in air are given by

$$k_z = \sqrt{k^2 - k_x^2 - k_y^2} = k \cos \theta \tag{5.7}$$

and

$$k_r = \sqrt{k_x^2 + k_y^2} = k \sin \theta. \tag{5.8}$$

In Eqs. (5.7) and (5.8) θ is the angle the incident ray makes with the optical axis. Changing coordinate systems, the integration becomes

$$\begin{aligned} I_{\text{det}} &= 2\pi \int_{\text{det}} |U_T + U_B|^2 k_r \, dk_r \\ &= 2\pi k^2 |U|^2 \int_{\theta_1}^{\theta_1 + \Delta} [R_T^2 + R_B^2 + 2R_T R_B \cos(2n_f kt \cos \theta_f)] \sin \theta \cos \theta \, d\theta, \end{aligned} \tag{5.9}$$

where θ_f is the incident angle the ray makes with the optical axis in the film.

It is assumed in Eq. (5.9) that the illumination is uniform with angle. It follows that the $|U|^2$ term may be removed from the integral. The third term in the integrand, $2R_T R_B \cos(2n_f kt \cos \theta_f)$, is the interference term. The sum and difference of the reflectivites R_T and R_B define the maxima and minima of the reflectivity curves. Since we are only interested in the interference term, the terms R_T^2 and R_B^2 will be dropped from the subsequent analysis.

For a given wavelength, the third term in the integrand of Eq. (5.9) can be directly integrated by using the paraxial approximation, $\cos \theta = 1$, and Snell's law, ($\sin \theta = n_f \sin \theta_f$) to change the integration variable

5.3 Film Thickness Measurements

from the angle of the focused beam in air, to the angle of the focused beam in the film. We then obtain the result

$$I_{\text{det}} = 4\pi k^2 n_f R_T R_B |U|^2 \int_{\theta_{1f}}^{\theta_{1f}+\Delta_f} \cos(2n_f kt \cos \theta_f) \sin \theta_f \, d\theta_f$$

$$= K \cos[n_f kt(\cos \theta_{1f} + \cos(\theta_{1f} + \Delta_f))] \frac{\sin[n_f kt(\cos \theta_{1f} - \cos(\theta_{1f} + \Delta_f))]}{n_f kt}. \quad (5.10)$$

In Eq. (5.10), the constants in front of the integral have been replaced by a single constant K. The interference term has a sinusoidal variation described by the cosine term in Eq. (5.10), modified by an envelope that falls off with increasing thickness t. There is, as would be expected, a strong similarity between Eq. (5.10) and the expression for the depth response of the correlation microscope obtained in Chapter 3.

In the VAMFO system, different detector elements are used to detect rays at different angles in the focused beam, i.e., the different components of the angular spectrum. In practice, a finite number of detectors are used, each with an acceptance angle Δ_f. The angle Δ_f is typically small so that Eq. (5.10) reduces to

$$I_{\text{VAMFO}} = K \Delta_f \cos[2n_f kt \cos \theta_{1f}] \sin \theta_{1f}. \quad (5.11)$$

For $\Delta_f \to 0$, the signal received is proportional to the acceptance angle Δ_f. It will be observed that a set of interference fringes is obtained from the microscope which for a given film thickness varies with the angle of incidence. By using a set of detectors placed at different angles, the angular spectrum can be determined and the thickness of the film measured.

In the CARIS system, the full angular range of the focused beam is accepted by each detector, so that the output is integrated over the full range of illumination angles. In this case $\theta_{1f} = 0$ and $\Delta_f = \theta_{0f}$, where $\sin \theta_{0f} = N.A./n_f$ the maximum angle subtended by the focused beam in the film. A spectrum analyzer consisting of a prism or grating, is used to produce separate beams of different wavelengths which impinge on different detectors. With these substitutions Eq. (5.10) becomes

$$I_{\text{CARIS}} = A \cos[n_f kt(1 + \cos \theta_{0f})] \frac{\sin[n_f kt(1 - \cos \theta_{0f})]}{n_f kt(1 - \cos \theta_{0f})}, \quad (5.12)$$

where we have normalized by dividing by $(1 - \cos \theta_{0f})$, and A is a constant.

Equation (5.12) gives the signal intensity as a function of wavelength ($k = 2\pi/\lambda$); the detected signal varies sinusoidally with both the film thickness and the wavelength. With the use of a spectrum analyzer and multiple detectors, the fringe pattern can be determined and, hence, the film thickness measured.

In the parallel beam limit both the CARIS and VAMFO techniques have the same response. The sensitivity of the VAMFO technique increases with increasing angle, provided the spread in angles at each detector is kept small. If a large range of focusing angles is used on each detector element, the fringe amplitude will be reduced. The sensitivity of the CARIS technique, on the other hand, decreases when high numerical aperture lenses are used. To improve the resolution and sensitivity, the wavelength must be reduced into the ultraviolet region of the spectrum. In a focused beam system, if the film thickness is greater than the depth of focus of the lens, coherent interference will not occur and the fringe amplitude will also be reduced.

Another way of thinking about the same phenomenon is that each angle in the focused beam will produce an interference pattern. The maxima and minima of the various patterns will, however, be slightly offset due to the variation in angle. When the offset patterns are averaged, the net result is an interference pattern with reduced fringe amplitude. The amplitude or envelope of the interference fringes in Eq. (5.12) is given by

$$I = A \frac{\sin[n_f k t (1 - \cos \theta_{0f})]}{n_f k t (1 - \cos \theta_{0f})}. \tag{5.13}$$

Equation (5.13) is plotted in Fig. 5.16 for a lens with numerical aperture N.A. = 0.55 and a film with refractive index $n_f = 1.5$.

The first zero of fringe amplitude in Eq. (5.13) occurs at a film thickness given by $n_f k t (1 - \cos \theta_{0f}) = \pi$, or solving for t

$$t = \frac{0.5\lambda}{n_f(1 - \cos \theta_{0f})}. \tag{5.14}$$

Equation (5.14) gives the approximate film thickness at which no interference fringes are produced. This region should be avoided in film thickness measurement tools. The easiest way to avoid this problem is to select a wavelength region at which fringes are still produced and make measurements in this range. The interference fringes will also undergo a π phase shift at the minimum amplitude point. This phase shift causes the fringes to reverse order, which must be accounted for in the measurement algorithms. An example of this phenomenon is given in Fig. 5.17. The figure shows experimental data from a 2.56-μm-thick silicon dioxide on

5.3 Film Thickness Measurements

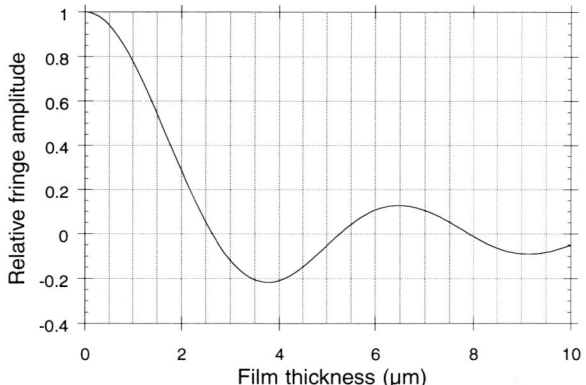

Figure 5.16 Graph of the interference fringe amplitude as a function of film thickness for a thin-film interference microscope.

silicon film made with a CARIS system. Figure 5.17(a) shows interference fringes produced by a low magnification $5\times/0.07$ $N.A.$ lens; Fig. 5.17(b) are interference fringes of the same sample produced using a $50\times/0.55$ $N.A.$ objective. The variation in fringe amplitude for different wavelengths and as a function of the numerical aperture of the focusing lens is clearly observable in this data. The higher aperture lens produces fringes with less contrast and which undergo a minimum in amplitude at approximately 550 nm.

Other Techniques In addition to the classical CARIS and VAMFO techniques, other methods of measuring film thickness are available on the CSOM and interference optical microscope. In the CSOM or interference microscope the height of an opaque film can be measured by observing the displacement of the peak of the depth response curve as the microscope scans over an edge.

If the film is transparent and sufficiently thick, the thickness can be measured directly from the $V(z)$ curves. In this case the depth response signal passes through two maxima, corresponding to the reflection of the light from the top and bottom surfaces of the film. The distance between the two intensity peaks can be measured and the film thickness calculated after correcting for the refraction of the light in the film as discussed in section 5.2.3.[49,50,51]

If the sample, however, is only one or two wavelengths thick, interference between the top and bottom surfaces can lead to dramatic changes

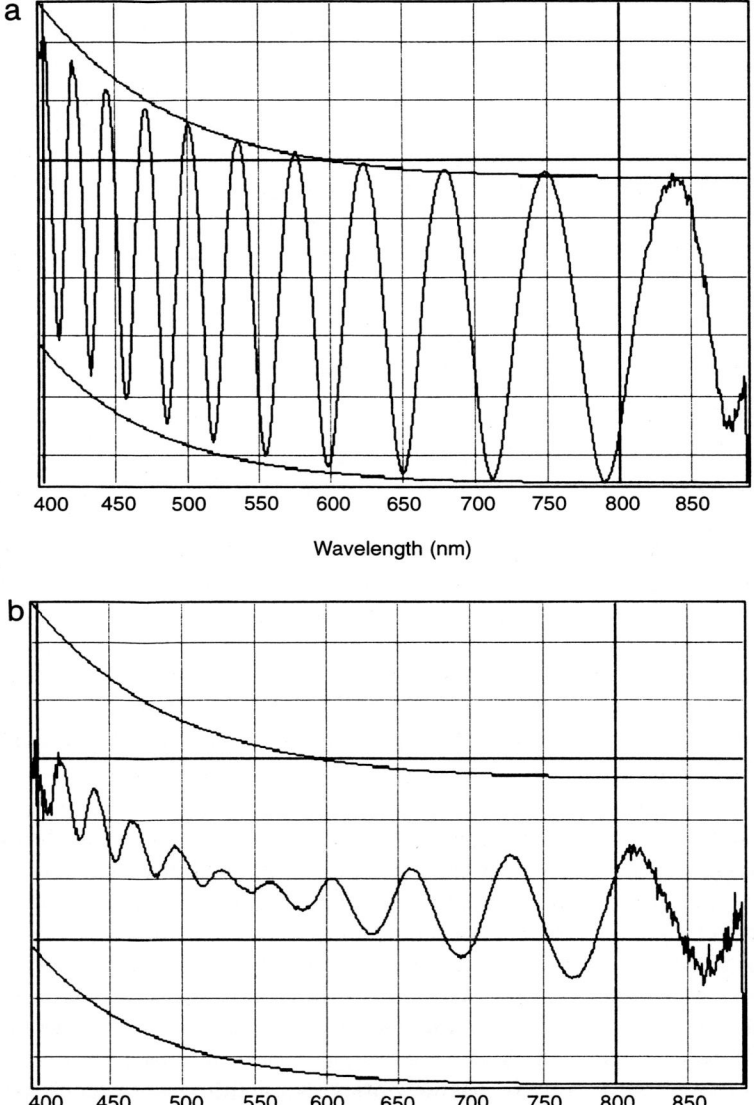

Figure 5.17 Experimental results showing the interference fringes from a 2.56-μm-thick silicon dioxide on silicon film made with a CARIS system using an (a) $N.A. = 0.07$ lens and (b) $N.A. = 0.55$ lens.

in the shape of the $V(z)$ curve. The shape of the curve will depend on the reflection coefficients of each surface of the film and the phase change through the film as discussed in Chapter 3. Broadband illumination can help to decrease the interference, an important consideration when a confocal microscope is being used for accurate metrology. For this reason, microscopes employing broadband illumination are better suited for metrology on a wide range of unknown samples.

In the next section we will discuss how the phase information can be used to make film thickness measurements with an interference microscope.

5.3.2 Film Thickness Measurements with the Mirau Interference Microscope

The interference microscope is better suited to the task of height measurement over a step in an opaque material like metal than a CSOM because a direct measure of phase is available when the data are processed. The method of obtaining the phase data in a Mirau correlation microscope (MCM) was described in Chapter 4. For films less than $\lambda/2$ thickness, the height is given by

$$z_s = \frac{\phi(x,y) + \phi_s}{k(1 + \cos\theta_0)}. \tag{5.15}$$

The term ϕ_s in Eq. (5.15) is a phase change that depends on the material properties of the sample. In practice, to accurately measure heights on the Mirau interference microscope using the Fourier transform technique, the material must be identical on both sides of the phase step so that ϕ_s = constant. In this case z_s will be directly related to $\phi(x,y)$ through Eq. (5.15). If both the height and material properties of the sample change, the height difference can be determined from the displacement of the envelope of the correlation function, or alternately ϕ_s can be calibrated on a sample with a known step height made of the same materials.

Chim and Kino have employed this technique to measure the step heights in a phase-shift mask used for optical lithography.[52] Phase-shift masks were developed to enable optical microlithography equipment to print smaller dimensions without reducing the exposure wavelength.[53,54,55] The mask measured by Chim and Kino had a design height difference between the inner and outer regions of Δh = 365 nm. Figure 5.18 shows a three-dimensional reconstruction of the mask obtained in the MCM. The height difference between the central and outer regions of the mask was measured to be 369 nm, very close to its design value.

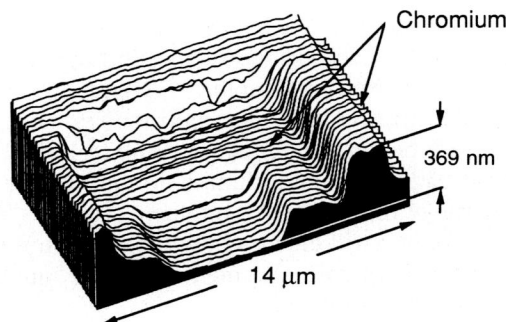

Figure 5.18 Three-dimensional reconstruction of the phase shift mask measured by Kino and Chim. [From S. S. C. Chim and G. S. Kino, "Measurement of a phase-shifting mask with the Mirau correlation microscope," *Integrated Circuit Metrology, Inspection and Process Control VI*, M. T. Postek, editor, SPIE **1673**, 266–271 (1992), with permission.]

5.4 Biological Imaging

One of the fastest growing fields for confocal scanning microscopes is biological imaging. In this area confocal microscopy has been used to study the brain,[56] nerve cells,[57,58,59,60,61] gene mapping,[62] the oxidative metabolism of cells in the cornea,[63] and in a host of other applications. It is extensively used as an experimental technique because it produces exceptionally clear images of specimens in which the depth discrimination property of the CSOM allows optical sectioning of living tissue.[64] In fluorescence imaging the sectioning ability eliminates glare from out-of-focus planes that can obscure the fine details in the image.

Imaging biological specimens using the confocal microscope is an entire subject unto itself. Several excellent books have been written covering observational techniques for different types of specimens and the use of mounting materials, fluorescent dyes, and imaging methodologies.[65,66,67] We will not attempt to duplicate this wealth of material here. In this section we will concentrate on some of the instrumental considerations for biological imaging. We will briefly discuss lenses and system aberrations in the context of brightfield and fluorescent imaging.

5.4.1 Brightfield and Phase Imaging

Brightfield imaging with a CSOM uses the full numerical aperture of the lens to illuminate the sample. The reflected light is collected by the same objective, again using the full aperture, and is focused onto a pinhole

5.4 Biological Imaging

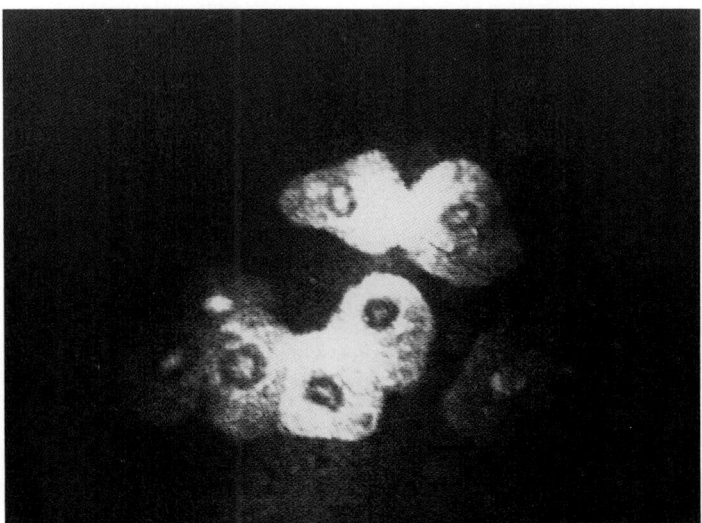

Figure 5.19 RSOM image of the superficial epithelial cells of a rabbit cornea. [From G. Q. Xiao, G. S. Kino, and B. R. Masters, Scanning **12**, 161–166 (1990). Copyrighted and reprinted with the permission of SCANNING, and/or the Foundation for Advances of Medicine and Science (FAMS), Box 832, Mahwah, New Jersey 07430, USA.

detector without passing through a phase plate or other optical components that modify the signal. As with semiconductors, the primary advantage of the brightfield imaging with the CSOM is the shallow depth response. The shallow depth response enables the microscope to optically section living tissue, and to observe weakly reflecting samples without having the image obscured by reflections from out-of-focus planes.

Ophthalmology One area in which this advantage is useful is in ophthalmology, or imaging of the eye.[68,69,70] Imaging of the cornea of the eye represents a unique challenge to optical microscopy. The lens of the eye is, by its nature, transparent. Images of the different cell layers within the cornea and lens are thus easily obscured by the light reflected from the first surface of the eye lens, where the index change between air and the eye is greatest. The ability of the RSOM to image the interior of the eye is illustrated in the following figures. Figure 5.19 is an RSOM image of the superficial epithelial cells of a rabbit cornea. These cells are located on the front surface of the eye.[71,72] Figure 5.20 is an RSOM image of the ocular lens fibers of a rabbit eye.[73] The fibers in this image are located about 200 μm below the surface of the lens. The separation between the

Figure 5.20 RSOM image of the ocular lens fibers of a rabbit eye. [From B. R. Masters "Confocal microscopy of ocular tissue," *Confocal Microscopy*, T. Wilson, editor (Academic Press, 1990), with permission.]

fibers in the image is about 2 μm. Submicron transverse bands within the cells are also visible in this image. In a standard optical microscope, it would not be possible to obtain this set of images because the reflection from the front surface of the eye would overwhelm the images of the cells.

In Vivo Imaging The images of the rabbit eye were made on an eye that had been excised, or removed from the rabbit. For diagnostics on human subjects, however, we obviously cannot remove the eye from the patient. Having the object under observation attached to a living organism leads to several unique imaging challenges.[74] The main challenge for *in vivo* imaging in larger organisms is the movement of the cells or tissue being imaged due to heartbeat, respiration, or other involuntary motions.[75,76,77,78] In addition, the high-energy UV photons used in many fluorescence imaging applications can potentially damage living tissue due to phototoxicity. Both of these limitations suggest that a high-speed scanning mechanism is needed for *in vivo* imaging. Confocal scanning laser microscopes have been used with limited success for *in vivo* imaging, but the bulk of the published work has used a Nipkow disk scanning microscope.

The motion of the specimen relative to the microscope can be reduced by using a surface contact or planarizing objective lens. These objectives have a planar front surface that is placed in contact with the specimen and thus limits movement of the sample out of the focal plane. Another type of objective used to image living specimens is the dipping cone

5.4 Biological Imaging

objective.[79] These lenses have a cone-shaped tip with a concave or flat front surface that is concentric with the axial focal point of the lens. The focus of the lens is changed by moving the lens elements within the objective so that the sample beneath the tip need not be disturbed.[77]

Other issues which arise are those due to spherical aberration and the orientation of the sample. When imaging features more than a few microns below the surface, an adjustable objective lens is needed to correct for spherical aberration as the microscope focuses into the specimen.[80] Lastly the objective should be accurately oriented perpendicular to the tissue planes under observation. Alignment of the objective in any orientation other than directly perpendicular to the tissue surface results in an oblique cross section of the sample which is sometimes difficult to interpret.

Phase Imaging Additional information on the structure and processes in biological samples can be obtained if the phase as well as the amplitude of the light is detected. In an early application of this principle, Brakenhoff used a phase-sensitive microscope to study a mouse germ cell in the pachythene stage of reduction.[81] Using phase imaging, thin strands of chromosomal material of an imbedded germ cell were successfully visualized. The main purpose of the research, however, was to investigate the performance of the microscope itself. They observed that optical path differences of the order of one hundredth of a wavelength were detectable, although instabilities in the two-path interferometer ultimately limited the phase measurement capability of the instrument. Phase imaging has also been used to study the attachment sites of cells to a substrate as discussed in Chapter 3.

5.4.2 Fluorescence Imaging

Fluorescence imaging is probably the most widely used application of the CSOM.[82,83,84] Fluorescence imaging is of great importance because fluorescent particles can be used as markers to follow biological activity. In this application, the CSOM's ability to remove glare from out-of-focus but still fluorescing planes in the sample dramatically improves the detail of the images. An example of the improvement is shown in Fig. 5.21. The figure shows two images of a human rib bone stained with brilliant sulphaflavine. Figure 5.21(a) is a standard optical microscope image, while Fig. 5.21(b) is an RSOM image of the same area. Both images were made using a 40×/1.3 *N.A.* oil immersion lens. The difference between the two images is dramatic. Fine details, which cannot be seen in the standard microscope image, are clearly visible in the RSOM image.[85]

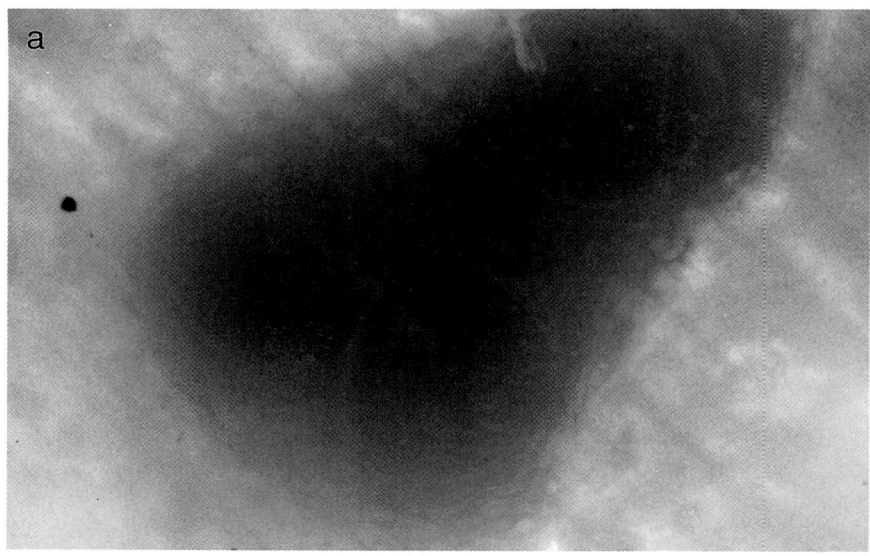

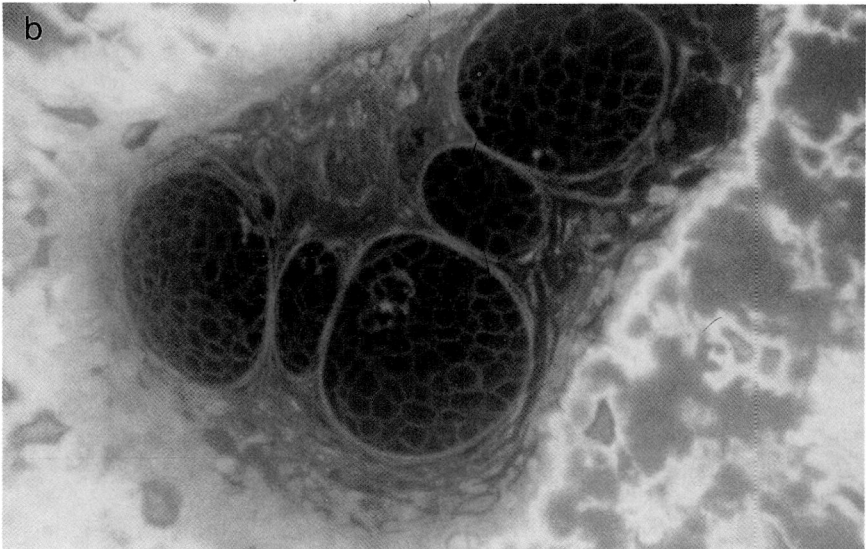

Figure 5.21 Two images of a human rib bone stained with brilliant sulphaflavine: (a) a standard optical microscope image, (b) an RSOM image. [Courtesy: Dr. G. Q. Xiao.]

5.4 Biological Imaging

Producing bright fluorescence images with this level of visible detail in a CSOM is limited by the bleaching of the fluorophores and by the ability of the microscope to collect the light that has been emitted by the molecules. Diffraction-limited excitation by more than a few milliwatts of radiation will bleach most fluorophores. When using strongly fluorescent dyes, better images are obtained by integrating multiple scans at low excitation power than with one scan at full power.[86]

For weakly fluorescent samples, a laser-based CSLM is normally needed. In this case, the light efficiency and power available from the mercury vapor or xenon light source in a Nipkow disk system are normally too low to yield a good image. In addition, the wavelengths available from an argon laser are well suited to fluorescent imaging, and it is easier to compensate for chromatic aberration in a system that uses only two wavelengths. The laser system is also able to concentrate the collected image light onto one highly sensitive detector, such as a photomultiplier tube, which improves the detection efficiency of the instrument.[87] In summary, the Nipkow disk scanning microscope and CSLM are complementary tools for fluorescence imaging. The Nipkow disk scanning microscope is suitable only for use with strongly fluorescing samples, while the CSLM can be used with either weakly or strongly fluorescing samples, although bleaching can be a problem with its strong light source. In practice, the implication is that the CSLM is normally the microscope of choice in this application.

Factors that limit the performance of a CSOM in fluorescence imaging applications include light efficiency, lens aberrations, and scattering by the sample. Many microscopes incorporate design features to counteract these limitations. We shall discuss a few of these here.

Light Efficiency The best conventional fluorescent microscopes are able to collect only approximately 5% of the fluorescence emitted from each resolution volume in the sample.[88] Factors that affect the efficiency include the fact that the solid angle subtended by the objective lens collects only a small proportion of the emitted light, the reflection losses within the lens itself coupled with those due to beamsplitters and other filters in the optical train, and the efficiency of the detector. In laser scanning confocal microscopes, the situation is even worse than with a standard microscope because the light must be focused through a pinhole. In these instruments, overall collection efficiencies may drop to less than 0.2%.[89] Attention must therefore be paid to maximizing the light collection ability of each component in the instrument. The highest possible numerical aperture lenses should be used to increase the amount of light collected. In addition, a large separation of excitation and fluorescent wavelengths will enable the dichroic mirror to efficiently reflect the image to the detector. This need for a large separation must be balanced by the ability to correct axial chromatic aberration of the objective lens.

Lens Aberrations Lens aberrations will also contribute to the degradation of the fluorescent image. Chromatic aberration has the strongest affect on the image. If the illumination wavelength and the fluorescent wavelength do not share a common focal point, the detector pinhole will exclude the fluorescent light emitted from the most intensely illuminated portion of the sample. If the microscope uses separate illumination and detector pinholes, this problem can be mitigated by repositioning the detector pinhole to correct for the chromatic aberration of the lenses. For beam scanning or disk scanning microscopes, the objective lenses themselves must be color corrected at both the excitation and fluorescent wavelengths.[89]

Spherical aberration and scattering of the illumination beam by thick samples will also affect the imaging performance by preventing the light from being focused to a small spot at the pinhole. The spherical aberration component should be measured for each lens by using the depth response of the microscope as outlined in Chapter 3.

The scattering component is related to both the number of scattering centers in the sample and the point spread function of the objective lens. A direct measurement of the point spread function using small fluorescent particles is difficult but has been accomplished by Shaw and Rawlins.[90] Brakenhoff and his colleagues used a different approach. They scanned a fluorescent grid under the microscope and calculated the PSF from the received image.[91] Such measurements are hampered by the poor signal-to-noise ratio of the image due to the small number of fluorescent molecules in the particle or line being imaged and bleaching effects in the small volume occupied by the fluorophores.

5.4.3 Two-Wavelength and Two-Photon Fluorescence Imaging

The fluorescent imaging capabilities of optical microscopes have been extended by the development of two-wavelength and two-photon fluorescent imaging. Two-wavelength imaging enables researchers to track metabolic activity in different sections of a cell that cannot be stained with the same dye. It is also used to monitor two different fluorescent markers in a cell to observe the relationship between different cell processes. Two-photon fluorescence imaging uses two red photons to excite a fluorophore that is normally excited by one green photon. It virtually eliminates the problem of bleaching and provides intrinsic three-dimensional resolution

5.4 Biological Imaging 315

because of the intensity-squared dependence of the two-photon absorption.

Two-Wavelength Microscopy The design of a two-wavelength microscope is considerably more complicated than the design for a single-wavelength system. First, two or more excitation wavelengths are needed because different fluorescent dyes may not respond to a single pump wavelength. The simplest method is to use an argon laser which is simultaneously lasing at several wavelengths. With this approach, however, it is difficult to adjust the relative power of the lasers at the different wavelengths. Since different dyes bleach at different rates, obtaining sufficient excitation power for a good image at one wavelength may bleach the second dye on the sample. Many of the commercial microscopes overcome this limitation by using multiple laser sources. Most two-wavelength fluorescent microscopes also use multiple dichroic filters and two or more detectors. Since it is difficult to construct one dichroic filter that can efficiently separate four different wavelengths (two excitation and two fluorescent), multiple detection paths are needed to eliminate the crosstalk or bleeding of the two fluorescent signals. These multiple detector paths, however, can have a significant impact on the detection efficiency of the microscope.

Two-Photon Imaging Bleaching of fluorescent dyes after a relatively short time is a major problem with fluorescence imaging which is not solved with the use of a confocal microscope. When a tissue sample is being examined, the beam is focused to a small focal spot, but on its way to this spot it passes through other regions of tissue, which it bleaches. As much light is absorbed by these intervening regions as at the focal plane.

In two-photon microscopy a fluorescent molecule such as Indo-1 which is normally excited by one UV or green photon is excited instead by two red photons of double the wavelength. The two red photons from a high peak power pulsed laser will produce the same energy as one UV photon and thus excite fluorescence.

This method of excitation has several advantages.[92] First, two-photon microscopy provides intrinsic three-dimensional resolution without the use of pinholes. The reason is linked to the two-photon absorption process. Since the two photons must be absorbed together, there is an intensity-squared dependence of the fluorescent emission on the excitation power, so that the fluorescence originates from a tightly confined region around the focal point. The confinement of the excitation to an area around the focal point minimizes the damage caused by photobleaching and photo-damage associated with short-wavelength illumination. Furthermore, since the definition will depend on the square of the intensity of the illuminating beam, *no pinhole is required in front of the detector* to obtain definition comparable to that of a confocal microscope.

Other advantages of this approach include the fact that conventional visible light optics can be used in both the illumination and collection systems, even if the fluorescence is in the UV range. Typical excitation wavelengths are of the order of 700 nm, and the fluorescent wavelength is typically less than 400 nm. In addition, the absorption and scattering of light by most living cells are significantly less in the red than in the UV. Thus most of the power in the focused beam reaches the focal point. The relative transparency of biological specimens at the longer wavelengths also permits deeper sectioning than is possible with UV light.

To concentrate sufficient light at the focal point for two-photon imaging, very high power densities of the order of TW/cm^2 are needed. These power densities can be generated by combining diffraction-limited focusing of a single laser beam with the enhanced power levels of a short pulsed laser. The pulsed Ti:sapphire laser is an ideal source for this purpose. It is tunable over the range 700–1000 nm and able to produce pulses with peak excitation intensities 10^6 times greater than those in a typical confocal microscope. To prevent sample damage, a low duty cycle ($\sim 10^{-5}$) is used to maintain the average input power at less than 10 mW, only slightly greater than used in conventional confocal microscopy. This light can be coupled into the scanners of a standard CSLM to produce the images.

Two-photon imaging has been used to study the dynamics of intracellular free calcium activity in rat cells.[92] In this application the use of the dye Indo-1 reduced the uncertainty of the calcium measurements caused by variations in the indicator concentration within the cells. Webb and his colleagues at Cornell University have also used two-photon microscopy to image vital DNA stains in developing cells and embryos such as sea urchins and to apply optically induced micropharmacology using charged bioeffector molecules.[93]

5.5 Conclusion

Confocal scanning microscopy has developed into an essential tool for solving many imaging problems in microscopy, particularly semiconductor and biological imaging. In semiconductor metrology, CSOMs hold the promise of improving the speed and precision of optical measurements. CSOMs are the only tools available capable of making both the critical dimension and overlay measurements at the speeds that today's advanced manufacturing processes require. CSOMs are also integral to most biological imaging laboratories. There they make it possible to optically section materials for detailed observations of cells and subsurface processes *in vivo* and help to produce exceptionally clear fluorescent images.

As with other imaging systems, further applications for these instruments will be found as they become available to a wider range of research scientists. New applications will be limited only by the imagination of researchers yet to enter the field.

References

1. A. Boyde, "Bibliography on confocal microscopy and its applications," Scanning **16**, 33–56 (1994).
2. H. Becker, "High resolution metrology for submicron masks and wafers," Microelectronic Manufacturing and Testing 11–13 (April 1988).
3. J. T. Lindow, S.D. Bennett, and I. R. Smith, "Scanned laser imaging for integrated circuit metrology," *Micron and Submicron Integrated Circuit Metrology*, Kevin M. Monahan, editor, SPIE **565**, 81–87 (1985).
4. T. Wilson, J. N. Gannaway, and C. J. R. Sheppard, "Scanning optical microscopy of semiconductor devices," *Scanned Image Microscopy,* E.A. Ash, editor, 227–232 (Academic Press, 1980).
5. N. Smith, "Ready of the half-micron process," European Semiconductor, 10–11 (November 1990).
6. D. A. Toy, "Confocal Microscopy: the ups and downs of 3-D profiling," Semiconductor International, 120–123 (April 1990).
7. T. Wilson and C. J. R. Sheppard, *Theory and Practice of Scanning Optical Microscopy* (Academic Press 1984).
8. M. T. Postek and D. C. Joy, "Microelectronics dimensional metrology in the scanning electron microscope," J. of Res. of the Nat. Bureau of Stand., **92**(3), 205–228 (1987).
9. C. W. T. Knight, "The future of manufacturing with optical microlithography," *Optics and Photonics News*, **1**(10), 11–13 (1990).
10. 1986 Annual Book of ASTM Standards, Designation E 456-83a, American Society for Testing and Materials, Philadelphia, USA (1986).
11. C. P. Kirk and D. Nyyssonen, "Modeling the optical microscope images of thick layers for the purpose of linewidth measurement," *Optical Microlithography IV,* Harry L. Stover, editor, SPIE **538**, 179–187 (1985).
12. C. P. Kirk, "A study of the instrumental errors in linewidth and registration measurements made with an optical microscope," *Integrated Circuit Metrology, Inspection and Process Control,* Kevin M. Monahan, editor, SPIE **775**, 51–59 (1987).
13. D. Nyyssonen and B. Larrabee, "Submicrometer linewidth metrology in the optical microscope," *J. of Res. of the Nat. Bureau of Stand.*, **92**(3), 187–204 (1987).
14. O. Hignette, J. Woch, and L. Gotti, "Large bandwidth deep-UV microscopy

for CD metrology," *Integrated Circuit Metrology, Inspection and Process Control IV,* William H. Arnold, editor, SPIE **1261**, 79–90 (1990).
15. N. Smith and R. Gale, "Advances in optical metrology for the 1990s," *Integrated Circuit Metrology, Inspection and Process Control IV,* William H. Arnold, editor, SPIE **1261**, 104–113 (1990).
16. D. Nyyssonen and C. P. Kirk, "Optical microscope imaging of lines patterned in thick layers with variable edge geometry: theory," JOSA A **5**, 1270–1280 (1988).
17. C. P. Kirk, D. S. Moore, and J. C. C. Nelson, "Analysis of linewidth measurement techniques for the purpose of automation," *Integrated Circuit Metrology II,* Kevin M. Monahan, editor, SPIE **480**, 33–39 (1984).
18. T. R. Corle, G. Q. Xiao, G. S. Kino, and N. S. Levine, "Characterization of a real-time confocal scanning optical microscope," *Integrated Circuit Metrology, Inspection and Process Control III,* Kevin M. Monahan, editor, SPIE **1087**, 91–145 (1989).
19. K. M. Monahan and J. P. Benchop, "Object contrast in the confocal microscope and applications to lithographic metrology," *Integrated Circuit Metrology, Inspection and Process Control IV,* William H. Arnold, editor, SPIE **1261**, 72–78 (1990).
20. B. R. Stallard and Y. Bukhman, "Use of a confocal scanning laser microscope for the measurement of submicrometer critical dimensions," *Integrated Circuit Metrology, Inspection and Process Control,* Kevin M. Monahan, editor, SPIE **775**, 60–67 (1987).
21. M. Dusa, G. Xiao, F. Menagh, E. Rauch, W. G. Gouin, and G. Mirth, "Application specific microscopy for half micron metrology," *Integrated Circuit Metrology, Inspection and Process Control VII,* Michael T. Postek, editor, SPIE **1926**, 72–83 (1993).
22. P. H. Singer, "CD measurement meets the submicron challenge," Semiconductor International, 64–68 (November 1989).
23. T. R. Corle, "Submicron metrology in the semiconductor industry," Solid State Electronics **35**, 391–402 (1992).
24. T. R. Corle, J. T. Fanton, and G. S. Kino, "Distance measurements by differential confocal optical ranging," Appl. Opt. **26**, 2416–2420 (1987).
25. T. R. Corle, "Studies in confocal scanning optical microscopy," Ph.D. Dissertation, Department of Applied Physics, Stanford University, Stanford, California, USA (June 1989).
26. I. R. Smith, "Dimensional metrology for partially etched photomasks in iterative-etch fabrication processes," Opt. Eng. **31**, 1723–1725 (1992).
27. M. C. Gupta and S. T. Peng, "Diffraction characteristics of surface-relief gratings," Appl. Opt. **32**, 2911–2917 (1993).
28. M. P. Davidson, K. M. Monahan, and R. J. Monteverde, "Linearity of coherence probe metrology: simulation and experiment," *Integrated Circuit Metrology, Inspection and Process Control V,* William H. Arnold, editor, SPIE **1464**, 155–176 (1991).
29. T. R. Corle, L. C. Mantalas, T. R. Kaack, and L. J. LaComb, Jr., "Polarization enhanced imaging of photoresist gratings in the real-time scanning optical microscope," Appl. Opt. **33**, 670–677 (1994).

30. M. G. Moharam and T. K. Gaylord, "Three-dimensional vector coupled-wave analysis of planar-grating diffraction," JOSA **73**, 1105–1112 (1983).
31. N. Garcia, "Exact calculations of P-polarized electromagnetic fields incident on grating surfaces: surface polariton resonances," Opt. Comm. **45**, 307–310 (1983).
32. J. Q. Lu, A. A. Maradudin, and T. Michel, "Enhanced backscattering from a rough dielectric film on a reflecting substrate," JOSA B **8**, 311–318 (1991).
33. D. T. Nguyen and M. L. Rustgi, "Diffraction of plane waves from a multilayered film with a grating surface," JOSA B **9**, 1850–1856 (1992).
34. G. Q. Xiao, T. R. Corle, and G. S. Kino, "Real-time confocal scanning optical microscope," Appl. Phys. Lett. **53**, 716–718 (22 August 1988).
35. N. S. Levine, T. R. Corle, R. T. Mumaw, C.-H. Chou, and G. S. Kino, "Multilevel CD/Overlay metrology using a real-time confocal scanning optical microscope," Microelectronic Engineering **11**, 669–674 (1990).
36. T. F. Hasan, D. S. Perloff, and C. L. Mallory, *Semiconductor Silicon/1981*, H. R. Huff, R. J. Kriegler, and T. Yakesihi, editors, The Electrochemical Society, **866**, Pennington, NJ (1981).
37. D. Yen, *Integrated Circuit Metrology*, D. Nyyssonen, editor, SPIE **342**, 73 (1982).
38. S. Kimura and T. Wilson, "Confocal scanning dark-field polarization microscopy," Appl. Opt. **33**, 1274–1278 (1994).
39. R. W. Wijnaendts-van-Resandt, "Semiconductor metrology," *Confocal Microscopy* T. Wilson, editor (Academic Press, 1990).
40. E. Betzig, J. K. Trautman, J. S. Weiner, T. D. Harris, and R. Wolfe, "Polarization contrast in near-field scanning optical microscopy," Appl. Opt. **31**, 4563–4568 (1992).
41. A. Starikov, D. J. Coleman, P. J. Larson, A. D. Lopata, and W. M. Muth, "Accuracy of overlay measurements and mark asymmetry effects," Opt. Eng. **31**, 1298–1309 (1992).
42. N. S. Levine, T. R. Corle, R. T. Mumaw, C.-H. Chou, and G. S. Kino, "Multilevel CD/overlay metrology using a real-time confocal scanning optical microscope," Microelectronic Engineering **11**, 669–674 (1990).
43. P. Troccolo, N. Smith, and T. Zantow, "Tool and mark design factors that influence optical overlay measurement errors," *Integrated Circuit Metrology, Inspection and Process Control VI*, Michael T. Postek, editor, SPIE **1673**, 148–156 (1992).
44. D. J. Coleman, P. J. Larson, A. D. Lopata, W. A. Muth, and A. Starikov, "On the accuracy of overlay measurements: tool and mark asymmetry effects," *Integrated Circuit Metrology, Inspection and Process Control IV*, William H. Arnold, editor, SPIE **1261**, 139–161 (1990).
45. F. Reizman and W. Van Gelder, "Optical thickness measurement of SiO_2-Si_3N_4 films on silicon," Solid State Elect. **10**, 625–632 (1967).
46. W. A. Pliskin and R. P. Esch, "Refractive index of SiO_2 films grown on silicon," J. Appl. Phys. **36**, 2011–2013 (1965).
47. A. Rosencwaig, J. Opsal, B. L. Willenborg, S. M. Kelso, and J. T. Fanton, "Beam profile techniques for dielectric measurements," Appl. Phys. Lett., **60**, 1301–1303 (1992).

48. W. A. Pliskin and S. J. Zanin, "Film thickness and composition," in *Handbook of Thin Film Technology*, L. I. Maissel and R. Glang, editors (McGraw-Hill, 1970).
49. W. Galabraith, "The optical measurement of depth," Quart. Journal of Micro. Science **96**, 285–288, (1955).
50. E. M. Glaser, "Snell's law: the bane of computer microscopists," Journal. of Neuroscience Methods **5**, 201–202 (1982).
51. L. Majlof and P.-O. Forsgren, "Confocal microscopy: important considerations for accurate imaging," Methods in Cell Biology **38**, 79–95 (1993).
52. S. S. C. Chim and G. S. Kino, "Measurement of a phase-shifting mask with the Mirau correlation microscope," *Integrated Circuit Metrology, Inspection and Process Control VI*, Michael T. Postek, editor, SPIE **1673**, 266–271 (1992).
53. M. D. Levenson, N. S. Viswanathan, and R. A. Simpson, "Improving resolution in photolithography with a phase-shifting mask," IEEE Trans. on Elect. Dev. **29**, 1828–1836 (1982).
54. T. Tereawa, N. Hasegawa, T. Kurosaki, and T. Tanaka, "0.3-micron optical lithography using a phase shifting mask," *Optical/Laser Microlithography II*, Burn J. Lin, editor SPIE **1088**, 25–33 (1989).
55. M. D. Levenson, "Wavefront engineering for photolithography," Physics Today 28–35 (July 1993).
56. U. Dirnagl, A. Villringer, and K. M. Einhaupl, "In-vivo confocal scanning laser microscopy of the cerebral microcirculation," Journal of Microscopy **165** Pt. 1, 147–157 (1992).
57. K. Carlsson, P. E. Danielsson, R. Lenz, A. Liljeborg, L. Majlot, and N. Åslund, "Three-dimensional microscopy using a confocal laser scanning microscope," Opt. Lett. **10**, 53–55 (1985).
58. K. Carlsson and N. Åslund, "Confocal imaging for 3-D digital microscopy," Appl. Opt. **26**, 3232–3238 (1987).
59. G. J. Brakenhoff, J. S. Binnerts, and C. L. Woldringh, "Developments in high resolution confocal scanning light microscopy (CSLM)," *Scanned Image Microscopy* E.A. Ash, editor, 183–200 (1980).
60. K. Carlsson, P. Wallen, and L. Brodin, "Three-dimensional imaging of neurons by confocal fluorescence microscopy," Journal of Microscopy **155** Pt. 1, 15–26 (1989).
61. J. N. Turner, J. W. Swann, D. H. Szarowski, K. L. Smith, D. O. Carpenter, and M. Fejtl, "Three-dimensional confocal light microscopy of neurons: fluorescent and reflection stains," Method in Cell Biology **38**, 345–367 (1993).
62. T. K. Tang, C.-J. C. Tang, T.-C. Tsou, and C.-W. Wer, "Applications of confocal fluorescent microscopy on modern cell biology and human gene mapping." Scanning **14** Suppl. II, II-27 (1992).
63. B. R. Masters, "Noninvasive redox fluorometry: how light can be used to monitor alterations of corneal mitochondrial function," Curr. Eye Res. **3**, 23–26 (1984).
64. G. M. Bernacca and R. Wilkinson, "Confocal microscopy of calcified heart valves," USA Microscopy and Analysis, **5** (January 1994).
65. *The Handbook of Biological Confocal Microscopy,* James Pawley, editor (IMR Press, 1989).

66. Confocal Microscopy, T. Wilson, editor (Academic Press, 1990).
67. S. J. Wright, V. E. Centonze, S. A. Stricker, P. J. DeVries, S. W. Paddock, and G. Schatten, "Introduction to confocal microscopy and three dimensional reconstruction," Methods in Cell Biology, **38**, 1–45 (1993).
68. Scanning–Special issue on ophthalmology, **16** (Sept.-Oct. 1994)
69. R. H. Webb, G. W. Huges, and F. C. Delori, "Confocal laser opthalmoscope," Appl. Opt. **26**, 1492–1499 (1987).
70. J. H. Massig, M. Preissler, A. R. Wegener, and G. Gaida, "Real-time confocal laser scan microscope for examination and diagnosis of the eye *in vivo*," Appl. Opt. **33**, 690–694 (1994).
71. G. Q. Xiao, G. S. Kino, and B. R. Masters, "Observations of the rabbit cornea and lens with a new real-time confocal scanning optical microscope," Scanning **12**, 161–166 (1990).
72. J. V. Jester, W. M. Petroll, R. M. R. Garana, M. A. Lemp, and H. D. Cavanagh, "Comparison of in vivo and ex vivo cellular structure in rabbit eyes detected by tandem scanning microscopy," Journal of Microscopy **165** Pt.1, 169–181 (1992).
73. B. R. Masters, "Confocal microscopy of ocular tissue," *Confocal Microscopy,* T. Wilson, editor (Academic Press, 1990).
74. J. I. Prydal and P. N. Dilly, "In vivo confocal microscopy of the cornea and tear film." Scanning **17**, 133–135 (1995).
75. K. C. New, W. M. Petroll, A. Boyde, L. Martin, P. Corcuff, J. L. Leveque, M. A. Lemp, H. D. Cavanagh, and J. V. Jester, "In vivo imaging of human teeth and skin using real-time confocal microscopy," Scanning **13**, 369–372 (1991).
76. W. M. Petroll, H. D. Cavanagh, M. A. Lemp, P. M. Andrews, and J. V. Jester, "Digital image acquisition in vivo confocal microscopy," Journal of Microscopy **165** Pt. 1, 61–69 (1992).
77. W. M. Petroll, J. V. Jester, and H. D. Cavanagh, "In vivo confocal imaging: general principles and applications," Scanning **16**, 131–149 (1994).
78. S. Myrdal and M. Foster, "Time-resolved confocal analysis of antibody penetration into living, solid tumor spheroids," Scanning **16**, 155–167 (1994).
79. C. J. Koester, "Scanning mirror microscope with optical sectioning characteristics: applications in ophthalmology," Appl. Opt. **19**, 1749–1757 (1980).
80. V. Sarafis, "Biological perspectives of confocal microscopy," in *Confocal Microscopy*, T. Wilson, editor (Academic Press, 1990).
81. G. J. Brakenhoff, "Imaging modes in confocal scanning light microscopy (CSLM)," Journal of Microscopy **117** Pt. 2, 233–242 (1979).
82. W. B Amos, J. G. White, and M. Fordham, "Use of confocal imaging in the study of biological structures," Appl. Opt. **26**, 3239–3243 (1987).
83. G. J. Brakenhoff, H. T. M. van der Voort, E. A. van Spronsen, and N. Nanninga, "Three-dimensional imaging in fluorescence by confocal scanning microscopy," Journal of Microscopy **153** Pt. 2, 151–159 (1989).
84. L. Loew, "Confocal microscopy of potentiometric fluorescent dyes," Methods in Cell Biology **38**, 195–209 (1993).
85. G. Q. Xiao, "Confocal optical imaging systems and their applications in microscopy and range sensing," Ph.D. Thesis, Stanford University, Stanford, California, USA (December 1989).

86. J. G. White, W. B. Amos, and M. Fordham, "An evaluation of confocal verses conventional imaging of biological structures by fluorescence light microscopy," J. Cell Biol. **105**, 41–41 (1987).
87. A. Boyde, S. J. Jones, M. L. Taylor, L. A. Wolfe, and T. F. Watson, "Fluorescence in the tandem scanning microscope," Journal of Microscopy **157** Pt. 1, 39–49 (1990).
88. D. R. Sandison and W. W. Webb, "Background rejection and signal-to-noise optimization in confocal and alternative fluorescence microscopes," Appl. Opt. **33**, 603–615 (1994).
89. K. S. Wells, D. R. Sandison, J. Strickler, and W. W. Webb, "Quantitative fluorescent imaging with laser scanning confocal microscopy," *The Handbook of Biological Confocal Microscopy,* J.B. Pawley, editor (IMR Press, 1989).
90. P. J. Shaw and D. J. Rawlins, "The point spread function of a confocal microscope: its measurement and use in deconvolution of 3-D data," Journal of Microscopy **163** Pt. 2, 151–165 (1991).
91. H. T. M. van der Voort, G. J. Brakenhoff, C. G. A. M. Janssen, J. A. C. Valkenburg, and N. Nanninga, "Confocal scanning laser fluorescent and reflection microscopy: measurements of 3-D image formation and applications in biology," *Scanning Imaging Technology,* Tony Wilson and Ludwig Balk, editors, SPIE **809**, 138–143 (1987).
92. D. W. Piston, M. S. Kirby, H. Cheng, W. J. Lederer, and W. W. Webb, "Two photon excitation fluorescence imaging of three dimensional calcium-ion activity," Appl. Opt. **33**, 662–677 (1994).
93. W. C. Holton, "Under a microscope: confocal microscopy casts new light on the dynamics of life," Photonics Spectra 78–84 (February 1995).

APPENDIX A

Vector Field Theory for Depth and Transverse Resolution of a CSOM

We derived a scalar theory in Chapter 3 for the depth and transverse response of a confocal microscope. In this Appendix, we will derive these results by using the full vector field theory of Richards and Wolf.[1] We shall also show that the depth response $V(z)$ derived from the vector field theory is identical to that obtained from scalar theory. However, there are slight differences between the results obtained by the two methods for the transverse response. The derivation is very similar to that used in the scalar theory. In the vector theory, however, the lens is assumed to be illuminated by a rectilinear quasi-monochromatic plane wave linearly polarized in the x direction. In addition, any aberrations introduced by the lenses or other optical components are neglected.

A.1 The Depth Response

For any point (r,z,φ), expressed in cylindrical coordinates, with the origin at the focal point ($z = 0$), Richards and Wolf have shown that the Cartesian components of the electric and magnetic fields, denoted by **e** and **h**, respectively, are given by the relations

$$e_x(r,z,\varphi) = -jA(I_0 + I_2 \cos 2\varphi) \qquad (A.1)$$

$$e_y(r,z,\varphi) = -jAI_2 \sin 2\varphi \qquad (A.2)$$

$$h_x(r,z,\varphi) = \frac{-jAI_2 \sin 2\varphi}{\eta} \qquad (A.3)$$

$$h_y(r,z,\varphi) = \frac{-jA(I_0 - I_2 \cos 2\varphi)}{\eta}, \qquad (A.4)$$

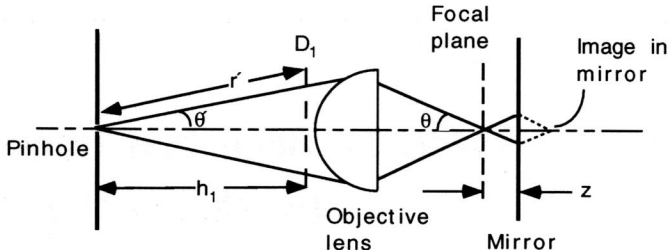

Figure A.1 Schematic of the focusing system treated in the theory.

where

$$I_0(r,z,\varphi) = \int_0^{\theta_0} P(\theta) \cos^{1/2}\theta (1 + \cos\theta) J_0(kr\sin\theta) e^{-jkz\cos\theta} \sin\theta\, d\theta, \tag{A.5}$$

and

$$I_2(r,z,\varphi) = \int_0^{\theta_0} P(\theta) \cos^{1/2}\theta (1 - \cos\theta) J_2(kr\sin\theta) e^{-jkz\cos\theta} \sin\theta\, d\theta. \tag{A.6}$$

In these equations J_0 and J_2 are Bessel functions of the first kind, $k = 2\pi/\lambda$ is the wave number, λ the wavelength, θ_0 is the aperture angle of the lens, and η is the impedance of free space. Richards and Wolf's expressions have been modified to express them in MKS units, with the assumption that all fields vary as $\exp j\omega t$, and to explicitly include the pupil function of the objective lens $P(\theta)$. The expressions given here are written in cylindrical coordinates rather than in the spherical coordinates of the original paper. In addition, we have assumed that the numerical aperture of the beam on the pinhole side of the objective is very small so that the exciting field at the pupil plane of the lens can be regarded as uniform in amplitude.

To derive the depth response of a confocal optical microscope from this result, we consider the system shown in Fig. A.1, with the pinhole on the left-hand side of the objective and, for simplicity, illumination through the pinhole. We first assume that the field incident upon the objective is reflected by a perfect plane reflector, such as a perfect mirror, located at the focal plane, $z = 0$, of the objective. The transverse variation of the reflected field $e_x^R(r',\theta',\varphi)$ at the pupil plane D_1, a distance h_1 from

A.1 The Depth Response

the pinhole, has a phase variation identical to that of the incident field $e_x^I(r',\theta',\varphi)$, which is regarded as uniform in amplitude. It follows that

$$e_x^R(r',\theta',\varphi) = P^2(\theta)e_x^I(r',\theta',-\varphi), \tag{A.7}$$

where (r',θ',φ) are the spherical coordinates of a point in the plane D_1 with the center of the coordinate system at the pinhole. The pupil function appears squared in Eq. (A.7) because the beam passes through the lens twice. As in Chapter 3, unprimed coordinates are used for fields on the pinhole side of the objective and unprimed on the sample side.

When the mirror is moved a distance z from the focal plane, the image of the focused spot in the mirror moves a distance $2z$. This implies that the fields in the back focal plane of the lens that give rise to rays converging on the focal point at an angle θ pick up an additional phase shift $2jkz\cos\theta$. It thus follows that

$$e_x^R(r',\theta',\varphi;z) = P^2(\theta)e_x^I(r',\theta',-\varphi;0)e^{-2jkz\cos\theta}. \tag{A.8}$$

The field $e_x^R(r',\theta',\varphi;z)$ is the excitation field at the pupil plane D_1 of the objective for a defocused lens. Richards and Wolf's equations can now be used to calculate the vector fields at the pinhole. If the system is properly aligned an infinitesimal pinhole will transmit only the on-axis fields. In this case $r = 0$, so that Eqs. (A.1) through (A.8) reduce to

$$e_x^R(z) = -\eta h_y^R(z) = -jA \int_0^{\theta_0'} \cos^{1/2}\theta'(1 + \cos\theta') e_x^R(r',\theta',\varphi;z) \sin\theta'\, d\theta'$$

$$= -jA \int_0^{\theta_0'} \cos^{1/2}\theta'(1 + \cos\theta') P^2(\theta') \tag{A.9}$$

$$e_x^I(r',\theta',-\varphi;0)e^{-2jkz\cos\theta} \sin\theta'\, d\theta'.$$

In Eq. (A.9), the integration is over the pinhole lens with angular coordinates θ'.

Equation (A.9) can be written entirely in terms of objective coordinates θ using the sine condition for a perfect lens,

$$\sin\theta = M\sin\theta', \tag{A.10}$$

and the transformation

$$\cos\theta\, d\theta = M\cos\theta'\, d\theta'. \tag{A.11}$$

In Eq. (A.11) M is the magnification of the optical system, and we have assumed for simplicity that the refractive index at the sample is unity.[2] With these substitutions, and the assumption that on the pinhole side,

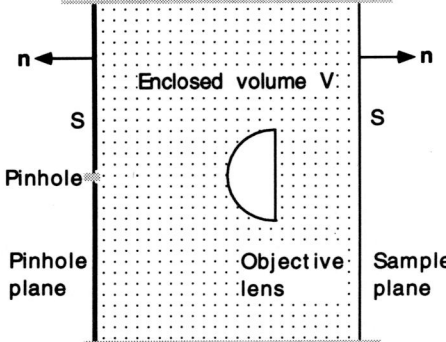

Figure A.2 Schematic illustrating the volume of integration and the enclosing surfaces.

where the angles are very small, $e_x^I(r',\theta',-\varphi;0)$ is uniform and of unit value, Eq. (A.9) becomes

$$e_x^R(z) = \eta h_y^R(z) = \frac{-2jA}{M^2}\int_0^{\theta_0} P^2(\theta)e^{-2jkz\cos\theta}\cos\theta\sin\theta\,d\theta. \quad (A.12)$$

Equation (A.12) gives the depth response of the CSOM. It can be normalized by writing

$$V(z) = \frac{e_x(z)}{e_x(0)} = \frac{h_y(z)}{h_y(0)} = \frac{\int_0^{\theta_0} P^2(\theta)e^{-2jkz\cos\theta}\cos\theta\sin\theta\,d\theta}{\int_0^{\theta_0} P^2(\theta)\cos\theta\sin\theta\,d\theta}. \quad (A.13)$$

Equation (A.13) is a nonparaxial vector solution for the depth response of the CSOM and is identical to the scalar result of Eq. (3.5).

A.2 Transverse Response

Because of the effects of polarization it is necessary to use a somewhat different approach than that used in the scalar theory to determine the transverse response by vector field theory. Richards and Wolf's theory gives the form of the fields transmitted to the sample through the objective lens. Knowledge of these fields enables us to use a vector form of the reciprocity theorem, or Green's theorem,[3] to find the fields excited at the pinhole by reflection of a focused beam from a point on the sample.

We consider the fields inside a volume V bounded by the sample surface and the plane of the pinhole, as shown in Fig. A.2. Let the fields

A.2 Transverse Response

reflected from the object be $\mathbf{E}^R(x,y,z)$, $\mathbf{H}^R(x,y,z)$. These fields must obey Maxwell's equations

$$\nabla \times \mathbf{E}^R = -j\omega\mu\mathbf{H}^R, \tag{A.14}$$

and

$$\nabla \times \mathbf{H}^R = j\omega\varepsilon\mathbf{E}^R. \tag{A.15}$$

Now consider the fields excited by a current $\mathbf{J}(x,y)$ per unit area. The fields associated with this current, $\mathbf{E}^G(x,y,z)$, $\mathbf{H}^G(x,y,z)$, also obey Maxwell's equations

$$\nabla \times \mathbf{E}^G = -j\omega\mu\mathbf{H}^G, \tag{A.16}$$

and

$$\nabla \times \mathbf{H}^G = j\omega\varepsilon\mathbf{E}^G + \mathbf{J}. \tag{A.17}$$

It is convenient to use the vector identities

$$\nabla \cdot (\mathbf{E}^G \times \mathbf{H}^R) = \mathbf{H}^R \cdot \nabla \times \mathbf{E}^G - \mathbf{E}^G \cdot \nabla \times \mathbf{H}^R, \tag{A.18}$$

and

$$\nabla \cdot (\mathbf{E}^R \times \mathbf{H}^G) = \mathbf{H}^G \cdot \nabla \times \mathbf{E}^R - \mathbf{E}^R \cdot \nabla \times \mathbf{H}^G, \tag{A.19}$$

to relate the fields to the volume integrals. We subtract Eq. (A.19) from Eq. (A.18), substitute from Eqs. (A.14)–(A.17), and integrate over a volume bound by the sample surface and the pinhole plane. It then follows, by employing Gauss's theorem, that

$$\int_V \mathbf{J} \cdot \mathbf{E}^R \, dV = \int_S (\mathbf{E}^G \times \mathbf{H}^R - \mathbf{E}^R \times \mathbf{H}^G) \cdot \mathbf{n} \, dS, \tag{A.20}$$

where V is the volume enclosed by the sample surface and the plane of the pinhole, S the enclosing surface, $\mathbf{n}$ the outward normal from the volume at these surfaces, and it is assumed that the fields fall off rapidly enough so that there is no contribution to the surface integrals at infinity.

We are interested in determining the reflected E field, polarized in the x direction at an infinitesimal pinhole. To find this quantity, we take $\mathbf{J}$ to be of the form $\mathbf{J} = \mathbf{a}_x\delta(x',y',z')$, where $\mathbf{a}_x$ is the unit vector in the x direction, it is assumed that the infinitesimal pinhole is located at $r' = 0$, and δ is the Dirac delta function. In this case, Eq. (A.20) becomes

$$E_x^R = \int_S (\mathbf{E}^G \times \mathbf{H}^R - \mathbf{E}^R \times \mathbf{H}^G) \cdot \mathbf{n} \, dS. \tag{A.21}$$

By using the Green's function technique, we have now derived a formula for the field E_x^R at the pinhole.

Since the pinhole is infinitesimal in size and the tangential E fields on a perfect conductor surrounding the pinhole must be zero, it will be seen that the only contribution to the integral in Eq. (A.21) is from the sample surface. The fields $\mathbf{E}^G$ and $\mathbf{H}^G$ are identical to the fields transmitted to the sample by the illuminating source and are given by Eqs. (A.1)–(A.4). If the reflectivity of the sample is $R(x,yz)$, it follows from Eq. (A.21) that $\mathbf{E}^R \cdot \mathbf{n} = R(x,y,z)\mathbf{E}^G \cdot \mathbf{n}$ and $\mathbf{H}^R \cdot \mathbf{n} = -R(x,y,z)\mathbf{H}^G \cdot \mathbf{n}$, so that in normalized form

$$E_x^R = -2 \int_S R(x,y,z)(\mathbf{E}^G \times \mathbf{H}^G) \cdot \mathbf{n} \, dS. \qquad (A.22)$$

Assuming that the reflector is a point (a delta function) located a distance r from the axis, and substituting from Eqs. (A.1)–(A.4), for the fields excited at the sample by the objective lens, it will be seen that

$$E_x^R(r) = K[I_0^2(r) - I_2^2(r)], \qquad (A.23)$$

where K is a constant and I_0 and I_2 are given by Eqs. (A.5) and (A.6). This is the exact nonparaxial form of the vector field theory for the transverse response of the confocal microscope. This theory can be extended fairly easily to calculate the cross-polarized fields associated with the I_2 term. These fields are normally small and only finite off axis.

Comparing this result to the scalar theory at $z = 0$, the term I_2^2 is zero at $r = 0$ and remains very small compared to I_0^2 at all values of r, so it can usually be neglected. In this case the amplitude PSF of the CSOM, Eq. (A.23), becomes in normalized form

$$E_x^{R^2}(r) = \left[\frac{\int_0^{\theta_0} J_0(kr \sin\theta)(1 + \cos\theta) \sin\theta \cos^{1/2}\theta \, d\theta}{\frac{2}{3}(1 - \cos^{3/2}\theta_0) + \frac{2}{5}(1 - \cos^{5/2}\theta_0)} \right]. \qquad (A.24)$$

We can compare this equation to the other forms of the amplitude PSF derived in Chapter 3. For an infinitesimal pinhole, the exact scalar theory of Eq. (3.83) reduces to the normalized form

$$\Phi_0^{R^2}(r) = \left[\frac{\frac{3}{2} \int_0^{\theta_0} J_0(kr \sin\theta) \sin\theta \cos^{1/2}\theta \, d\theta}{1 - \cos^{3/2}\theta_0} \right]^2, \qquad (A.25)$$

where $\Phi_0^{R^2} = 1$.

A.2 Transverse Response

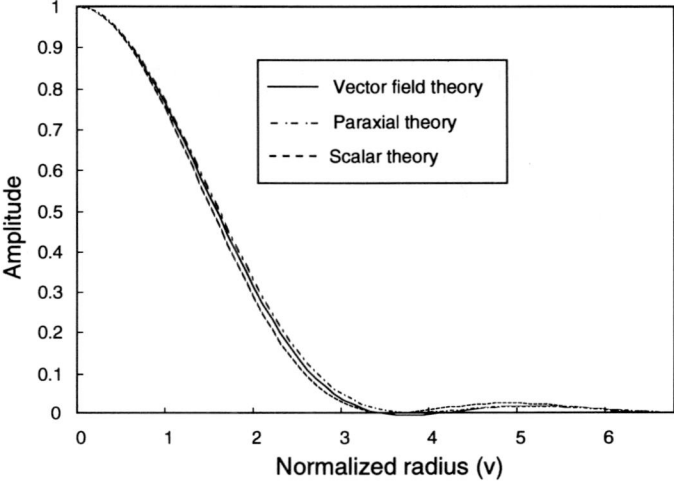

Figure A.3 Plots of the normalized PSF vs. normalized radius $v = knr \sin \theta_0$ for $\sin \theta_0 = 0.95$, from Eqs. (A.23), full line; (A.25), dashed line; and (A.26), dash and dot.

The equivalent paraxial normalized scalar form of the amplitude PSF is

$$h^2(r) = \left[\frac{2 \int_0^{\theta_0} J_0(kr \sin \theta) \sin \theta \cos \theta \, d\theta}{\sin^2 \theta_0} \right]^2 = \left[\frac{2 J_1(kr \sin \theta_0)}{kr \sin \theta_0} \right]^2. \quad (A.26)$$

Equations (A.24)–(A.26) are identical in the paraxial approximation. We have calculated the PSF for a lens with $\sin \theta_0 = 0.95$ from the full vector theory of Eq. (A.23), rather than Eq. (A.24). We have compared this result to the PSFs calculated from Eqs. (A.25) and (A.26) in Fig. A.3 as a function of the normalized radius $v = knr \sin \theta_0$. The differences between the scalar and vector theories are normally too small to be of concern. It will be noted, however, that the PSF of Eq. (A.23) goes slightly negative due to a phase change in I_2 in the region where $I_0 \approx 0$ and I_2 is finite. Interestingly enough, the scalar paraxial theory is in slightly better agreement with the full vector theory of Eq. (A.23) than the "exact" scalar theory as far as sidelobe levels and FWHM are concerned, but the vector theory and the "exact" scalar theory are in perfect agreement for $V(z)$. At all events, even for this large a numerical aperture, the FWHMs are all within 3% of each other.

References

1. B. Richards and E. Wolf, "Electromagnetic diffraction in optical systems II. Structure of the image field in an aplanatic system," Proc. Roy. Soc. **A 253**, 358–379 (1959).
2. M. Born and E. Wolf, *Principles of Optics*, 6th edition (Pergamon, 1983).
3. S. M. Mansfield, "Solid immersion microscopy, Ph. D. Dissertation, Department of Applied Physics, Stanford University, Stanford, California, USA (March 1992).

Index

A
Abbé, 17, 19
Aberration, 37, 47, 73, 93, 150, 153, 193, 245, 314
   astigmatism, 163
   chromatic, 52, 70, 93, 164, 314
   coma, 163
   spherical, 52, 163, 180, 194, 200, 202, 311, 314
Accuracy, 73, 110, 165, 279, 296
Acousto-optic cell, 33, 75, 113, 239, 262
Airy function, 12, 39, 169, 187
Airy pattern, 17, 74, 87, 90, 175
Amplitude apodization, 164
Analyzer, 29, 44, 92, 107, 118, 248, 288
Annular phase filter, 227
Aperture stop, 16
Arc lamp, 14, 45, 70, 90, 193, 254, 282
Atomic force microscope, 123, 129

B
Back focal plane, 9, 16, 19, 29, 75, 227, 248, 325
Basic requirements, 33, 67
Beam expander, 69, 71, 174
Beam scanning, 33, 75, 83, 97, 314
Beamsplitter, 45, 73, 84, 95, 118, 196, 197, 200, 202
Bilateral scanning, 78, 99
Biological imaging, 13, 41, 308, 316
Birefringent optical material, 29, 248

Brewster angle, 165
Brightfield imaging, 26, 84, 308
Brightness, 25, 70
Broadband illumination, 90, 164, 191, 296

C
Calibration, 294
Cloud plots, *see* Cross-sectional image
Coherence length, 147, 189
Coherence time, 14
Coherent detection, 108, 110, 214
Coherent illumination, 14, 108, 179, 182
Coherent interference, 70, 90, 156, 296, 304
Commercial microscopes, 68, 79, 95, 315
Condenser lens, 16, 17, 38, 90, 175, 227
Confocal, 33, 108
Confocal scanning laser microscope, 30, 68, 79, 97, 168, 173, 313
Constant Angle of Reflectance Spectrometry, 300, 303
Cornea, 97, 309
Correlation length, 192, 189
Critical dimension, 278, 280
Critical illumination, 16, 48
Cross-sectional image, 3, 38, 52, 281, 282, 284, 290
Cutoff mode, 206, 215
Cutoff radius, 208

331

D

Darkfield imaging, 26
Deconvolution, 109, 245
Dense arrays, 286
Depth of focus, 77, 152
Depth response
  confocal scanning optical microscope, 17, 35, 72, 93, 323, 326
  fluorescent sample, 159, 160
  interference microscope, 49, 51, 52, 189, 192, 230, 269
  other microscopes, 108, 113, 236, 238, 303, 305
  plane reflector, 149, 151, 160
  point reflector, 155
  real-time scanning optical microscope, 87, 88, 171
  solid immersion microscope, 214
Detector pinhole, 30, 74, 77, 80, 111, 314
Dichroic beamsplitter, 79
Dichroic filter, 29, 315
Dichroic mirror, 69, 100, 314
Dielectric tip, 209
Differential amplitude imaging, 125, 83, 260, 265
Differential contrast imaging, 260
Differential depth response, 234, 236, 266
Differential interference contrast, 30, 247, 260, 262
Differential phase images, 83, 261, 265
Dipping cone objective, 310
Directional coupler, 83, 84
Dual detectors, 232

E

Edge response, 40, 178, 180, 205
Efficiency, 74, 80, 83, 87, 92, 97, 101, 123, 206, 313
Electro-optic phase modulator, 113, 160, 167, 172, 173, 233, 236, 238
Entrance pupil, 74
Envelope function, 3, 49, 114

Experimental results, 161, 194, 215, 219, 238, 271, 284, 290, 298
Eyepiece, 9, 91, 93

F

Ferro-electric liquid crystal, 252
Fiber optics, 83, 109, 123
Field lens, 93, 95
Film thickness measurements, 284, 304
Fluorescence imaging, 27, 79, 125, 157, 246, 308, 311
Focus-exposure wafers, 295
Focus position, 52, 73, 171, 238, 239, 269, 282, 296
Focus tracking, 238
Fourier transform technique, 269, 307
Fraunhofer diffraction theory, 169, 170
Fringe spacing, 268

G

Galvanometer mirror, 33, 75, 77, 81
Gouy shift, 128

H

Halo, 229
Heterodyne interferometer, 226, 239, 261

I

Illumination, 16, 25, 27, 43, 89, 102, 114, 163, 177, 227, 234, 248, 288, 303, 314
  spatially coherent, 13, 39, 69, 95, 108, 229
  spatially incoherent, 13, 38
Imaging characteristics, 16, 21, 39, 40, 177, 196
Immersion lenses
  oil, 13, 72, 118
  water, 13, 72
Impulse response, 11

Index

Infinitesimal pinhole, 149, 178
Integrated circuit inspection, 278
Integrating bucket technique, 266
Interference fringes, 3, 47, 152, 189, 192, 300, 303, 304
Interference microscopes, 3, 32, 110, 195, 266, 305
   4Pi, 107, 113, 246
   confocal scanning optical, 67, 110, 229, 230
   Linnik, 45, 115, 230
   Mach-Zehnder, 110
   Michelson, 3, 111, 113
   Mirau, 47, 116, 196, 202, 230, 307
   Tolanski, 119
Interference term, 112, 191, 218, 231, 303
Interferogram, 112, 113, 233, 266, 271
Intermediate optical system, 73
In vivo imaging, 310

K

Kerr effect, 131
Kirchhoff approximation, 170
Köhler illumination, 16, 48, 90, 119

L

Lambert's law, 25
Laser, 24, 69, 70, 89, 108, 126, 295, 313, 315
Linearity, 279, 286
Linear systems theory, 11
Line spread function, 20, 21, 180
Linewidth measurement, 282
Lithography, 13, 278, 307

M

Macroscope, 78
Magneto-optical storage, 131, 136, 219
Magnification, 9, 10, 13, 150, 212
Metrology, 68, 70, 93, 164, 278, 279, 295, 300
Modulated carrier, 269

N

Narrowband source, 51, 192
Near-field scanning optical microscope, 3, 68, 120, 206
Negative phase-contrast, 228
Neutral density filters, 81, 90
Nipkow disk, 33, 42, 84, 86, 186, 288, 313
Nipkow disk fabrication, 89
Nomarski, 30, 125, 128, 247
Normalized pinhole radius, 172, 173, 188
Numerical aperture, 9, 10, 13, 18, 114, 169, 265, 268

O

Objective, 9, 13, 29, 33, 39, 71, 93, 75, 102, 151, 176, 268, 310, 324
Objective pupil, 91, 93, 95, 150, 257, *see also* Pupil function
Ophthalmology, 309
Optical beam-induced currents, 279
Optical fiber, 33, 83, 123, 207, 263
Optical isolator, 89, 91, 92, 96, 101, 288
Overlay measurements, 278, 295, 298

P

Partially coherent illumination, 176
Pellicle beamsplitters, 198
Phase contrast microscopy, 29, 226, 230
   ac Zernike, 234
Phase detection, 54, 191, 232, 307
Phase imaging, 54, 113, 152, 226, 234, 246, 266, 311
Phase information, 46, 51, 114, 192, 225, 231, 232, 233, 266, 267, 307
Phase-shift mask, 307
Pinhole
   diameter, 74, 87, 120, 171, 210
   optimum pinhole, 87, 165, 166, 188
   size, 87, 167, 183
   spacing, 88

Pinhole lens, 73, 150, 166
Pipette, 122, 209
Plasmon resonance, 125
Point detector, 39, 73, 77, 83, 178, 258
Point spread function, 10, 109, 157, 185, 202, 217, 229, 238, 329
    amplitude point spread function, 12, 15, 39, 196, 217
    intensity point spread function, 10, 12, 14, 39, 175, 188, 218
Pointing stability, 70
Polarization, 29, 92, 96, 118, 131, 169, 248, 252, 288, 294
Polarization-enhanced imaging, 288, 291
Polarization isolation, see Optical isolator
Polarizer, 29, 90, 92, 96, 118, 248, 288, 294
Polarizing beamsplitter, 74, 90, 92, 94, 111
Positive phase-contrast, 228
Precision, 279
Propagation constant, 264
Pupil function, 11, 18, 150, 151, 164, 324
Pupil plane detector, 258

Q

Quadrature phase imaging, 233
Quarter-wave plate, 32, 92, 118, 288

R

Radiating dipole, 210
Range resolution
    fluorescent microscope, 159
    interference microscope, 52, 53, 193
    plane reflector, 35, 52, 162, 169, 171
    slit microscope, 160
    transmission microscope, 104
Rayleigh criterion, 23, 179
Rayleigh-Sommerfeld diffraction theory, 11, 170
Rayleigh two-line criterion, 182
Real-time Scanning Optical Microscope, 43, 86, 152, 168, 172, 183, 186, 220, 226, 248, 286, 288
Reference beam, 54, 112, 190, 192, 229, 242, 246, 267
Reference surface, 47, 117
Reflection coefficient, 164, 173, 197, 249, 301
Repeatability, 279

S

Sample scanning, 33, 73
Sample tilt, 165
Scanning acoustic microscope, 41, 56
Scanning electron microscope, 57, 279, 281
Scanning tunneling microscope, 59, 123
Schwarzchild lens, 120
Secondary objective, 90, 91, 94
Seidel coefficients, 163
Sidelobe, 12, 16, 37, 88, 102
Signal-to-noise, 29, 41, 74, 92, 110, 258, 314
Silicon nitride beamsplitter, 198, 201
Sine condition, 25, 150, 325
Single point resolution, 13, 39, 178
Slit microscope, 97, 101, 160
Snell's law, 201, 302
Solid immersion lens, 3, 133, 212
Solid immersion microscope, 3, 120, 133, 212
Sparrow criterion, 24, 40, 156, 159, 180
Spatial filter, 32, 69, 77
Spatial frequency, 17, 18, 19, 22, 51, 190, 270
    maximum, 18
    response, 14, 40, 118, 192
Speckle, 15, 88, 97, 152, 179
Spectroscopy, 131
Split detector, 83, 254, 257, 263
Spot size, 17, 70, 87, 102, 120, 123, 137, 202, 218, 262, 294
Stage scanning, 68, 73
Standard optical microscope, 7, 13,

26, 33, 39, 40, 79, 175, 225, 247, 257, 280, 298, 310
Stigmatic focusing lens, 137, 213
Struve function, 181
Surface roughness, 165

T

Tandem scanning reflected light microscope, 43, 84
Tapered fiber, 207
Telecentric, 75, 79, 81, 248
Television, 86
Temporal interference, 192
Temporally coherent, 14
Thin films, 54, 56, 165, 300
Time-averaged point spread function, 186
Tool-induced shift, 299
Transfer function, 17, 18, 21, 40, 72, 160, 245
Transfer optics, 75, 77
Transmission microscope, 34, 104
Transverse resolution, 39, 51, 55, 72, 74, 97, 113, 175, 177, 183, 189, 195, 196, 203, 216, 219, 326

Tube length, 9, 10, 72, 87
Tube lens, 93, 94
Two-mode optical fiber, 263
Two-photon
 fluorescence, 314
 imaging, 315
Two point detectors, 258
Two-wavelength imaging, 314

V

Variable Angle Monochromatic Fringe Observation, 300, 303
Vignetting, 76

W

Wafer-induced shift, 299
Waveguide, 123, 207, 284
Wavelength filters, 90
Wollaston prism, 30, 248, 252

Z

Zernike, 29, 225, 226, *see also* Phase contrast microscopy
Zero crossing, 234